Securing Industrial Control Systems

Mohammad Ashiqur Rahman •
Syed Bahauddin Alam • Kishor Datta Gupta •
Roy George • Sunzida Siddique •
Kazuma Kobayashi

Securing Industrial Control Systems

Advanced Strategies and Technologies

 Springer

Mohammad Ashiqur Rahman
Florida International University
Miami, FL, USA

Kishor Datta Gupta
Clark Atlanta University
Atlanta, GA, USA

Sunzida Siddique
Daffodil International University
Birulia, Bangladesh

Syed Bahauddin Alam
University of Illinois Urbana-Champaign
Urbana, IL, USA

Roy George
Clark Atlanta University
Atlanta, GA, USA

Kazuma Kobayashi
University of Illinois Urbana-Champaign
Urbana, IL, USA

ISBN 978-3-032-03017-7   ISBN 978-3-032-03018-4   (eBook)
https://doi.org/10.1007/978-3-032-03018-4

This Springer imprint is published by the registered company Springer Nature Switzerland AG
The registered company address is: Gewerbestrasse 11, 6330 Cham, Switzerland

If disposing of this product, please recycle the paper.

Contents

Acronyms

AGVs	Automated Guided Vehicles
AI	Artificial Intelligence
APTs	Advanced Persistent Threats
BMS	Building Management System
CEH	Certified Ethical Hacker
DCS	Distributed Control System
DDoS	Distributed Denial of Service
FTP	File Transfer Protocol
HDOs	Health Delivery Organization
HIDS	Host-based Intrusion Detection System
HMI	Human Machine Interface
HTTP	Hypertext Transfer Protocol
HVAC	Heating, Ventilation, and Air Conditioning
IACS	Industrial Automation and Control Systems
IAM	Identity and Access Management
ICs	Industry Control System
IDS	Intrusion Detection System
IEDs	Intelligent Electronic Devices
I/O	Input/Output
IoT	Internet of Things
IP	Internet Protocol
IPS	Intrusion Prevention System
IT	Information Technology
MES	Manufacturing Execution System
MFA	Multi-Factor Authentication
MSPs	Managed Service Providers
NAT	Network Address Translation
NIDS	Network Intrusion Detection System
OPC	Open Platform Communications
OT	Operational Technology
PAC	Programmable Automation Controller

PIDS   Perimeter Intrusion Detection System
PLC   Programmable Logic Controller
RAM   Random Access Memory
ROM   Read-Only Memory
RTU   Remote Terminal Unit
SCADA   Supervisory Control And Data Acquisition
SSH   Secure Shell
SSL   Secure Sockets Layer
TCP   Transmission Control Protocol
TLS   Transport Layer Security
VM   Virtual Machine
VPN   Virtual Private Network
VSA   Virtual System Administrator

Chapter 1
Comprehensive Overview of Industrial Control Systems ICS: Evolution, Components, and Security Challenges

Abstract Industrial control systems (ICS) are the backbone of the world's modern infrastructure and are used to power manufacturing, critical infrastructure, including utilities and building systems. They combine sensors, control, connectivity, and human interface to automate, monitor, and adjust industrial processes. This chapter analyzes the basic elements of ICS, its use in multiple sectors, and, especially, the particular aspects of their requisite safe, secure, and efficient operation, which have a tangible effect in the real world.

Keywords Industrial control systems · Automation · Infrastructure · Sensors · Controllers · Network communication · Security · Safety · Industrial processes · Critical infrastructure · Human-machine interface

1.1 Introduction

Industrial control systems (ICS) are the operational plumbing used in industries including water treatment, manufacturing, and energy. In various technologies, including building and industrial automation, these systems control and monitor end-devices, including sensors, controllers, and networked devices, to perform both human and automated control [137]. ICS is now increasingly closely associated with legacy IT systems that are already in place. These systems are frequently connected via Internet Protocol (IP), which adds new capabilities but also exposes them to new cybersecurity threats. However, a complete digital failure of ICS can have tangible consequences, such as a factory closing or the destruction of an environment. Not so much in the case of essential infrastructure, where there is no downtime allowed, but each of these must be paid off for a lower overall brute-force attack performance with regard to availability or security. Furthermore, ICS systems emphasize availability through the use of redundant, failover, or fault-tolerant elements [10]. Predictive maintenance using machine-learning-based algorithms will enable early detection of component wear or degradation—which translates to less unscheduled downtime and a longer-lived asset. Such a pro-active

M. A. Rahman et al., *Securing Industrial Control Systems*,
https://doi.org/10.1007/978-3-032-03018-4_1

attitude enhances the security of service and is beneficial for asset control in a cost-effective way. However, as businesses' ICS become more interconnected with their corporate IT environment and the broader Internet of Things (IoT), these values are more exposed than ever before to a multitude of cybersecurity risks [125]. In the intersection of OT and IT, risk must be managed to minimize the risk associated with the plethora of vulnerabilities in legacy systems and ensure safe operating environments.

1.2 Examples of Industrial Control Systems

ICS are utilized to manage, monitor, or automate the major processes of different industrial sectors. One of the prevalent ICS is a supervisory control and data acquisition (SCADA) system consisting of PLCs, RTUs, communication networks, and HMI software that facilitates monitoring and controlling remote control processes [56]. Such SCADA systems can be able to read field device data in intelligent ways, such that the operators can command the industrial operations remotely or can combine human judgment with real-time situational perception [45].

SCADA has general use in managing electric power grids, water and sewage treatment plants, and oil and gas pipelines monitoring [8]. This all contributes to continuous process refinements, fault indication, and decreased on-site manual inspection. But SCADA systems are susceptible to corporate networks and the Internet upon which to communicate, and therefore the attack surface increases [116, 122]. Inadequate credentialing, insecure network topology, and inadvertent system firmware (DoS) and privilege escalation to turn off services.

The 2021 Oldsmar water treatment plant attack is a reflection of the critical necessity of hardened cyber architecture, including network segmentation, anomaly detection, and secure updates to preclude penetration by malicious SCADA systems.

Building Management System Another type of ICS is that of BMS (Building Management System) and is utilized for building control of the environment; these systems can range from the air conditioning of an office-block, to airport ventilation systems, to commercial premises, and in some instances, the medical suite-buildings' in hospitals. A building management system AMS functions as an illustration is BMS system ranging from HVAC, lighting, security, firefighting, to energy, all managed via control ports with sensing feedback from the sensors themselves. Through this, energy and recycling management are optimally used for environmental and operational benefits. Threat to cybersecurity is not confined to SCADA systems but BMS [125]. A compromised BMS can have serious consequences like temperature manipulation in a facility's critical areas, blocking fire alarms, or unauthorized access. The security of BMS should be protected against information authentication, firmware validation, vulnerability tracking, and so on. As more and more buildings become smart and connected, cybersecurity must become part of their basic design and operational paradigms.

1.3   Core Components of Industrial Control Systems

Industrial control systems, which impose the connection of the physical machinery and the computer-based control and monitoring, enable for operation of the industrial system in a more efficient, reliable, and automated way in real time. Such systems are based on a single core set of components with their inherent capabilities that determine the responsiveness, the stability, and the security of ICS. These are the fundamentals needed to build, run, and defend secure industrial infrastructure.

1. **Control Loops**
   The basic building blocks of process control are control loops. These feedback loops monitor process variables in real time (temperature, pressure, flow, etc.) with sensors and compare them to setpoints to trigger corrective action through actuators. Controllers loop the system back by applying control algorithms that keep the system at equilibrium despite dynamic loads or environmental conditions. The accuracy of control loop design has a very direct impact on throughput, energy consumption, and product quality.

2. **Human-Machine Interfaces**
   Human-Machine Interfaces (HMI) are intervening devices between human operators and control layers that are automated. Their design is of utmost significance—not aesthetically only, but cognitive ergonomics-wise. A good HMI dissociates raw sensor signals into understandable visual cues, which enable humans to easily detect anomalies, to intervene in abnormal occurrences, and to perform tuning. In a high-risk extreme situation, such as a chemical or nuclear plant, having the ability to understand and respond to an HMI can be the difference between a small-scale event and a catastrophic disaster.

3. **Sensors**
   Sensors are the ICS sensors and translate physical movement to an electrical reaction that can be quantified. The reliability of their controller decisions, calibration dependability, and resistance to environmental influences determine the dependability of the control decisions. A faulty sensor will, particularly in a closed-loop, propagate faulty signals, which will destabilize entire processes. For this, sensor reliability plays a pivotal role in upholding the process integrity and safety.

4. **Actuators**
   Actuators convert control signals into physical movement: they can open or close valves, drive at variable speeds, or turn relays on and off. Their fast response time, good accuracy, and longer lifetime are key to efficient process control and quick emergency responses.

5. **Controllers**
   Programmable Logic Controllers (PLCs) and Distributed Control Systems (DCS) are the base computational devices of ICS. They handle input streams, manage logic flows, and produce input commands with strict timing deadlines. The reliability of such systems, whose components have certain failure rates, has been addressed due to the various stages of STC they must traverse and the

importance of these characteristics (deterministic and fault containment) for system availability and safety certification.

6. **Maintenance and Asset Management Tools**
   Advanced ICS architectures bring the first line of maintenance intelligence to the intelligent edge, including predictive analytics, asset tracking, and condition monitoring. They employ telemetry and historical information to predict failure, analyze resource allocation, and provide total cost of ownership reduction. Predictive maintenance is a boon for asset-heavy industries where asset operational reliability is critical.

7. **Remote Diagnostic Capabilities**
   Detect and respond with remote diagnostics and system health monitoring that help reduce downtime—especially with dealerships that are spread out. Nonetheless, secure remote access is vital due to inherent weaknesses if not correctly authenticated or segmented. Thus, diagnostic tools must incorporate secure communication protocols and real-time anomaly detection.

8. **Layered Network Topologies**
   The ICS communications shall be controlled by a hierarchical network structure (modeled according to the OSI (Open Systems Interconnection) architecture). Each layer (from physical signaling to application-level protocols) offers two levels of functions: abstraction and control, fostering scalability and cybersecurity by logical isolation. For example, separating supervisory control from field-level communications can control the spread of failure and the blast radius of cyber incidents.

These components are the cyber-physical part of ICS. They are required to cooperate in order to ensure a secure operation of crucial infrastructure, and inefficiencies in aspects like sensor veracity, or network separation can affect performance, safety, or reliability. Thus, to design, operate, or secure industrial environments, a systems-level understanding of how engineering constraints are incorporated in future Internet technologies and operational ecosystems is necessary.

1.4 Cybersecurity Incidents in Industrial Control Systems

Industrial Control Systems manage operations in critical infrastructure like energy, water, manufacturing, and transport, and these must function correctly in real time. And the fact that previously ICS were air gapped, and increasingly networked ICS are being linked to IT networks (and across the Internet) increased the ICS attack surface, if not made attacks on modern ICS possible, as discussed in more detail below. But, unlike most of the other IT hacks, SCADA/ICS attacks could be physiologically crippling to physical infrastructure, lives, and the country due to the possibility of physical disruption and financial loss. Attacks on industrial control systems can be used to destroy power plants, poison water supplies, or disable safety features [54]. No ifs, ands, or buts about it: even within an industrial context, the threat landscape is changing and industrial-specific cybersecurity measures are more and more required, as testified by the latest incidents [5]. Another analysis of the risk will need to be intentionally potential in the domain of ICS and would encompass the financial-impact-analysis: where risk for the financial dependability: that would be in the form of asset-process failures, cyber attacks, problems of field and control-room automation, etc.; along these lines of own this Probabilities of Occurrence for the same and the possible financial loss [132]. It is a follow-on monitoring and regular risk reporting (e.g., audit, event log) to enable the governor oversight to rationalize ICS controls and to financial risk model. Internal controls are structured by the COSO Model—policy, segregation of duties, and process controls are present in all ICS operations to minimize financial exposure; periodic system audits are performed with third-party service agencies to detect failures in controls (error and fraud controls) and security, privacy, and compliance.

1. **2015/2016: Attacks on Ukraine's Power Grid**
   Synchronized cyberattacks targeted Ukraine's power grid SCADA systems in 2015 and 2016. The 2015 attack remotely disabled circuit breakers, impacting hundreds of thousands of people. The 2016 attack employed the industrious malware, directly manipulating grid communication protocols. These attacks, among the first to cause substantial physical disruption, marked a new era of geopolitical threats to national infrastructure by bridging cyber and physical domains.

2. **2017: Triton Malware in Saudi Arabia**
   HatMan was a game-changer in the evolution of the cyber threat. Triton was intentionally designed not to attack process control; it was meant to tamper with Safety Instrumented Systems (SIS), which serve as a system to ensure the well-being in case of a necessary shutdown. By hacking Schneider Electric's Triconex SIS controllers, hackers were trying to prevent emergency shutdown systems from activating at a petrochemical plant in Saudi Arabia. An early detection from a deployment snafu notwithstanding, the sabotage to safety functions represented an escalation in adversarial tactics, from economic disruption to human fatalities.

3. **2019: Norsk Hydro Aluminum Plant**
   A ransomware assault on Norwegian aluminum [36] maker Norsk Hydro halted production lines, affected operations around the world, and suffered losses of more than $70 million. The assault, which involved the LockerGoga ransomware, encrypted essential systems, and the company took its network offline and has since been operating mainly manually. This incident was a typical illustration of why manufacturing companies, with their significant investment in aging ICS assets, are still low-hanging fruit for attackers who are out to cause extreme damage and capture the value in the operating downtime and the urgency of recovery.

4. **2020: Cyber Intrusion in Israeli Water Installations**
   There were numerous efforts to penetrate Israeli water-treatment and pumping facilities. There were almost no immediate operational consequences, but the attacks demonstrated how a state or a proxy actor could target public health systems by interfering with the dosing of chemicals. These incidents acted as an early warning that water infrastructure—often a little-funded portion of cybersecurity—might get swept up in hybrid warfare tactics.

5. **2021: Ratcheting up on critical infrastructure**

   (a) *Oldsmar, Fla. Water Utility*
       In February 2021, an intruder remotely accessed a city's water treatment plant's SCADA system and tried to increase the amount of sodium hydroxide (caustic soda) above safe levels. A watchful operator intervened before the alterations went into effect. This event underscored poor authentication processes and shone a light on the importance of segmentation between IT and OT networks.

   (b) *Colonial Pipeline Ransomware*
       Colonial Pipeline, a major fuel supplier to the East Coast, shut down after the DarkSide group unleashed a ransomware attack. While the operational technology systems were left unaffected, the initiative to temporarily suspend pipeline operations was due to the fact that there was no information available to perturb lateral movement within the network. It illustrated the link between business systems and physical infrastructure and the devastating ripple effects of cyberattacks on economic stability.

   (c) *JBS meat processing disruption*
       Ransomware forced JBS, one of the largest meat processors in the world, to temporarily halt operations in North America and Australia. The incident upended food supply lines and showed that cyberthreats are not limited to utilities but also include the broader industrial world, from agriculture to logistics.

   (d) *Israeli Hospital Systems*
       The Israeli hospitals were hit by the Black Shadow hacking group and were left without access to patient records and medical systems. This example is not limited to ICS, but rather indicates the extensive crossover between cyber and physical safety threats in an operational context, particularly in cases where human lives are at stake.

6. **2022: Risks of Supply Chain and Software Integrity**

   (a) *SolarWinds SUNBURST Backdoor*
       The SolarWinds breach, which was discovered in 2020, extended into 2022 as government departments and private companies found that Russian hackers had been persistently present in their networks over many months. The

attackers implanted a backdoor, SUNBURST, into Orion software updates, making clear the vulnerability of trusted software pipelines. Although IT systems were its main target, the incident hastened the uptake of zero-trust architectures and the best practice of secure by design for ICS networks.

(b) *Kaseya Ransomware Incident*
The malicious actors had exploited a bunch of security holes in Kaseya's VSA for MSPs to deliver "ripple effect" ransomware to hundreds of MSPs' downstream customers. The attack highlighted the compound risk of third-party dependencies—a pressing issue for industrial manufacturers with single vendor-managed, integrated platforms.

7. **2023: Beyond the Factory Floor Cyber Threats**

(a) *Blizzard DDoS Attack*
The DDoS attack on Blizzard Entertainment highlights the growing threat of malicious actors flooding digital services. It also underscores the importance of incorporating resiliency measures like traffic filtering and redundant connectivity into all mission-critical systems, including those in industrial settings.

(b) BlackCat Ransomware found on Reddit
The BlackCat group's attack on Reddit, though not targeting industrial systems, demonstrates increasingly sophisticated ransomware-as-a-service (RaaS) tactics, techniques, and procedures (TTPs) relevant to ICS defenders, particularly regarding lateral movement and data manipulation.

These incidents represent an escalation in the type of threats faced by Industrial Control Systems—from ransomware that takes advantage of easy intrusion vectors to sophisticated, nation-state-supported cyber-physical attacks. The distinction between IT and OT security is further breaking down. With adversaries increasingly strategic and dangerous, protecting ICS environments requires more than traditional patching and perimeter defences. A heavy defense-in-depth techniques—network segmentation, anomaly detection, secure development, and human supervision—are mandatory. If we don't adjust, industry efficiency is no doubt going to be affected, and that's not to mention threats to public safety and national security.

1.5 Types of Industrial Control Systems

Industrial Control Systems (ICS) are fundamental technologies accelerating the automation, monitoring, and control of industrial processes. These systems are tailored for guaranteed operational performance, safety, and efficiency in manufacturing, energy, water treatment, transportation, and other industries. ICS belong into several different architectures addressing different control requirements or system complexity and scalability requirements. It is important to differentiate these different types to adequately support robust industrial ecosystems as well as to alleviate CPPS risks.

1.5.1 Programmable Logic Controllers

PLC (Programmable Logic Controller) is a ruggedized, real-time control system designed for discrete and continuous industrial automation. Although they were originally intended to substitute an OEM's relays, they now control sophisticated tasks in harsh conditions—dust, moisture, shock and vibration, and electric noise.

PLCs run user defined code written in the IEC 61131-3 languages such as Ladder Logic, Structured Text, and Function Block Diagram. Both low-level control of the machine and high-level automation of the process are accomplished in these languages. PLCs, which are interfaced and integrated via SCADA and DCS, are also providing real-time data to SCADA and DCS for monitoring, controlling, and diagnostics.

Reliability is a focus with watchdog timers, redundant I/O, and error-checking to always maintain uptime. Predictive maintenance is part of the new generation PLCs due to the fact that sensor data analytics and AI approaches are part of the data processing and a proactive failure prediction, along with performance optimizing is possible.

Communication interoperability: Connectivity is realized using protocols such as Modbus, EtherNet/IP, PROFIBUS, etc., which makes the linkage with HMIs, sensors, actuators, and IT systems very straightforward. PLCs and IoT in Industry 4.0: In the Industry 4.0 technology movement, PLCs enable edge computing and IIoT with cloud access and analytics-based automation.

The modular design of PLCs makes the system flexible in size. Upgraded diagnostics, remote configuration options, and cybersecurity capabilities, such as access control and secure communication, help ensure security against cyber threats.

In the end, PLCs are essential in industrial automation because of their deterministic, robust, and versatile properties that are of fundamental importance to secure, effective, and reliable industrial infrastructures in smart manufacturing and cyber-physical systems.

1.5.1.1 Simple Logic Example

```
PROGRAM MAIN
VAR
    A: BOOL; (* Input A *)
    B: BOOL; (* Input B *)
    Y: BOOL; (* Output Y *)
END_VAR
```

```
 8  (* Logic: Y = A AND B *)
 9  IF (A AND B) THEN
10      Y := TRUE;
11  ELSE
12      Y := FALSE;
13  END_IF;
14  END_PROGRAM
```

Note The entire pseudocode and algorithms referred throughout this book can be found at: https://github.com/Sunzidasiddique1/ICS

The code represents a simple PLC ladder logic example written in Structured Text (ST). It defines a program called `MAIN` that operates based on two boolean inputs, `A` and `B`, and produces an output `Y`.

Three variables are declared:

- A: Input boolean variable
- B: Input boolean variable
- Y: Output boolean variable

The program checks if both inputs `A` and `B` are `TRUE` (i.e., both are 1). If this condition is met, it sets the output `Y` to `TRUE`. Otherwise, `Y` is set to `FALSE`. This code simulates a basic logical AND operation, where the output `Y` is `TRUE` only when both inputs `A` and `B` are `TRUE`.

1.5.2 Distributed Control Systems

A Distributed Control System (DCS) ensures safe and stable operation in industries like oil and gas, chemical manufacturing, power generation, and water treatment. Designed for continuous, reliable, real-time plant control, a DCS distributes control intelligence across individual controllers, field devices, and operator stations, connected via a redundant, high-speed network. This decentralized structure enhances system robustness and fault tolerance, enabling scalable, fine-grained control of interconnected processes.

A typical DCS comprises a hierarchy of layers, such as field control units (FCUs), supervisory controllers, Human-Machine Interfaces (HMI), and a central engineering workstation. FCUs also consist of real-time control loops manipulating variables using sensor and actuator references [68], in which supervisory layers harmonize systems, trends, and decision-making. Such a hierarchical organization provides for local control and central control, it keeps latency low, it enhances service availability, and it makes it easier to isolate faults.

DCS provides deterministic control of the control loop for the steady states with limited process drift. In addition, their nonstop running ability is ideal for mission-critical applications. The system has failover configured in the hardware and software design, including dual power supplies, network paths, and processor failover to maintain system integrity. Interconnectivity is essential, and modern DCS can work with a number of open communication protocols (such as Modbus, FOUNDATION Fieldbus, Profibus, OPC UA) for smooth integration with legacy and third-party systems, too. It also supports embedded control applications such as PLC, SCADA, and ERP. There are two important factors at play: security and regulatory compliance. Vital national infrastructure employs DCS with multi-layered defence against both cyber and physical threats, namely network segregation, role-based access control, secure communication, and frequent audit. DCS for regulated industries also logs process data related to compliance standards like ISA-95, NERC CIP, or Good Manufacturing Practice (GMP) for auditing or traceability.

1.5.2.1 Pseudocode for Distributed Control Systems

```
# Initialize DCS components
fieldDevices = []
sensors = []
actuators = []
localControllers = []
centralControlRoom = initialize_central_control_room()
communicationNetwork = initialize_communication_network()
HMIs = []

while DCS_is_operational:

    # Real-time data acquisition from field devices and sensors
    for sensor in sensors:
        sensorData = collect_data_from_sensor(sensor)
        store_in_localController(sensorData)

    # Data processing by local controllers
    for localController in localControllers:
        processedData = process_data(localController)

        # Control actions based on processed data
        for actuator in actuators:
            controlSignal = generate_control_signal(
processedData)
            apply_control_signal(actuator, controlSignal)

    # Data communication to central control room
    for localController in localControllers:
        send_data_to_central(localController)

    # Real-time monitoring and supervision by operators
    display_real_time_data(HMIs)

    # Alarm and notification for critical conditions
    if critical_condition_detected():
        trigger_alarm()
```

```
36        notify_operators()
37
38    # Integration with other technologies
39    if integration_required():
40        integrate_with_data_analytics()
41        integrate_with_cloud_computing()
42
43    # Redundancy and fault tolerance
44    if fault_detected():
45        switch_to_backup_system()
46
47    # Wait for the next cycle
48    wait_for_next_cycle()
49
50 # Shutdown DCS components
51 shutdown(fieldDevices, sensors, actuators, localControllers)
52 shutdown(centralControlRoom, communicationNetwork, HMIs)
```

Listing 1.1 Distributed Control System

The algorithm simulates a DCS operation, beginning by initializing system processes, including field devices, sensors, actuators, local controllers, the control room, communication lines, and HMIs. During installation, the system collects live sensor data, which local controllers process to generate actionable results. Control signals drive actuators to adjust processing parameters based on this data. The processed data is then transmitted to the central control room, where HMIs display it for supervisor oversight. The system identifies critical events, triggering alarms and alerting operators to initiate responsive actions. Integration with technologies like data analytics and cloud computing enhances performance. Redundancy, fault tolerance, and monitoring maintain system health and trigger fail-over mechanisms when needed. This cycle repeats until the system is deactivated and reset.

1.5.3 Supervisory Control and Data Acquisition

SCADA (Supervisory Control And Data Acquisition) systems monitor and control geographically distributed industrial and infrastructure processes, such as electricity grids, water distribution networks, transportation, and production plants. By using sensing, communication, and control technologies, SCADA enables visible operations and maintenance, data-driven decision-making, and reliable and safe systems.

Unlike Distributed Control Systems (DCS), which excel in continuous processes, SCADA focuses on remote supervision, data acquisition, and control. SCADA systems transmit control signals, like set point changes, between a control center and remote supervised stations across long distances, making them suitable for various data exchange needs. Designed for supervisory control and centralized data monitoring, SCADA accommodates data exchange with varying latency, from high internal latency to event-triggered communication. This makes SCADA ideal for applications requiring both local autonomy and centralized coordination, including power transmission, industrial infrastructure, pipelines, and municipal infrastructure.

SCADA system design prioritizes reliability and availability through redundant communication paths, failover servers, and hot-swap hardware to ensure continuous operation during hardware failures or network outages. Advanced fault detection and diagnostics facilitate rapid anomaly identification and resolution, preventing system-wide failures.

With the increasing convergence of OT and IT systems and cloud connectivity, security is a growing concern for previously isolated SCADA systems. Robust cybersecurity measures, such as encryption, network segmentation, intrusion detection, and role-based access controls, are essential. Frameworks for security are provided by standards like NIST SP 800-82 and IEC 62443.

Modern SCADA systems integrate analytics and interoperability, connecting to cloud computing and IIoT for real-time analysis, predictive maintenance, and automated reporting. These capabilities optimize asset utilization, support operational planning, and adapt SCADA to Industry 4.0 demands.

SCADA enhances critical systems management by providing real-time visual field operation data, minimizing operator errors, and enabling swift, data-driven responses. This transforms SCADA systems from passive data aggregators into active decision support platforms, vital for industrial automation and digital transformation. As infrastructure becomes smarter and more connected, SCADA's role in managing operations through high-quality imagery leads to fewer human errors and faster decision-making in dynamic environments.

Figure 1.1 illustrates a SCADA setup, including control servers, communication routers, HMIs, and data historians. Various communication protocols and topologies are employed for efficient data exchange between control centers and field sites. Large systems often adopt sub-control servers, reducing primary control server load. This implementation ensures robust remote management, troubleshooting, and control, contributing to operational efficiency and responsiveness.

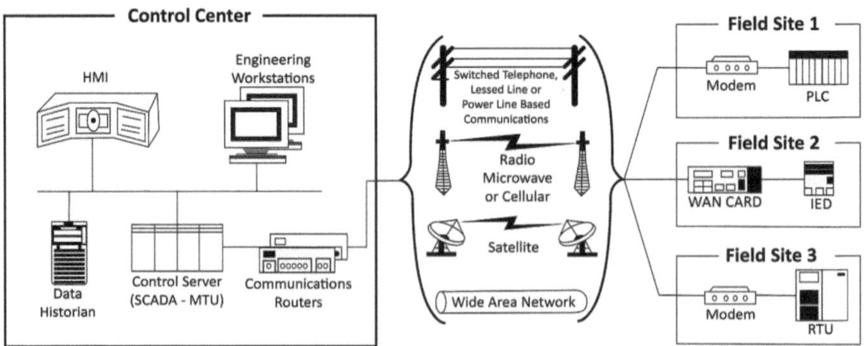

Fig. 1.1 SCADA control system (Reprinted from [137])

1.5.3.1   Pseudocode for Supervisory System

```
# BEGIN SupervisorySystem

# Initialize components
connections = initialize_connections(PLCs, RTUs, sensors)
HMI_software = initialize_HMI()
configure_communication_protocols()

while SupervisorySystem_is_operational:
    # Real-time data collection
    dataCollection = collect_data(PLCs, RTUs, sensors)
    store_in_SCADA_database(dataCollection)

    # Display and monitoring
    display_data_on_HMI(dataCollection)
    update_HMI(processSchematics, maintenanceSchedules,
    diagnosticData, logisticalInfo)

    # Control and interaction
    if operatorCommand_received():
        send_command(PLCs, RTUs, actuators)

    # Redundancy and fault tolerance
    if check_server_status() == "FAULT":
        switch_to_redundant_server()

    # Disaster recovery
    if disaster_detected():
        activate_disaster_recovery_plan()

    # Wait for the next cycle
    wait(updateInterval)

# Shutdown components
shutdown_connections()
shutdown_HMI()

# END SupervisorySystem
```

Listing 1.2 Supervisory control system

A supervisory system in an ICS operates by first establishing connections between PLCs, RTUs, and sensors, and configuring HMI software for user interaction, along with a communication protocol for reliable data transfer. Users then connect via the established protocol. Once running, the system automatically polls real-time data from field devices (PLCs, RTUs, etc.) and stores it in the SCADA database for processing and retrieval. This data is integrated into the HMI to display schematics, schedules, diagnostics, and logistics, providing a practical monitoring tool. The system can also translate HMI commands into instructions for PLCs, RTUs, and actuators to control processes. Live badminton servers operate as failovers, ensuring immediate reconnection if one fails. A disaster plan mitigates system failure risks. This process is continuous. Finally, the system is systematically powered off, safely stopping all components and connections.

1.5.4 Remote Terminal Units

RTUs link the field sensors to the central control systems in industrial automation. They are used to collect, process locally, communicate, and operate from remote sites. RTUs offer real-time values, control industrial processes at a distance, and alert in the event of a fault. Redundancy boosts system dependability. Secure RTUs seamlessly connect real-world processes to the digital world of control systems in the industrial sector. Positioned between the physical and digital layers of industrial applications, RTUs enhance performance and visibility. RTUs are usually used in applications that require devices for remote control and monitoring, including oil and gas pipelines, water treatment facilities, and electric grids. They are meant to be used in rough conditions to keep them rugged and reliable. RTUs support a number of communications protocols and can be installed on a range of industrial networks. Figure 1.2 was from [142].

Figure 1.2 shows an RTU's control system, providing an antenna holding "Radio Transmitter," "Receiver," and the "Main RTU," connected to the "Modem" shows the SPO data send and receive. A "Central Bus" also includes "Power Supply," "Central Processing," and "Volatile" and "Non-Volatile" memory architectures. Types of signals are also depicted. The system is powered by a "240V AC Source," so it is safe to assume ideal electricity. The system is field configurable for industrial or automation-specific applications with "Optional Serial Ports" (RS-232/RS-422/RS-485) and an "Optional Operator Station/Programmable Logic Controller Programming Terminal" with a "Spare RS-232 Port" for data communication and control. In general, it is a full RTU control system process.

Remote Terminal Unit Components

RTU is also referred to as an integral part of the industrial automation system, which is used to connect field devices and the central control system. RTUs are designed to operate in extreme environments and provide real-time control and monitoring of the processes. Here are some key components of an RTU and how they all work together to deliver reliable functioning, top performance, and excellent functionality.

1. **Microprocessor:** The microprocessor is the center of an RTU's composition, to execute the instructions, process the data, and make decisions based on the pre-programmed logic. Real-time data processing and timely response to other devices and control systems demand high speed and performance of the

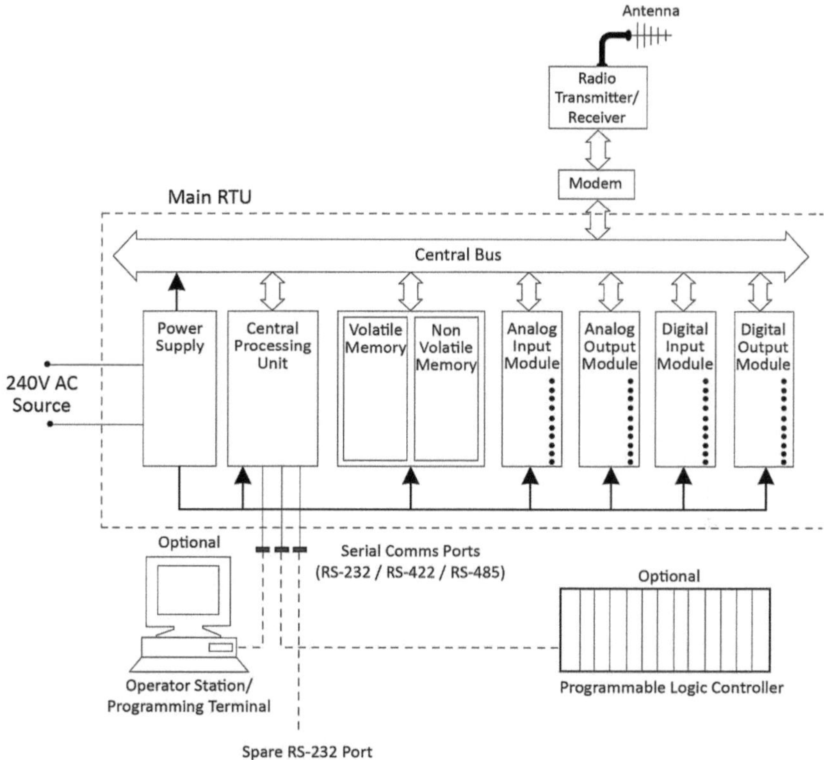

Fig. 1.2 RTUs control system

microprocessor. The RTU is equipped with a strong microprocessor, so the ability
to perform complex tasks without sacrificing performance (especially in time-
dependent applications).

2. **Input/Output (I/O) interfaces:** RTUs include different I/O interfaces to communicate with sensors and actuators. These may be analog or digital inputs
   for receiving inputs and analog or digital outputs for sending control signals to
   actuators. I/O interfaces allow an RTU to monitor and control physical processes
   such as components and machinery which operate between the system field level
   (in which fieldbus networks are used for transfer of data to the SCADA system)
   and the Human-Machine Interface (HMI) [159].
3. **Communication Modules:** The capability for communicating is a feature of
   the RTU through which data can be exchanged with the central systems,
   other RTUs, and remote monitoring centers. They can operate over different
   communication protocols, including Modbus, Distributed Network Protocol 3
   (DNP3), Profibus, and Ethernet. These modules' versatility is critical when
   bypassing various systems and subsystems in the industrial environment. In

addition, good communication is important for integrated remote control and monitoring.

4. **Memory:** RTUs also have memory in the form of Random Access Memory (RAM) and Read Only Memory (ROM). RAM is for temporary storage during processing, and ROM contains important facts: firmware, configuration data, and the RTU's programming. The memory also affects how much data an RTU can process in a local manner, so that it becomes less reliant on communicating with the central control system and works more efficiently.

5. **Local Data Processing and Logic:** A lot of RTUs can do data processing locally, so they can run control logic and perform data processing before sending it to a central location. This feature can be particularly important when real-time control protocols have to be implemented, e.g., alarms triggered as a response to sensor data. Through the reduction of communication with the central system, the local processing accelerates the directives and decisions from the RTU significantly.

6. **Power supply:** RTUs can be powered by different sources (such as AC or DC power supplies), depending on usage, also by batteries or renewable energy alternatives. The power supply is required to be durable and dependable, especially in faraway or extreme conditions where an electricity supply may not be readily available. Reliable power is crucial and is a necessity for the continuous operation of the RTU, particularly in adverse conditions.

7. **Housings and Enclosures:** RTUs are encased in robust housings, which shield them from weather while safeguarding vulnerable hardware. These housings are often tested and rated to withstand both water, dust, and temperature variation, so that they can be used in extreme conditions. Here get a lot of ratings, like IP67, which may suggest how weatherproof the device is, giving a hint about how tough it is out on the field.

8. **Redundancy Backup:** In applications where system uptime is critical, advanced RTUs are equipped with redundancy features. These may include secondary power sources or alternative communication methods to prevent system failures. Redundancy is a key requirement in industrial environments for which RTUs are designed, as any failure can lead to work stoppages, safety risks, or financial losses.

9. **Security Aspects:** Due to the control and monitoring nature of RTUs for industrial processes, it is crucial to maintain the privacy and protection. Encryptions, authentications, and firewalls can be implemented in an RTU in order to protect from unauthorized access and hacking. Since RTUs are essential to critical systems, compliance with the cybersecurity procedures is needed to protect the RTU and the sensitive data, and thus further maintain the security of the system.

1.5.4.1 Pseudocode for Remote Terminal Unit

```
# BEGIN RemoteTerminalUnit

# Initialize RTU components
communicationModule = initialize_communication()
sensors = initialize_sensors()
actuators = initialize_actuators()
localStorage = initialize_local_storage()

while RTU_is_operational:
    # Read data from sensors
    for sensor in sensors:
        sensorData = read_sensor(sensor)
        store_in_localStorage(sensorData)

    # Process sensor data and control actuators
    for sensorData in localStorage:
        processedData = process_sensor_data(sensorData)
        for actuator in actuators:
            if action_required(processedData):
                activate_actuator(actuator)
            else:
                deactivate_actuator(actuator)

    # Send sensor data to central control system
    for sensorData in localStorage:
        send_to_control_system(sensorData, communicationModule)

    # Clear local storage after data is sent
    clear_localStorage()

    # Wait for the next cycle
    wait(cycleTime)

# Shutdown RTU components
shutdown_communicationModule()
shutdown_sensors()
shutdown_actuators()
shutdown_localStorage()

# END RemoteTerminalUnit
```

Listing 1.3 Remote Terminal Unit (RTU) algorithm

Explanation of the Pseudocode

1. **Initialization:**

   - The RTU is fully powered on the communication module, sensors, actuators, and local storage.

2. **Main Operational Loop:**

- *Reading Sensor Data:* The RTU reads the data from the attached sensors and buffers it in its local storage temporarily.
- *Data Processing and Actuator Control:* Process the data sensed to determine what action is needed. Based on the processed data:

  - If action is necessary, the relevant actuator is triggered.
  - Otherwise the actuator is disabled.

- *Data Shared:* After processing, the sensor data will be transmitted to the central control system through the communication module.
- *Storage Management:* In order to avoid overflow, local storage is flush once data is transferred out.
- *Cycle Time:* After the system waits for a duration, it restarts the loop.

3. **Shutdown:**

- All parts are safe to turn off once the RTU is out of operation.

This pseudocode represents a methodically engineered prescription of controlling the RTU fundamental actions battery and power structures to operate effectively and harmoniously in an industrial control system.

1.5.5   *Programmable Automation Controllers*

Part of the popular **PAC family of products** and offering an industrial controller in a compact package available at a value-oriented price—**PAC 3000 Series** targeting system integrators and OEMs. It is especially suitable for use in machine vision, PLC and PAC-based control systems, and robotics and other industrial automation applications. These controllers offer the rich software features of a PC and the reliability of a PLC to provide a new control platform that excels in high-level control applications.

PACs are designed to drive even the most challenging applications, such as data manipulation, signal processing, communication, and complex control. PAC architectures, as defined by **Automation Research Corporation (ARC)**, support multi-domain integration, intelligent development environments, flexible engineering tools, open connectivity standards, and secure external corporate network access.

It is necessary to clarify the terms *PAC systems* and *PAC modules*. After all, PAC systems rely on purpose-specific modules dedicated to specific roles such as analog and digital I/O (input/output), relays, counters, serial communications, servo/stepper controller, and timers, but PAC modules are specialized entities in their own right.

Some of the options can include surge suppression, diagnostic indicators, real-time clocks, and watchdog timers, which further improve system reliability and operational safety.

Programmable Automation Controllers Advantages

1. **Analog Control Capabilities:** PACs are good for controlling analog values, which is something that is useful for precise control over things like temperature and pressure. This feature is important for applications with high precision and stability.

2. **Open Architecture:** PACs are open architecture, and hence are less monolithic than the traditional, and founder of the PAC concept, PLCs, as they can interface easily with different devices and networks, such as human-machine interfaces (HMIs), other PACs/PLCs, and fieldbuses. This open approach simplifies the design, integration, and sustainment of systems.

3. **Costs:**PACS typically pay for themselves within a few years due to hardware cost savings and low maintenance expenses, especially when avoiding costly radiology IT contracts. They are generally more durable and less prone to corruption or failure than standard computers.

4. PAC **Compact and Hard-wearing:** PACs are compact and tough, enabling them to perform in small areas and industrial environments. Their rugged, durable construction ensures long-term operation even under harsh conditions like temperature extremes, vibration, and mechanical shock.

5. **Increased connectivity:** PACs provide a choice for more communication protocols (Ethernet, EtherCAT, field buses, wireless). With these available connectivity options, the PACs can easily be connected to a variety of devices and systems, providing reliable, transparent communication across great distances.

6. **Scheduled Cyclic Model:** Controllers adhere to a fixed pattern to deliver time-deterministic and reliable services for control functionalities, which is critical for the stability and safety of industrial processes.

7. **Scalability:** PACs are highly scalable; one can add or remove I/O ports from PACs based on their system's needs. This flexibility is important for the control, data acquisition with sensors and actuators.

8. **Flexible:** PACs are flexible in controlling operations, such as start/stop control, move control, and process control. What's special about them and what makes them so useful is that they are very programmable.

9. **Increased Reliability:**As compared to relay-based conventional control systems, PACs are more reliable due to their advanced ICS features, which protect against electrical noise and extreme temperatures, making them suitable for harsh environments.

10. **Improved Response Time:** Considering computation time and fixed components, PAC relays offer a faster response than traditional relay systems, enabling real-time control, which is highly desirable in many applications.

11. **Higher Precision:** PACs are recognized for their precision, especially in process automation and robotics. Their integrated diagnostics also improve system performance and troubleshooting capability.

12. **Enhanced Safety:** In addition to complying with strict safety standards, PACs enhance safety through emergency shutdowns, machine guarding, and safety monitoring.

13. **Remote Access:** Modern PACs have a remote access feature, where on-site operators and maintenance staff can monitor and access or debug the system remotely, increase the operational flexibility, and minimize the system downtime.
14. **Data Logging:** PACs log performance data for historical analysis, preventative maintenance, predictive maintenance, and problem solving. This data-centric design minimizes downtime, optimizes system performance, and enhances long-term reliability.

1.5.5.1 Programmable Automation Controller Pseudocode

```
# BEGIN Programmable Automation Controller

# Initialize PAC components
communicationModule = initialize_communication()
sensors = initialize_sensors()
actuators = initialize_actuators()
controlAlgorithm = initialize_control_algorithm()
dataLoggingModule = initialize_data_logging()
userInterfaceModule = initialize_user_interface()

while PAC_is_operational:
    # Read data from sensors
    for sensor in sensors:
        sensorData = read_sensor(sensor)
        processedData = process_data(sensorData,
    controlAlgorithm)

        # Control actuators
        for actuator in actuators:
            if action_required(processedData):
                activate_actuator(actuator)
            else:
                deactivate_actuator(actuator)

    # Log data for analysis
    for dataPoint in processedData:
        log_data(dataPoint, dataLoggingModule)

    # Update user interface
    update_user_interface(userInterfaceModule, processedData)

    # Communicate with central system
    send_to_control_system(processedData, communicationModule)

    # Wait for the next cycle
    wait(cycleTime)

# Shutdown PAC components
```

```
38  shutdown_communication()
39  shutdown_sensors()
40  shutdown_actuators()
41  shutdown_control_algorithm()
42  shutdown_data_logging()
43  shutdown_user_interface()
44
45  # END Programmable Automation Controller
```

Listing 1.4 Programmable Automation Controller (PAC) algorithm

The pseudocode describes a Program for Automation Controller (PAC) in an industrial control system with initialization, real-time operation, and safe shutdown in focus. During startup, the system sets up communication modules, sensors, actuators, control and data logging algorithms, and a user interface. The main control loop reads sensor inputs, processes them with control algorithms to determine actuation, turns actuators on and off, logs data for remote analysis, updates a screen interface with real-time action status, and communicates processed data to a central control system. This cycle is repeated at a selected cycle rate to provide continuous operation. Finally, all the PAC components are appropriately deactivated in the shutdown, preventing in all a safe termination of the operations [10].

1.5.6   Intelligent Electronic Devices

Intelligent electronic devices (IEDs) are advanced microprocessor based electronic devices. Such devices (regulators, protection relays, circuit breakers, controllers, and fault recorders) play a fundamental role in the recently introduced automation of power grids and interact with other devices that belong to the network. It is a significant revolution in the operation of power systems, it can realize a complex of monitoring, protection, m and communication functions necessary for the operation of power systems efficiently.

Properties and features of modern IEDs Of relevance, with respect to modern IEDs is the raised computing power that is built in, as well as the communication ability that is provided. They are certified for IEC 61850, Distributed Network Protocol 3 (DNP3), and Modbus, guaranteeing them to fit into any standardized substation automation system. These features allow IEDs to implement the complex applications of signal recording and event recording, waveform recording, and power quality monitoring. These are the innovations that will allow utilities to observe and track deviations, perform post-event analyses, and implement predictive maintenance for the grid to make it more reliable and more efficient.

IEDs are a key component in smart grid systems, providing distributed intelligence to the closed-loop system. This reduces the need for human intervention and enables automated decision-making through substation automation processes. IEDs can perform logic-based control, trigger alarms, and transmit real-time data to Supervisory Control and Data Acquisition (SCADA) systems, allowing the grid

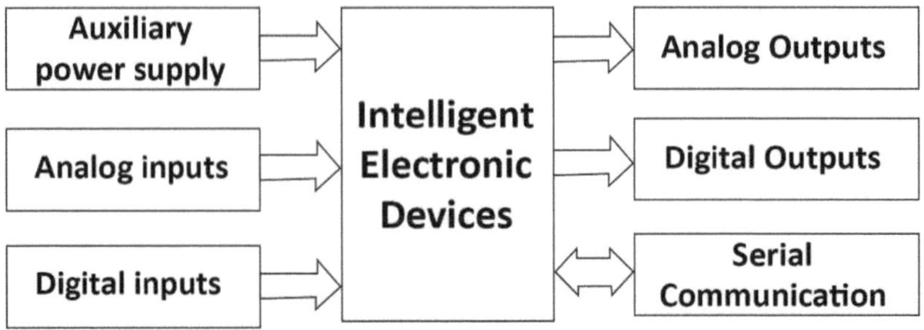

Fig. 1.3 Block diagram of IED (Reprinted from [55])

to operate smarter, more reliably, and with greater flexibility and responsiveness. Decentralized active control also enables more flexible and robust power management.

They are also an important part of the future grid, such as enhanced energy management systems (EMS) and distribution management systems (DMS), where features like adaptive load control and FLISR can be provided. These features are fundamental for the stability and security of the grid, particularly under dynamic load changes and disturbances. In a general sense, IEDs constitute the key devices in the transition toward full automation and digital interconnection of power systems, where the physical, electrical, and digital advanced control worlds are interconnected.

Figure 1.3 depicts the typical block diagram of an IED [55]. The block diagram includes auxiliary power supply, analog inputs, digital inputs, analog outputs, and digital outputs.

The auxiliary power supply delivers DC and AC standby power for IEDs. Analog inputs are for sensor data, ; digital inputs receive commands or status sense on some lines. The digital outputs may have several forms, and the analog outputs are converted by software transducers to signals. IEDs provide the ability to communicate real-time data as well as the ability to have multiple ports that can be connected by electrically operated relays, control switches, or the like, and have wiring to provide the ability to communicate data in near real-time. Adaptability (IDEs are modular and good communicators). There are other benefits that they have:

1. **Monitoring:**
   IEDs measurement and event sorting at the substation online to information models synchronizing devices to prevent a jerk transfer of power (set delta vector point). IEDs coordinate the playback of PTP-detected events across devices to reproduce ultra-accurate sub-second events. They will be watch voltage, current, and temperature in the circuit in no time! Analysis of data using sophisticated algorithms highlights those anomalies and predicts failures in advance to hinder operations.

2. **Data Access:** Local remote IEDs terminologies used to provide information about power system networks. They store pattern histories and ICS as real-time representation data vaults. Data may be locally analyzed for performance problems and for maximizing maintenance plans. IEDs can be linked to SCADA control systems, resulting in a single control features presentation of key grid parameter ICS, alarm, and event details throughout the grid. Accepted, the power system may have a health dashboard with a remote real-time data pull.

3. **Testing and Analysis:** IEDs can perform various of test, report, fault diagnose, and programmable logic control. They provide a means for examining the behavior and performance of the system in a more detailed manner.

```python
import random
import time

class IED:
    def __init__(self):
        self.sensors = ["Sensor1", "Sensor2"]
        self.communication_module = "Module1"
        self.intelligent_algorithms = lambda x: x * 1.4

    def collect_data(self, sensor):
        return random.uniform(0, 100)

    def process_data(self, data):
        return self.intelligent_algorithms(data)

    def send_data(self, data):
        print(f"Intelligent data sent to central system: {data}")

    def run(self):
        while True:
            data = [self.collect_data(sensor) for sensor in self.sensors]
            intelligent_data = [self.process_data(d) for d in data]
            for idata in intelligent_data:
                self.send_data(idata)
            time.sleep(1)

# Instantiate and run the IED system
ied = IED()
ied.run()
```

Listing 1.5 IED example code

The changes made to the pseudocode include updating the In and Out sections to reflect actual examples: the input is a list of sensors (e.g., Sensor1, Sensor2), and the output is the intelligent data sent to the central system. These updates better align with the operational flow of the system, where sensors are collected, processed, and the intelligent data is sent to the system. The gets operator is used for assignment to make the pseudocode clearer and more readable. This change improves understanding for readers by making the data transformation process more explicit. It also helps ensure consistency between the system model and its algorithmic representation.

As an example of the output, when the code runs, it would display:

```
Intelligent data sent to central system:  43.56
Intelligent data sent to central system:  78.34
```

1.6 Digital Outputs

Digital outputs are essential elements in today's industrial control systems that allow devices to report status and drive control of actions. These outputs can be switched by three types of contacts: solid state, normally closed, and normally open. Where digital input is sensing or receiving information, digital output is action or transmission of information in response to specific conditions, like predefined logic or system states. In order to guarantee reliable system functionality, it is important to know the load and function properties of these outputs.

IEDs make use of different kinds of digital and analog outputs to perform various monitoring and control, and protection tasks. One of the important advantages of IEDs is that to combine a variety of functions into one device, which provides more flexibility and simplicity in system design. For example, IEDs often include digital outputs that are combined with relays, control switches, or other interfaces, and, as a result, the system architecture is simplified while responses are faster with enhancements in system performance and reliability.

(a) **Auxiliary Power Supply:** IEDs have rugged auxiliary power supplies, which maintain reliable operation despite power interruptions. They operate in a wide DC voltage range from 15 to 150 V for any nom. voltage or switch to AC power (110–140 V), providing continuous operation in critical applications. This capability provides power systems with much greater stability and dependability, particularly in regions of the world where the grid is unreliable or blackouts are common.

(b) **Analog inputs:** Conventional relays use current and potential transformers for analog input only; nevertheless, with IEDs, this function is also extended by connecting to a variety of sensors. This direct interface permits online acquisition from different types of sensors, like pressure or temperature sensors. Allowing the request of such data enables the IEDs to make complex system-wide data-driven decisions.

(c) **Digital Input** IEDs have enhanced flexibility using digital input that allows communication of a wider variety of status data or command data. These inputs are suitable for potential-free contacts, so any metallic grounding is not a problem, as is the case with other older designs. This additional flexibility allows it to be easily integrated into different control systems and communication protocols, which in turn can improve the coordination of larger and more complex networks.

(d) **Analog Outputs:** While receiving data through their analog outputs, IEDs can also effusively send data. Based on transducers, IEDs produce an output that can be active or passive contingent upon the implementation. Passive outputs need external power to go offline, but active outputs can give control signals by themselves. This flexibility is what makes IEDs capable of communicating to actuators, displays, and controllers, so that the message can be sent out in the best way for the application requirements.

1.7 Functionality and Control

1. **Process Control:** Process control is a fundamental component of Industrial Control Systems (ICS), ensuring stable operations across various sectors, including manufacturing, chemical plants, and power generation. The primary objective of process control is to maintain key variables—such as temperature, pressure, flow rates, and chemical composition—as close as possible to their reference setpoints. This stability is crucial for maintaining productivity, ensuring safety, and delivering high-quality output.

   Process control systems continuously monitor and adjust these variables through feedback loops, typically governed by control algorithms. Sensors and actuators work in conjunction to apply incremental corrections, keeping parameters within tight operational limits. The efficiency of a control process is often measured by its disturbance rejection capability, energy efficiency, and responsiveness to abrupt changes in plant conditions.

   Advanced Process Control (APC) systems, particularly those based on predictive models, can anticipate process upsets well in advance and apply corrective actions to avoid disruptions. These systems are designed for high availability and reliability, meeting the stringent requirements of critical industrial environments. Effective process control minimizes downtime, optimizes operational transitions, and enhances safety and efficiency. Additionally, enforcing robust cybersecurity measures within process control systems is increasingly vital to prevent breaches that could compromise both operational performance and plant safety.

   In practice, process control often addresses multiple objectives simultaneously, such as minimizing costs, reducing energy consumption, and maximizing performance. Furthermore, systems must be scalable to accommodate different production levels and adaptable to integrate additional control layers, including Supervisory Control and Data Acquisition (SCADA) systems. Consequently, an effective process control strategy must balance short-term operational goals with long-term sustainability.

1.7.1 Code Example of Process Control

Below is an example of a simple process control implementation that simulates
the control of a system's temperature. The system continuously monitors the
temperature and makes adjustments by controlling a heater or cooler, ensuring that
the temperature stays within a safe operating range.

```python
class ProcessControl:
    def __init__(self, desired_temperature):
        self.desired_temperature = desired_temperature

    def adjust_heater(self, current_temperature):
        if current_temperature < self.desired_temperature:
            return "Heater ON"
        else:
            return "Heater OFF"


# Example usage
pc = ProcessControl(desired_temperature=75)
current_temp = 70
action = pc.adjust_heater(current_temp)
print(f"Current Temp: {current_temp}, Action: {action}")
```

Listing 1.6 Process Control Class

1.7.2 Automated Control

Automated control systems are a crucial part of ICS (Industrial Control System)
to increase process efficiency and diminish the dependence on human control.
The above systems facilitate the autonomous migration of common tasks usually
executed by human workers. With automation, precise and consistent in-process
production control is achieved to optimize performance and repeatability.

The automated control systems make adjustments in real-time based on varying
inputs, and so the processes are "running" at optimal settings all of the time. So
not only are errors in human operation reduced, but operation is more reliable and
response times are faster. Automated systems can be programmed to detect the
anomalies and then adapt autonomously or trigger alerts as needed.

In addition, one of the greatest benefits of automating equipment is performing
predictive maintenance. Leveraging powerful analytics and machine learning, these
systems can anticipate failures before they happen, enabling predictive maintenance
and minimizing costs related to unplanned downtime and maintenance. Predictive
analytics supports the system to be reliable and sustainable in the long term, and to
lower the operating costs.

One of the main obstacles in automation systems are security; the system is
susceptible to cyber-attacks that can compromise or damage the system. Therefore,

security enhancement technologies, such as encryption and access control, must be implemented in ICS to prevent unauthorized access and further enhance their safety and security of the industrial processes.

Automatic control systems are crucial in manufacturing, energy production, and chemical synthesis. They enhance productivity, system reliability, and reduce operational risks. In the future, automation technology in bio-based processes will incorporate intelligent algorithms like machine learning for real-time process optimization and closer integration with digital systems such as ERP and cloud-based analytics.

1.7.3 Code Example of Automated Control

Automatic control is a component of contemporary Industrial Control Systems (ICS), which can use it to perform day-to-day tasks without human intervention. These systems automatically take care of the everyday actions, so that optimal operations can be avoided, the human is also reduced to a minimum error ratio, and equal values as much as possible on all systems are produced. Automatic control is especially valuable in systems involving high accuracy, high reliability, and stringent control requirements, such as industrial and energy-related (e.g., chemical process) applications. Here's some Python code:

```python
class AutomatedControl:
    def __init__(self):
        self.task_complete = False

    def perform_task(self):
        # Simulate task automation
        self.task_complete = True
        return "Task Completed"
```

Listing 1.7 Automated Control Class

Example Usage:

```python
automated_control = AutomatedControl()
task_status = automated_control.perform_task()
print(f"Automated Control Status: {task_status}")
```

Input:

- **automated_control.perform_task()** is called to simulate the task execution.

Output:

```
Automated Control Status: Task Completed
```

This example demonstrates a typical use case of the `AutomatedControl` class. The `perform_task` function encapsulates the task execution logic and returns a status indicating successful completion. By abstracting automation tasks within such classes, Industrial Control Systems (ICSs) can perform routine operations previously handled by human operators, thereby enhancing efficiency, consistency, and reliability.

The implication of automation on ICS is striking. ICS automates repetitive, time-consuming tasks and become more efficient while reducing labor cost and downtime on systems. Desirable as they are as well, it is also important to look at the problems, such as those regarding system reliability, error reporting, validation, and the like, with the same skepticism, as only thus automatic processes do what one expects of them and do not pose additional risks.

1.7.4 Regulation and Setpoint Control

ICS regulates system parameters to desired setpoints, maintaining stable operation and minimizing deviations. By processing the real-time data with such means, an efficient means of continually observing the environment (and checking limits with the environment) can be made. Upon a disturbance occurring, the control system acts to correct for the disturbance—e.g., by employing automatic control techniques—to attempt to bring the system back to a near-optimal condition for the intended environment. Process control, such as this, makes the process stable, and the potential for machines to break down or to operate at less than design capacity is reduced. Specifically, in other instances, setpoint control is also a critical enabler for robustness and safety in many industrial process applications.

1.7.5 Code Example of Regulation and Setpoint Control

Regulation and setpoint control are of prime importance in many ICS to ensure that values are kept within desired ranges. Being able to adapt the system operation to dynamic environmental changes is therefore both crucial to ensure stability and to improve performance. One of the simplest systems to maintain is setpoint control, where a desired process value, called the setpoint, is specified and the actions of the system are decided upon according to whether the actual process variable differs from this value.

The following snippet is for a simple setter control loop implemented in a simple class, which compares the real value of a process to the setpoint. The system gives corrective measures as a result of this comparison.

```python
class SetpointControl:
    def __init__(self, setpoint):
        self.setpoint = setpoint

    def regulate(self, actual_value):
        if actual_value < self.setpoint:
            return "Increase"
```

```
 8            elif actual_value > self.setpoint:
 9                return "Decrease"
10            else:
11                return "Maintain"
```

Listing 1.8 Setpoint Control Class

Example Usage:

```
1  setpoint_control = SetpointControl(setpoint=100)
2  actual_value = 95
3  action = setpoint_control.regulate(actual_value)
4  print(f"Regulation Action: {action}")
```

Input:

- Setpoint: 100
- Actual value: 95

Output:

```
Regulation Action: Increase
```

Explanation:
The `SetpointControl` class defines a simple regulation mechanism based on the comparison between the actual process value and a target setpoint. The `regulate` method compares the actual value to the setpoint and returns one of three possible actions:

- **Increase**: When the actual value is less than the setpoint.
- **Decrease**: When the actual value exceeds the setpoint.
- **Maintain**: When the actual value is equal to the setpoint.

This simple control mechanism is foundational in many industrial applications, where precise adjustments are necessary to maintain system performance within predefined operational boundaries. By using such regulation mechanisms, ICS can minimize deviations, thereby ensuring process stability.

1.7.6   Feedback Control

Feedback control continuously measures process variables and adjusts them to match the desired setpoint, enhancing stability and responsiveness. Feedback control systems, commonly applied in temperature regulation, speed control, and other industrial automation tasks, enhance stability, accuracy, and responsiveness. Utilizing sensors to monitor variables and controllers such as PID (Proportional-Integral-Derivative), these systems can actively correct errors in real time, ensuring optimal operation even under disturbances.

The integration of setpoint and feedback control highlights the role of regulation within Industrial Control Systems (ICS), where maintaining processes within predefined limits is essential for both performance and safety.

1.7.7 Code Example: Feedback Control

Feedback control is a fundamental principle of control systems, designed to maintain system performance at its setpoint by dynamically managing outputs based on real-time error detection. This concept is widely implemented in applications ranging from industrial automation to robotics, where it stabilizes operations, enhances performance, and minimizes deviations from the desired state. At its core, feedback control enables the controller to continuously correct errors, thereby optimizing system behavior and improving accuracy. Below is a Python implementation of a simple proportional feedback control algorithm. In this case, the control output is determined by multiplying the error by a proportional gain factor (k_p).

```python
class FeedbackController:
    def __init__(self, kp):
        self.kp = kp

    def compute(self, setpoint, measured_value):
        error = setpoint - measured_value
        return self.kp * error
```

Listing 1.9 Feedback Controller Class

Example Usage:

```python
feedback = FeedbackController(kp=2.0)
setpoint = 100
measured_value = 90
control_output = feedback.compute(setpoint, measured_value)
print(f"Feedback Control Output: {control_output}")
```

Input:
Setpoint: 100, Measured value: 90, Proportional gain (kp): 2.0

Output:
Feedback Control Output: 20.0

This Python code illustrates the key components of a simple feedback control mechanism:

- The `FeedbackController` class is initialized with a proportional gain factor (kp).
- The `compute` method calculates the control output using the formula:

$$\text{Control Output} = \text{kp} \times (\texttt{setpoint} - \texttt{measured\_value})$$

- The error represents the difference between the setpoint and the measured value.
- In this example:
  - `setpoint` is `100`.
  - `measured_value` is `90`.
  - The proportional gain (kp) is `2.0`.
- The control output is calculated as `20.0`, which is printed as the result.

```python
class FeedbackController:
    def __init__(self, kp):
        self.kp = kp  # Proportional gain

    def compute(self, setpoint, measured_value):
        error = setpoint - measured_value
        control_output = self.kp * error
        return control_output

# Example usage
controller = FeedbackController(kp=2.0)
setpoint = 100
measured_value = 90
control_output = controller.compute(setpoint, measured_value)

print(f"Control Output: {control_output}")
```

Listing 1.10 Simple feedback control algorithm

This example demonstrates how feedback control adjusts the system's outputs based on the proportional relationship between the error and control action, thereby improving accuracy and responsiveness.

1.7.8 Feedforward Control

Feedforward control predicts external disturbances and compensates proactively, often combined with feedback control for better system resilience. Feedforward control serves as a counterbalance to feedback control since both are sometimes used within a single system: ... instead of reacting to a disturbance during a system's operation, feedforward control uses a model of the disturbance to compensate the output automatically. This technique proves to be very effective in chemical processing, power generation, and robotic control activities, where avoiding severe disruptions and delays is paramount toward preserving overall system stability and performance. On the other hand, feedforward control relies on precise disturbance modeling and is unable to deal with unknown errors, which is not very efficient in high uncertainty scenarios. As such, feedforward is supplemented with feedback control in order to deliver a more resilient and flexible system solution.

1.7.9   Code Example: Feedforward Control

Below is an example of a simple feedforward control algorithm in Python. This algorithm calculates the control output based on the input setpoint, multiplied by a gain factor.

```python
class FeedforwardController:
    def __init__(self, gain):
        self.gain = gain

    def compute(self, setpoint):
        return self.gain * setpoint
```

Listing 1.11   Feedforward controller class

Example Usage:

```python
feedforward = FeedforwardController(gain=1.5)
setpoint = 100
control_output = feedforward.compute(setpoint)
print(f"Feedforward Control Output: {control_output}")
```

Input:
Setpoint: `100`, Gain factor (`gain`): `1.5`
Output:

```
Feedforward Control Output: 150.0
```

This Python code demonstrates a feedforward control algorithm:

- The `FeedforwardController` class initializes with a gain factor (`gain`).
- The `compute` method calculates the control output using the formula:

$$\text{Control Output} = \texttt{gain} \times \texttt{setpoint}$$

- In this example:
- `setpoint` is `100`.
- The gain factor (`gain`) is `1.5`.
- The calculated control output is `150.0`, which is printed as the result.

1.7.10   Sequential Control

Sequential control ensures processes execute in the correct order, critical for batch processing, assembly lines, and synchronized operations. The primary purpose of sequential control is to ensure that each process control function is executed in the correct order, as required in systems such as manufacturing lines, batch processes, and automated assembly systems.

By dividing tasks into individual steps, sequential control ensures that each phase is completed before proceeding to the next. This structure is essential for systems that require precise timing and synchronization across different components. For example, in batch operations, sequential control guarantees that every chemical reaction or material blend occurs in the proper order. To prevent errors or mishaps, analog transmitters must also be calibrated accordingly.

One of the most valuable features of the sequential control model is its ability to automate task transitions, eliminating reliance on manual intervention and reducing the potential for human error. This leads to improved operational efficiency and more effective resource utilization. However, implementing an effective sequential control system can be challenging, as it requires thoughtful design to achieve accuracy while remaining adaptable to exceptions or new process conditions.

When unforeseen changes occur—such as equipment breakdown—the system must accommodate these variations. For instance, a chamber can be excluded from the process to maintain continuity. Ultimately, stabilizing the sequence under variable conditions requires a robust control system capable of handling interferences while preserving overall stability and consistency. Sequential control is a key component of modern automation systems, offering benefits such as consistency, higher productivity, and efficient resource utilization. However, its design requires careful consideration to address potential challenges and ensure adaptability across various scenarios. For example, in mass production environments, Sequential Control Systems (SCS) can be integrated with predictive maintenance algorithms to anticipate and prevent malfunctions before they disrupt the production process.

1.7.11 Code Example: Sequential Control

```python
class SequentialControl:
    def __init__(self):
        self.phase = 1

    def next_phase(self):
        self.phase += 1
        return f"Phase {self.phase} started"
```

Listing 1.12 Sequential control class

Example Usage:

```python
sequential_control = SequentialControl()
current_phase = sequential_control.next_phase()
print(f"Sequential Control: {current_phase}")
```

Input:None
Output:

```
Sequential Control: Phase 2 started
```

1.7.12   Safety and Emergency Shutdown Systems

Safety systems, such as Safety Instrumented Systems (SIS), protect personnel and equipment by initiating shutdowns or alarms during hazardous conditions. These systems are designed to avoid accidents by rapidly stopping critical operations when an unsafe situation is detected. At the core of this functionality is the Safety Instrumented System (SIS), which operates independently to initiate emergency shutdowns or other protective actions as necessary.

SISs assist in detecting and mitigating abnormal situations such as pressure spikes, temperature excursions, or equipment failures. When the system identifies a potential unsafe deviation from normal conditions, it initiates corrective actions such as shutting down machinery, isolating process sections, or triggering alarms. This quick response significantly reduces the risk of accidents, human injury, and equipment damage.

In high-risk sectors such as oil and gas, chemical processing, and power generation, safety and emergency control systems are critical. These systems must be tested and validated regularly to ensure proper operation during emergencies. Routine tests, including functional testing and fault-injection testing, help maintain system effectiveness and reliability so that it can perform when needed most.

Integrating safety systems into ICS not only safeguards system integrity in compliance with regulatory frameworks but also improves overall reliability by reducing human error and incorporating fail-safes. Although safety and emergency controls enhance operational security, their design and maintenance require a deep understanding of potential hazards and the ability to anticipate failure scenarios. This makes them indispensable in high-risk industrial environments.

1.7.13   Human-Machine Interface (HMI)

The Human-Machine Interface (HMI) is a key component of Industrial Control Systems (ICS), serving as the bridge between operators and complex industrial processes. HMIs allow operators to monitor system status, view sensor values, control processes, and visualize alarms in real time. These interfaces simplify control and monitoring, enabling faster execution and reducing human error. HMIs provide operators with clear insights into system behavior, allowing quicker decision-making, especially when processes deviate from normal conditions. Key HMI functions include monitoring sensor readings, operating equipment, and managing alarms to ensure continuous and safe operations.

Below is an example in Python using the `tkinter` library. This example demonstrates how an HMI can be created to refresh sensor data and control processes through an easy-to-use graphical.

```python
import tkinter as tk
from tkinter import messagebox
```

```python
import random

class HMIApp(tk.Tk):
    def __init__(self):
        super().__init__()
        self.title("HMI for Industrial Control System")
        self.sensor_label = tk.Label(self, text="Sensor Data:",
     font=("Arial", 14))
        self.sensor_label.pack(pady=10)
        self.sensor_display = tk.Label(self, text="0.0", font=(
    "Arial", 24))
        self.sensor_display.pack(pady=10)
        self.start_button = tk.Button(self, text="Start Process
    ", font=("Arial", 14),
        command=self.start_process)
        self.start_button.pack(pady=10)
        self.stop_button = tk.Button(self, text="Stop Process",
     font=("Arial", 14),
        command=self.stop_process)
        self.stop_button.pack(pady=10)
        self.update_sensor_data()

    def start_process(self):
        messagebox.showinfo("Process Control", "Process Started
    ")
        print("Process started.")

    def stop_process(self):
        messagebox.showinfo("Process Control", "Process Stopped
    ")
        print("Process stopped.")

    def update_sensor_data(self):
        sensor_value = random.uniform(20.0, 30.0)
        self.sensor_display.config(text=f"{sensor_value:.2f}")
        self.after(1000, self.update_sensor_data)

if __name__ == "__main__":
    app = HMIApp()
    app.mainloop()
```

Listing 1.13 HMIApp class for industrial control

Input: None
Output:
Sensor data is refreshed every second in the graphical interface, and a message box is displayed when the process starts or stops.

The following code provides a minimalistic example of an HMI built using the **tkinter** library. The application consists of the following components:

- The class **HMIApp** generates a graphical window that displays sensor data and provides buttons to control the process.

- The `start_process` and `stop_process` methods display message boxes to inform the operator about the process state.
- Sensor data is updated every second to provide real-time feedback to the operator.

These interfaces enable operators to manage and control industrial operations efficiently, reducing downtime and simplifying complex processes. The clear, high-resolution display allows real-time data monitoring and full process control through an intuitive interface, ensuring safer and more efficient operations. Additionally, the HMI can be integrated with other ICS components, such as trend analysis and predictive maintenance systems.

1.7.14 Remote Monitoring and Control

Remote monitoring allows centralized management of ICS processes, improving flexibility, response times, and operational efficiency. This capability offers significant operational benefits, including increased flexibility, cost reduction, and faster response times. Remote access allows operators to manage multiple sites from any location, enabling quick problem resolution and real-time adjustments. Furthermore, remote monitoring improves troubleshooting by providing system diagnostics without requiring physical presence. To ensure security and data integrity, all communications are encrypted to protect against network attacks.

1.7.15 Code Example: Remote Monitoring and Control

Remote monitoring and control have become indispensable functionalities in Industrial Control Systems (ICS), empowering operators to monitor and manage processes from centralized or off-site locations. This capability significantly enhances operational flexibility, enabling real-time decision-making and improving system response times, even in geographically dispersed environments. The ability to monitor system health and issue commands remotely reduces the reliance on on-site personnel, leading to both cost savings and operational efficiency. Furthermore, it allows for rapid adjustments in response to changing conditions, making ICS more adaptable to dynamic industrial environments.

The following Python code example illustrates how remote monitoring and control can be implemented within an ICS context:

```python
class RemoteControl:
    def __init__(self):
        self.system_status = "running"

    def monitor_system(self):
```

```python
        return f"System Status: {self.system_status}"

    def control_system(self, command):
        if command == "shutdown":
            self.system_status = "shut down"
        return f"System {self.system_status}"

remote_control = RemoteControl()
status = remote_control.monitor_system()
print(f"Remote Monitoring: {status}")
control_status = remote_control.control_system("shutdown")
print(f"Remote Control: {control_status}")
```

Listing 1.14 Remote control class

Input:
Command: "shutdown"
Output:

```
Remote Monitoring: System Status: running
Remote Control: System shut down
```

In this example, the RemoteControl class provides two key methods:

- monitor_system: This method monitors the system's current operational status and returns the system's state.
- control_system: This method allows the operator to issue commands (e.g., "shutdown") that modify the system's state, in this case transitioning the system to "shut down."

First, we check the status of the system, which returns as "running." Next, the operator issues a "shutdown" command, causing the system status to change from "shutting down" to "shut down."

Automatic monitoring and control play a crucial role in ensuring uninterrupted industrial operations, even when the operator is at a remote location. This feature is commonly embedded in Supervisory Control and Data Acquisition (SCADA) systems, enabling the system to be expanded with advanced monitoring, fault detection, and troubleshooting capabilities. These capabilities help reduce system downtime, increase reliability, and mitigate potential negative impacts on system operations due to the absence of on-site personnel.

Furthermore, with increasing complexity, data logging and advanced analytics can be utilized to gain deeper insights into system performance, enabling predictive maintenance and real-time decision-making.

1.7.16 Data Logging and Analysis

ICS collects, stores, and analyzes data for trend identification, anomaly detection, predictive maintenance, and compliance auditing. Advanced analytics—powered by machine learning and applied statistics—enable businesses to predict maintenance requirements, identify waste or inefficiencies, and enhance decision-making. This helps organizations increase system uptime, reduce maintenance costs, and ensure compliance with industry standards. Additionally, historical data logging supports audits and quality control by providing a comprehensive overview of system performance over time.

1.7.17 Code Example: Data Logging and Analysis

ICS continuously collects and stores information for monitoring, historical reference, troubleshooting, and optimization. This data includes sensor readings, actuator values, system events, alarms, and operational aggregates, which are recorded periodically. By maintaining structured logs, ICS facilitates trend analysis, anomaly detection, and predictive maintenance. Historical data, including troubleshooting and compliance-related information, is also critical for auditing purposes. In modern ICS architectures, this data is often processed via edge computing or sent to cloud-based platforms for further analysis.

The following code snippet demonstrates how to log sensor readings to cloud storage and later analyze the data to extract performance trends and detect potential failures:

```python
class DataLogger:
    def __init__(self):
        self.log = []

    def log_data(self, data):
        self.log.append(data)
        return "Data logged"

    def analyze_data(self):
        return f"Analyzing {len(self.log)} data points"

data_logger = DataLogger()
log_status = data_logger.log_data({"temperature": 70})
print(f"Data Logging: {log_status}")
analysis_result = data_logger.analyze_data()
print(f"Data Analysis: {analysis_result}")
```

Listing 1.15 DataLogger class

Input:
Data: "temperature": 70
Output:

```
Data Logging: Data logged
Data Analysis: Analyzing 1 data points
```

This code simulates data logging and analysis:

- The `DataLogger` class logs data points into a list.
- The `log_data` method adds data to the log.
- The `analyze_data` method analyzes the logged data.
- In this example, one data point is logged, and the analysis returns the count of data points.

1.7.18 Business Integration

ICS integrates with ERP systems to synchronize production and business processes, supporting real-time decision-making and predictive maintenance. This interface makes it possible to transfer data directly and in real time between the operational sphere and those responsible for making decisions. For example, ICS refreshes ERP with real-time inventory and production data, and ERP reschedules or order restocks based on demand projections. That way, departments like production, sales, and logistics can be better synchronized, which ultimately leads to more efficient processes. Plug-and-Play integration gives managers a real-time view of results so they can make informed decisions in advance. This also allows for advanced capabilities, such as predictive maintenance, where ICS data predicts when equipment will fail and decreases equipment downtime. By streamlining resource allocation, reducing waste, and keeping pace with market demands, the automation and productivity made possible by ICS-ERP integration is maximized. This integration powers the capability to work lean, flexible, with the goals of the organization in mind, and more effectively (making companies more competitive).

1.7.19 Code Example: Business Integration

Connecting Industrial Control System (ICS) equipment to Enterprise Resource Planning (ERP) systems is crucial for tying manufacturing operations to wider business goals. With the integration to the ERP system, it offers the possibility to have the production-process data being directly transferred to the ERP application and thus an automated integration between the shopfloor and the backoffice system from where one can optimize workflows and performance indicators and take real-time decisions. This convergence can narrow the gap between OT (operational technology) and IT (information technology), improving coordination and resource management, and creating overall business efficiencies.

The code snippet below illustrates the integration of production data with SAP (or any other ERP) to ensure the business and production-related efforts are aligned within the enterprise.

```python
class BusinessIntegration:
    def __init__(self, erp_system):
        self.erp_system = erp_system

    def sync_with_erp(self, production_data):
        return f"Synced with {self.erp_system}: {
    production_data}"

# Example usage
business_integration = BusinessIntegration(erp_system="SAP")
sync_status = business_integration.sync_with_erp({"
    units_produced": 500})
print(f"Business Integration: {sync_status}")

Output:
Business Integration: Synced with SAP: {'units_produced': 500}
```

Listing 1.16 BusinessIntegration class

Input:

Production Data: `"units_produced": 500`

Output:

```
Business Integration: Synced with SAP: {'units_produced': 500}
```

This code demonstrates the integration of ICS with an ERP system:

- The `BusinessIntegration` class initializes with an ERP system (e.g., SAP).
- The `sync_with_erp` method sends production data (such as the number of units produced) to the ERP system for synchronization.
- In this example, the system syncs the production data, reflecting the units produced in the SAP ERP system.

This integration supports the ability for production metrics to be visible and provide real-time data for more informed decision-making between all areas within the enterprise. Automation of synchronization between ICS and ERP systems contributes to elimination of manual data entry errors, an elevated level of operational efficiency, and consistent data in various departments.

1.7.20 Cybersecurity and Access Control

Cybersecurity ensures ICS integrity, preventing unauthorized access, breaches, and attacks using authentication, encryption, and intrusion detection. Because of the growing connections between ICS and external networks, it's critical to include strong cyber protection to guard against possible attacks and to preserve the integrity of the system.

Access controls, including MFA and role-based access control (RBAC), limit access to key system components so only approved individuals can access sensitive data and operations. Data-in-transit and data-at-rest are secured using encryption protocols in order to minimize the risk of data interception or tampering. In addition, Push-Intrusion Detection and Prevention Systems (PIDS/PIPS) are constantly scanning network traffic, trying to identify and remove invasive threat actions before they are able to execute. If these layered defenses are utilized properly, this can mitigate and manage cyber threats while providing the robustness and safety of ICS. As the attack surface expands and sophisticated threats to industrial systems continue to emerge, protecting these systems from vulnerabilities and exploits is essential.

1.7.21 Example Code: Cybersecurity and Access Control

SCO contains features to secure against unauthorized access, data loss, and cyber attacks. Due to the fact that these systems tend to manage important parts of critical infrastructures, the security of such system is highly important in order to guarantee proper operations and safety. Common cybersecurity practices are user authentication, role-based access control (RBAC), encryption, intrusion detection system (IDS), and network segmentation. Access control access of individuals to a specific location or site based on security requirements. Additionally, logging user activity and creating audit trails aid in post-incident analysis and compliance with common standards. Sample code provided below demonstrates a simple access control logic, focusing on user authentication, as well as permission evaluation to protect the system.

```python
class Cybersecurity:
    def __init__(self):
        self.access_granted = False

    def verify_credentials(self, user, password):
        if user == "admin" and password == "password123":
            self.access_granted = True
            return "Access Granted"
        else:
            return "Access Denied"

cybersecurity = Cybersecurity()
security_check = cybersecurity.verify_credentials(user="admin",
    password="password123")
print(f"Cybersecurity: {security_check}")
```

Listing 1.17 Cybersecurity class

   Input:
User: "admin", Password: "password123"
   Output:

```
Cybersecurity: Access Granted
```

This code demonstrates a basic cybersecurity and access control system:

- The `Cybersecurity` class verifies user credentials.
- The `verify_credentials` method grants access if the correct username and password are provided.
- In this example, the correct credentials are used, so access is granted.

1.8   Developments in Control Systems

The field of control systems has seen remarkable progress over the years. The development of control systems has not only influenced industrial processes but also contributed to advances in automation, robotics, and various engineering fields. Below is a summary of the key achievements and advancements in control systems over the years, as outlined in Table 1.1.

Table 1.1 Increasing developments in control systems

Year	Achievement
1769	James Watt's steam engine and governor were developed, marking the Industrial Revolution
1771	Richard Arkwright invents the first water-powered automated spinning mill
1800	Eli Whitney pioneers interchangeable parts manufacturing, initiating mass production
1868	J.C. Maxwell formulates a math model for steam engine governor control
1913	Henry Ford introduces mechanized assembly for automobile production
1927	H.S. Black conceives a negative feedback amplifier; H.W. Bode analyzes feedback
1932	H. Nyquist develops a stability analysis method for systems
1941	Creation of the first anti-aircraft gun with active control
1952	Numerical Control (NC) developed at MIT for machine tool axes
1954	George Devol creates the first industrial robot design
1957	Sputnik launch drives computer miniaturization, control theory advances
1960	Unimate robot (Devol's design) installed for machine tending
1970	State-variable models and optimal control developed
1980	Robust control system design widely studied
1983	Personal computers and control software were introduced
1990	Emphasis on automation in export oriented manufacturing
1994	Feedback control widely used in automobiles, demands robust systems
1997	The first autonomous rover Sojourner explores Mars
1998–2003	Advances in micro/nanotechnology; intelligent micromachines and nanomachines
2003-present	Rise of cyber-physical control systems with AI technology

Table 1.2 Comparison of IT vs ICS

Aspect	Information technology system (non-real-time)	Industrial control system (real-time)
Response requirement	Reliable, consistent	Immediate, time-critical
Data processing capacity	Substantial	Moderate
Tolerance for delay	Some allowance	Not tolerated
Criticality and security	Lower urgency, data privacy	Utmost importance, process safety
Purpose	Business operations, data management	Process automation, machinery control
Focus	Digital information management	Physical process control
Impact of failure	Data integrity, communication	Process disruption, safety hazards
Communication protocols	TCP/IP, common IT protocols	Industrial protocols (MODBUS, DNP3, etc.)
Real-time requirement	Not critical	Essential

1.9   Differences of IT System and ICS

IT networks lack the remote I/O systems used in ICS for real-time control. ICS focuses on reliability and monitoring, while IT emphasizes data and security. Table 1.2 compares their key differences.

1.10   ICS Control System Architecture

ICSs increasingly underpin modern enterprise by replacing traditional human-controlled systems to automate and optimize processes across a variety of industrial fields. Such systems are highly dependent on field devices, e.g., sensors, actuators, and motors, that provide real-time data and that can be remotely monitored and controlled. But as these systems have developed and they become part of more sophisticated environments, security weaknesses in core parts of ICS—such as HMIs and engineering tools have become apparent, putting ICS at risk of cyber attacks.

The 2010 Stuxnet attack manipulated the Programmable Logic Controllers was a game changer for cybersecurity awareness in ICS systems [163]. The attack demonstrated for the first time that hackers could disrupt essential industrial processes and resulted in heightened attention to vulnerabilities in ICS. Since then, a large number of researchers have identified security vulnerabilities in different parts of ICS devices, ranging from the fact that firmware updates are performed with lack of authentication (resulting in allowing anyone to install any firmware) to the poor practice of passwords (and then making that information available in the wild). With the rapid growth in the number of cyber attacks and threats, it becomes necessary not only to pay attention to the ICS design, but also to consider the history of security breaches and how the manufacturing process introduce vulnerabilities. And, as these are increasingly linked, the impact of cyberattacks could grow, further underlining the need for security to be a priority. Enterprises need to be more strategic in how they evaluate and secure their ICS, and the decision process should make security part of the lifecycle of these systems.

The security domain [96] of ICS is evolving, and it is important that companies should start to secure all levels of their control systems, from field devices through to the central control complex, to guarantee a secure operation of industrial equipments. This means implementing a safe communication standard along with monitoring and vulnerability management that is able to withstand new threats.

Figure 1.4 adapted from [133] highlights these concerns, which underline the urgent need for continuous improvements in secure ICS architectures and protocols.

Industrial Control Systems have a complex cyber attack surface [61] and threat profile. Figure 1.4 shows cyber attack surface of industrial control systems. They consist of several levels, such that users interact with user interfaces, HMI software, etc. Physical processes are controlled and monitored through software, as for example Control System Apps and Supervisory Control and Data Acquisition (SCADA). It also involves the field devices and pieces of hardware that interface directly to the physical process (Sensors, Actuators, RTUs). On the bottom is the

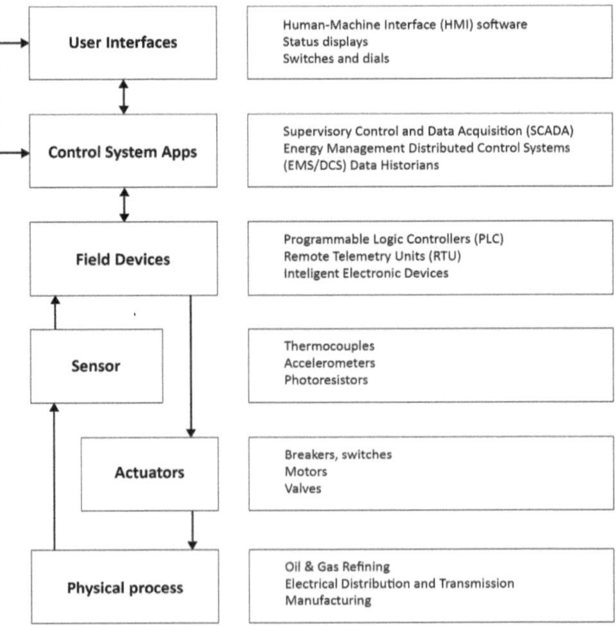

Fig. 1.4 Cyber attack surface

physical process, for which the control system manager (that is, us in this article) is responsible, which includes components such as breakers, switches, motors, valves, oil, gas refining, electric, and manufacturing.

1.11 Applications of Industry Control System

ICS is utilized in almost all industrial sectors and vital infrastructure, including water treatment, energy, transportation, manufacturing, and so on. Figure 1.5 was from shows in below:

ICS covers a wide range of technologies and applications for the monitoring and control of processes in critical infrastructure and industry. The following are some uses to which the invention may be put:

1. **Manufacturing:** The manufacturing sector depends on ICS, as they control machines, optimize processes, and ensure the quality of the manufactured

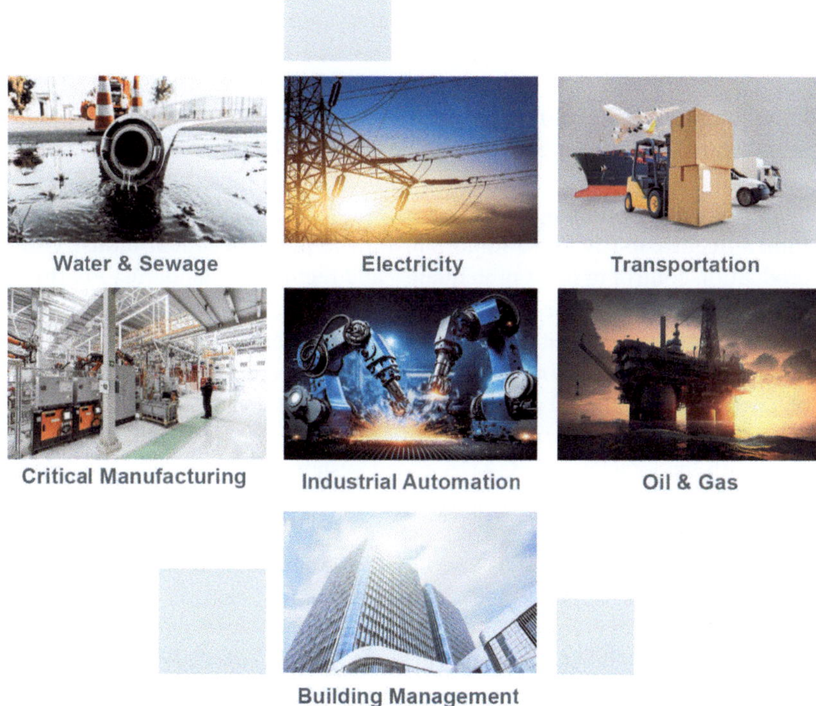

Fig. 1.5 Some application areas of ICS

goods. As part of ICS, SCADA enables end-users to monitor and control their equipment in real time, helping companies maximize operational performance, minimize downtime, and maximize productivity.

Energy Production: In the energy production field, ISC is important in operating renewable as well as fossil fuel power plants. And using SCADA systems, operators can monitor the safe and efficient operation of bearings of vital infrastructure, control the distribution of electricity, or manage the complex processes involved in an energy-generating facility.

2. **Solid and Water Waste Management:** For efficient water management and garbage (waste) monitoring, ICS are vital. By enabling monitoring and control of both wastewater and fresh water infrastructure, including sewers, treatment facilities, and solid waste operations, SCADA solutions enable to optimization of resource utilization as well as adherence to environmental regulations.

3. **Oil and Gas**: Industry system is crucial in oil and gas for data management, safety, and control of the process on different operations such as distribution, extraction, and refining. 5) SCADA Systems: By delivering real-time intelligence of critical processes, SCADA systems enhance performance and minimize hazards in a series of hostile environments.

4. **Transportation:** Transportation infrastructure, including air travel, railways, and traffic-managing systems, relies on ICSs for their safe and efficient operation. Transport networks of enhanced safety and reliability can be achieved by SCADA systems, which also enhance railway operation and monitor/control traffic.

5. **Food and Drink Manufacture:** ICs make the food and drink manufacturing industry more automated and efficient. SCADA permits accurate control of production processes, which in turn delivers consistent product quality, hygiene, and the optimal use of resources.

6. **Pharmaceuticals:** ICS are important for the pharmaceutical industry in drug production control to meet high quality and regulatory standards. Pharmaceutical products can be made more efficiently and to GMP standards using SCADA systems' extensive monitoring and control functions.

7. **Chemical Manufacturing:** To improve safety and efficiency in the production of chemicals, industrial control systems are needed to monitor and control chemical operations. Systematically, to minimize the risks and ensure compliance with the legislation, SCADA-based systems allow for an exact control of process's parameters, chemicals reaction's real-time monitoring and risks early detection.

8. **Building Automation:** ICS is also deployed in building automation for control and management of individual building systems such as lighting, security, HVAC, and access control. Integrating multiple systems into one central, control platform, SCADA systems help to minimize energy consumption while increasing occupant safety and comfort and supporting active management of natural and occupant-related functions [8].

9. **Aerospace and Defense:** Industrial control systems are also crucial in the aerospace and defense industries, such as for the management and monitoring

of aircraft, spacecraft, and military installations. Complex aircraft systems are assured of reliable operations through SCADA systems, which offer real-time telemetry data and fault diagnoses in addition to remote control to increase the chances of missions accomplished and safety [8, 26].

10. **Mining and Metals:** ICS as a part of Automation, Control and Monitoring levels are required by the mining and metals industry for mining and metal production automation, control and monitoring applications. SCADA systems improve production and profitability through the tight control of processing strategy, the exact control of mining operations and of the compliance with safety and environmental rules.

11. **Auto Industry:** ICS automatizes and controls many of the processes in the production of cars. Ultimately, ICS increases manufacturing plant productivity and car quality by enabling accurate, efficient, and consistent manufacturing of vehicles using assembly line robots or users that operate quality control systems.

12. **Agriculture:** In agriculture applications, ICS is used to deliver precision farming. ICS also optimizes farming practices including planting and harvesting, automates irrigation, and employs sensors and drones to keep an eye on crop conditions. ICS helps farmers achieve better agricultural yields, conserve resources, and make informed decisions by providing real-time data and control.

13. **Healthcare:** Contemporary healthcare systems are also increasingly having to include industrial control systems for their monitoring systems, medical gear, and process automation solutions.powered by industrial control systems. With control over temperature and humidity in the hospital environment and ensuring correct delivery of drugs and treatments, gaining a higher level of patient care, safety, and operational efficiency in hospitals with ICS.

14. **Environment Monitoring:** Industrial control system as necessary for environmental conservation by observing, managing air and water quality. ICS -based surveillance networks gather and analyze information on the existence and severity of pollutants, weather patterns, and ecosystem health in an effort to allow a preemptive response to minimize the danger to the environment or natural resources.

15. **Telecommunication:** ICS is also required for the management and control of telecommunication networks. ICS is essential for reliable and high-fidelity telecommunications services provision toward customers all over the world, ranging from ensuring seamless connection and network optimization to fault diagnosis and fast response in case of service disruption.

16. **Textile:** Many textile industry processes, such as weaving, dying, and finishing, can be automated and optimized with industrial controllers. In textile production processes, ICS contributes to optimum output, consistent quality, and efficient resource use by controlling material streams, surveillance of production parameters, and machine control.

17. **Pulp and Paper Mills:** Industrial management systems for quality control, process optimization, and energy conservation are employed in the pulp

and paper industry. ICS A single control system ICS manages the entire paper-making process—from wood preparation to finishing—to ensure that the process runs with maximum cost efficiency, produces high-quality end products, and minimizes the impact on our environment.

18. **Green Energy:** Industrial control systems play an important role in producing and distributing green energy generated from, e.g., wind or solar power plants. To enable the transition to clean and renewable energy, smart control systems are employed to significantly improve the process of energy production, to control the distribution of produced power, and to secure grid stability.

19. **Entertainment:** ICS systems are used in the entertainment industry to control and manage systems that are useful in theme parks, theaters, and entertainment venues. ICS enhances overall audience experience and boosts operating efficiency for the entertainment industry. That covers signals and lighting and sound effects, all the way to how rides are operated and shows are automated.

1.12 Evolution and Trends

ICT step into the ICS arena instead of analog technology, enabling more automation and remote monitoring aspects. Integration of IT and OT, and the free communication and data sharing, have laid the foundation for effective production. Cutting-edge ICS and AI techniques with predictive maintenance and manufacturing process optimization are changing the way industry looks at ICS. Cybersecurity remains a critical issue that, in combination with strong defenses against cyber attacks, ensures the consistent operation of industrial processes.

1. **Early Automation:** Automation of industries began in 1900s with gas and hydraulic systems. These primitive Automation [Automation systems] solutions make Manufacturing operations productive and efficient.

2. **Electromechanical Systems:** Electromechanical systems were developed in the middle of the twentieth century, very important advancements in industrial automation. The 1960s saw the introduction of the first Programmable Logic Controller (PLC) s , revolutionizing industrial control and monitoring functions.

3. **Digital Revolution:** The digital revolution in the 1970s and 1980s radically transformed ICS, enabling DCS and SCADA to offer more precise control, accurate data collection, and reliable systems through the transition from analog to digital operation.

4. **IT/OT Integration:** In the latter half of the twentieth century to early twenty-first century, a better networked system was developed with the integration of IT and OT.

5. **Industry 4.0 and IIoT:** The convergence of OT with IT was a milestone development in industrial automation and phenomena of the late 20th and early 21st centuries. This convergence facilitated the construction of networked, linked

systems and improved operational efficiency which facilitated the exchange and communication of data.

6. **Edge Computing:** With the advent of Industry 4.0 and the Industrial Internet of Things (IIoT), a whole new paradigm of automationindustry has emerged. Utilization of the IoT, to compile and study massive data inputs from devices as an input to preventive and predictive maintenance, and to optimize asset performance and minimize unexpected downtime, by the way of cloud computing, Data analyst ICS and real-time networking.

7. **Cybersecurity Orientation:** The focus of ICS is strengthened on the aspect of cybersecurity as the vulnerability of ICS increases with the expansion of network connectivity. Deploying measures like network segmentation, intrusion detection and mitigation and encryption in order to cope with attacks in cyberspace and to refine the confidentiality and integrity of essential industrial infrastructure [64].

8. **Artificial intelligence**
They use artificial intelligence (AI) and machine learning (ML) technologies to improve process optimization, anomaly detection, and equipment failure prediction, revolutionizing Industrial Automation . These state-of-the-art analytical instruments help increase ICS reliability, efficiency and maintenance practices.

9. **Remote and Cloud Monitoring:** The backbone of the Industrial Control technologies has become the remote access, data storage and analysis with the use of cloud based technologies. These capabilities enable industry decision makers to supervise what is happening in the plant from any location with an Internet connection, allowing for real-time visibility into what's happening. What's more, cloud storage provides scalability and flexibility, this means that a business can store large amount of data securely. In addition, cloud-based analyst ICS tools can harvest the data and manage swings of change of operational pattern, thus allowing for informed decisions and pro-active maintenance. In general, the remote and cloud monitoring can lead to more efficient operation, more informed decision-making, better utilization of resources, and more effective ongoing improvement of industrial processes.

10. **Augmented Reality and Virtual Reality):** AR and VR are ICS technology now that are changing how maintenance, instruction, and troubleshooting are performed within an organization. Augmented reality (AR) Through overlaying digital data on the real world, AR provides the workers with real-time data and instruction when they perform maintenance tasks. It optimizes the efficiency and accuracy of maintenance processes, minimizing downtime and maximizing asset reliability. VR technology also has users enter virtual worlds where they can experience lifelike training simulations, attend remote troubleshooting sessions, and much more. By integrating AR and VR in ICS operations, companies can enhance staff training, decrease human error, simplify troubleshooting process, and advance operational efficiency and safety.

11. **Sustainability:** ICS decrease energy consumption, reduce the generation of waste, and generally use resources more efficiently—key to any sustainabil-

ity strategy. Predictive ICS-based analytics and advanced control algorithms empower ICS to minimise energy consumption while maintaining peak operational efficiency. Also, through proper tuning of material flows and industrial progress, ICS can contribute to waste reduction in the form of trash and pollutants. Moreover, it directly contributes to be sustainability, because it allows the addition of renewable and environmentally friendly production to the ICS. Organizations can help reduce the potential environmental impact, comply with the law and enhance their reputation as a socially responsible business by adopting sustainable practices.

12. **Regulatory Compliance:** ICS should adhere to relevant regulations to ensure data security and privacy. Achieving a secure ICS environment requires robust protocols [17], practices, and policies, enforced using state-of-the-art methodologies and best practices such as those provided by NIST, IEC 62443, and GDPR. Organizational security measures must be implemented in accordance with applicable laws governing the handling of sensitive data and the transfer of relevant infrastructure, including network separation, access control, encryption, and intrusion detection systems (IDS). By complying with these standards, organizations can reduce their exposure to cybersecurity risks, protect themselves from potential attacks, and instill confidence among stakeholders and regulators.

13. **Human-Centered Design:** Industrial control systems need to be designed with concepts and interfaces that support the ease and safety of the operator. ICS By prioritizing operator needs and preferences through family design and operator-orientation, interfaces become more intuitive, providing clear and simple information for timely and well-informed decisions. This human-centered design enhances worker happiness, effectiveness, and security in industrial environments, ultimately improving performance and outcomes.

14. **Resilience and Disaster Recovery:** Redundancy and backup systems are required in industrial control systems to ensure operation even under emergency or breakdown. Because backup systems fail when the underlying hardware fails, or during a power failure or other disruption, they are an important element in the ICS design. Disaster recovery plans are also put in place so that systems can be quickly brought back online in the event of a catastrophic failure. Enterprises can safeguard their business and brand by: Reducing downtime, managing risk, and keeping things up and running by focusing on resilience and disaster recovery capabilities.

15. **5G Support:** Integrating 5G networks into ICSs enables ultra-low latency and real-time data transfer due to 5G's high channel bandwidth and low delay compared to previous wireless network generations, facilitating smooth operation. communication among the systems and tools present in the industry premises. This low-latency communication is especially favorable for real-time monitoring, control, and decision-making, and can significantly improve ICS performance, speed, and scalability. In addition, combined with 5G network capabilities, edge computing and IoT applications can improve industrial processes and enable greater innovation and scalability.

1.13   Prospective of Industry Control System

A number of important trends and developments are likely to shape future directions for ICS:

1. **Riding on innovations:** Future industrial control systems will exploit sensing, communication, and pervasive computing advancements to achieve more efficient and performing systems. This integration will enable true just-in-time interchange of data and decisions, fostering seamless communication between devices, systems, and processes. Moreover, advances in artificial intelligence and machine learning will enable better predictions, and as a result, we will better use our resources and need less offline time [144].

2. **Adopting a Data Revolution:** Processing data at scale will revolutionize industrial control, enabling data-driven decisions that reveal trends, patterns, and deeper business insights.

3. **Changing industry control:** The convergence of software and hardware will transform industrial control by merging virtual and real domains. This enables new reaction control strategies that blur the lines between digital and physical systems. Virtualization accelerates innovation and reduces product testing time by allowing testing on virtualized control systems.

4. **Move away from components:** Future industrial control systems (ICS) will leverage real-time sensing and correction to move beyond traditional control components and achieve ambitious goals. Adaptive control methods, responding in real time to varying conditions, will enhance system resilience and reliability. Sophisticated sensors and actuators will optimize each intricate process, minimizing waste generation.

5. **Toward changing in control:** Future ICS will interconnect diverse physical and information systems, in order to build integrated industrial ecosystems. Such a development will allow machine, robot and software systems to interact with one another in addition to humans. They will continue to create more flexibility and adaptability in industrial production. Smart decision-making and adaptive control techniques will be enabled by real-time data exchange and coordination facilitated by advanced networking technology.

6. **Autonomy and enterprise-wide resource integration:** In order to promote operational excellence, autonomy and enterprise-wide resource allocation will be both integrated into future versions of the ICS. Robust decision-making and adaptive control will be facilitated to differentiation efficiency and effectiveness of resource allocation in such a scenario. Advanced optimization algorithms which enable dynamic resource allocation across the organization, and to adjust production schedules, inventory levels, and energy consumption to changes in market conditions and demand.

7. **Extending control's reach:** The horizon for future ICS is to expand production, safety and efficiency multipliers in wider sectors by stretching control to unconventional systems. Control Automation centers will make critical systems in multiple areas (smart cities, health, transports) automatic and optimized".

Predictive maintenance, fault detection and adaptive control methods can be achieved utilizing advance control algorithms which significantly ameliorates the robustness and reliability of systems.

Critical defense technology will be empowered by future ICS—power command by control and autonomous systems. Control systems also will be critical in military applications such as unmanned platforms, command and control centers and surveillance systems. Situational Awareness and Battlefield Performance/Intelligence Technology A modular and interoperable vehicle control system and other advanced control algorithms will result in increased situational awareness and mission performance through the autonomy to make decisions and adjust to dynamically changing and uncertain operating environments.

8. **Enhancing threat detection and risk mitigation:** Future ICS will enhance threat detection and risk mitigation through microsystem centric control across industrial sectors [152]. Advanced sensors and analyt ICS tools will be connected to the control systems to detect and mitigate operational risk, physical hazard and Cybersecurity risk. A fast response to emerging perils will also be provided through real-time monitoring and anomaly identification, which will also reduce operational impacts, and ensure business continuity.

9. **Development of secure, dependable, and flexible networks:** Future ICSs would implement feedback-driven communication to form secure, dependable, and flexible networks for dynamically changing conditions. The actuators of control systems will have access to advanced networking technologies, such as *Software-Defined Networking* (SDN) and Network Function Virtualization (NFV), to be able to adjust dynamically to dynamic network conditions and security requirements. Reliable and secure connections among devices, systems and users can be enabled by the communication protocols that are based on feedback-driven real-time data transmission and coordination.

1.14   Security Challenges of Industry Control System

Five primary problems affect the ICS and have left it vulnerable to cyber attacks.

1. **IT/OT Convergence:**
   IT/OT convergence is a big challenge for industrial controls. Historically, different teams have been responsible for operating IT and OT systems in isolation from each other. This overlap between the two areas has become increasingly relevant as organizations' reliance on networked systems has increased and technology has advanced. Though IT/OT convergence may provide businesses with better supply chain integration, visibility, it also becomes easier for hackers to exploit vulnerabilities as it increases the attack surface. In addition, most companies have OT infrastructure that must be protected from cyberattacks.

measures could disrupt legitimate actions, leading to a loss of production or endangering safety. Consequently, they can be used in support of OT environments.

2. **Legacy Systems:** The abundance of legacy systems in industry is another considerable challenge for ICS. Because they were built decades ago with security as an afterthought, industrial control systems frequently do not have the features—like authentication and encryption—required to prevent cyberattacks.

3. **Remote Access:** As a lot of industrial control system are not able to provide enough access control so that exteriors breach doors to this vitals system without your permission. They also must contend with their own internal and external users who want to remotely access industrial assets for reasons such as maintenance needs. Third-party users can be especially difficult to support, since they can typically not use jump servers or other infrastructure (which can also be expensive and complex for admins). For businesses who do not have secure remote access, transparency and oversight of what is taking place in the environment also start to get lost, trickling its way to the safety and uptime of these environments. Nor do they have the equivalent of a control center to identify and respond to cyber problems.

4. **Patching:** Industrial environments do not often have windows for maintenance considering required system down times are intolerable. The systems themselves are vulnerable to a known attack that could have been prevented, which makes them especially hazardous.

5. **APT Attacks:** ICSs are regarded as targets by advanced persistent threat attacks and other polymorphic threats. APTs have their own custom tools to attack critical infrastructure systems, and their attacks are designed to be stealthy, which means the damage can develop over time and could be threatening to the critical infrastructure. It can be difficult to protect systems if a viable ICS security plan is not present.

1.15 Summary

Industrial control systems have transformed industries by increasing efficiency, dependability, and safety. Adopting technology such as AI increases adaptability and smart decision-making. In this age of technological sophistication, cybersecurity is critical. Robust protocols and aggressive risk management are in place to protect interconnected systems. Comprehensive training enhances the possibilities of technology. Global collaboration fuels creativity. Responsible innovation, sustainability, and ICS drive us to safer, more efficient, and more interconnected sectors.

Chapter 2
SCADA Systems in Industrial Control: Cloud Connectivity, Security Protocols, and Architectural Design

Abstract SCADA (Supervisory Control and Data Acquisition) systems are cornerstones of the modern industrial control, enabling the real-time observation and adjustment in the power generation, manufacturing, and water treatment infrastructures, among others. This chapter describes the development of SCADA systems and their connection to the modern interconnected and IoT-based environment as well. SCADA contains an in-depth review of components such as HMIs (Human-Machine Interfaces), RTUs (Remote Terminal Units), and PLCs (Programmable Logic Controllers), and how they enable data acquisition, system control, and all of the human intervention required. Furthermore, the chapter discusses SCADA communication protocols and security issues as well as the different software environments related to these systems. Through investigations of SCADA's system architecture and components, any reader will develop the ability to evaluate, use, or operate these control systems and determine the levels of productivity and safety of the industrial processes they control and maintain.

Keywords Supervisory Control and Data Acquisition · Real-time monitoring · IoT · Remote Terminal Units (RTUs) · Programmable Logic Controllers (PLCs) · Control systems · SCADA communication · Industrial security · Industrial software

2.1 Introduction

The software in the essential area of SCADA is even able to control industrial processes. SCADA provides organizations a way to make business decisions based on data gathered in real time from remote locations, to monitor and control equipment and conditions, and to ensure that processes are efficient, safe, and comply with appropriate standards. Its usefulness also explains why millions of units of this ICS are present worldwide in applications ranging from power distribution, water treatment, and manufacturing. Thanks to its hardware-software combinative capability, SCADA is able to gather and process the data in an efficient manner to take the form of useful information in a Human-Machine Interface

M. A. Rahman et al., *Securing Industrial Control Systems*,
https://doi.org/10.1007/978-3-032-03018-4_2

Fig. 2.1 General SCADA network

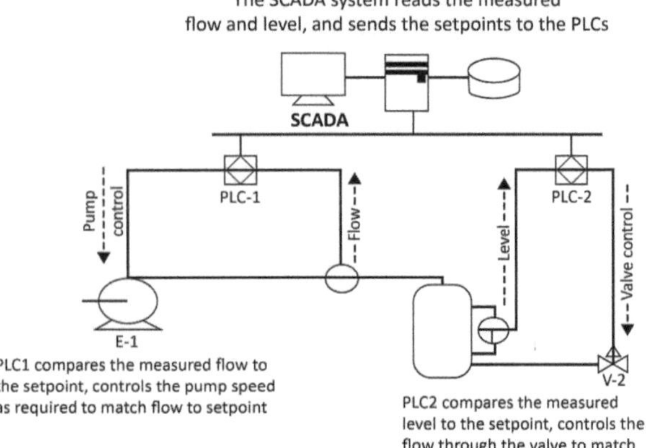

(HMI). It also provides safety with continuous monitoring, event tracking, and alerts. SCADA Is The Operating System Of The Modern Industrial World. At all levels of the enterprise, the massive computerization and automation revolution in industry involves SCADA.

SCADA is a widely used software system for industrial operations that leverages real-time data from remote sources to monitor and control equipment. This ICS offers flexibility and supports diverse functions, enabling efficient data collection and analysis through software-hardware integration. SCADA provides analytics via an HMI and ensures safety through event tracking and alerts, effectively optimizing production speed and safety by integrating data and control.

A Fig. 2.1 is referenced from [56].

Figure 2.1 illustrates a SCADA system work process with PLCs in the industry control system. The system controls and monitors industrial processes using a SCADA system connected to PLC1 and PLC2. These PLCs compare actual and desired values to control the process, operating two pumps (Pump1 and Pump2) via on/off switches. A tank with a level sensor transmits data to the SCADA system, which then sends setpoints to the PLCs to regulate liquid flow. This entire setup, including the flow control component, constitutes a distributed SCADA network.

2.2 Types of Supervisory Control and Data Acquisition Systems

These four kinds of SCADA systems monitor and manage industrial processes, giving operators the knowledge and tools they need to operate safely and efficiently [57]. SCADA systems have four types [57].They are:

1. **Monolithic SCADA systems**
2. **Distributed SCADA systems**

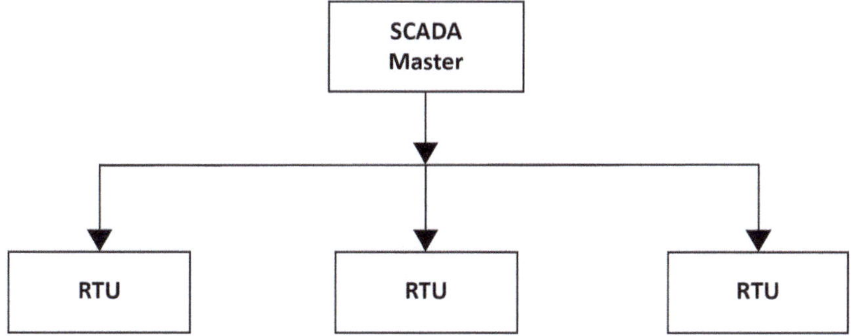

Fig. 2.2 Monolithic SCADA network (Reprinted from [57])

3. **Networked SCADA systems**
4. **IoT SCADA systems**

2.2.1 Monolithic SCADA

SCADA systems have evolved into networked systems. Initially centralized, data acquisition, control, and visualization were performed on a single server for small installations. Early SCADA systems used mainframe computers for fault-tolerant redundancy, with hardware compatible enough to run existing software.

Figure 2.2 illustrates a simple SCADA system with a master and two RTUs. RTUs are masters of the field of sensors and actuators. The master-RtU communication can be wired or wireless. The master generates data and gives orders, and RTUs consume operations driven by them. This configuration was adopted in telemetry and ICS for remote monitoring and control.

2.2.1.1 Algorithm: Monolithic SCADA System

```
\textbf{Initialize:}
    single_master_station
    rtus[] = array of Remote Terminal Unit (RTU) connections
    database = local_database
    hmi = single_workstation_interface

\textbf{Main Process:}
while system_running:
    for rtu in rtus:
        data = poll_rtu(rtu)   # Input: Data from RTU (e.g.,
temperature, pressure)
        processed_data = process_data(data)   # Process raw data
```

```
        store_data(database, processed_data)  # Output: Store
    processed data in the database
        update_hmi(processed_data)  # Output: Display data on
    the Human-Machine Interface (HMI)

    if operator_command_received:
        validate_command()  # Input: Operator command to modify
    system behavior
        send_command_to_rtu()  # Output: Send command to RTU
        log_action()  # Output: Log operator's action for audit

\textbf{Error Handling:}
if communication_failure:
    activate_backup_system()  # Output: Switch to backup system
    notify_operator()  # Output: Alert operator of
    communication failure
```

Listing 2.1 Monolithic SCADA system algorithm

Input:

- RTU Data: Temperature, pressure readings from RTUs
- Operator Command: Adjust the temperature of RTU 1 to 30 °C

Output:

- Processed Data: Stored in the local database and displayed on the HMI
- Command Execution: Adjusted temperature to 30 °C on RTU 1

The monolithic SCADA system algorithm starts with an initialization of basic components: master station, RTUs, local database, and HMI. The main loop repeatedly monitors the RTU data, processes the raw data, stores the data in a database, and refreshes the HMI with the processed data. If an operator command is received, it is authenticated, posted to the appropriate RTU, and recorded for auditing purposes. In the event of a communication failure, the system goes into a stand-by condition, and the operator is notified.

The system input is formed by data from the RTUs, such as temperature and pressure readings, and operator commands such as altered RTU settings. The output includes the processed data output on the HMI and recorded in the database, and the reaction to operator commands (such as temperature change).

2.2.2   Distributed SCADA

The introduction of small Master Terminal Unit (MTU) computers and the proliferation of Local Area Networks (LAN) caused a revolution in Supervisory Control and

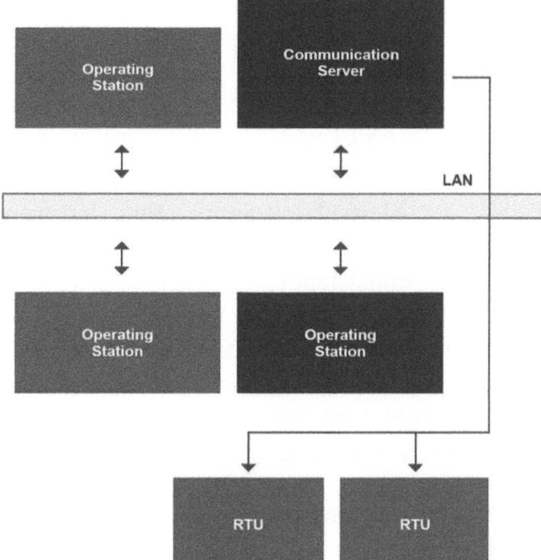

Fig. 2.3 Distributed SCADA network (Reprinted from [57])

Data Acquisition (SCADA) implementation. Distributed SCADA systems followed, with control functions spread across multiple systems interlinked via a LAN. This design provided good use of resources and efficiency. Net communication of real-time data was improved due to simultaneous data transfer via the LAN. The different pieces of these systems are distributed across multiple servers or so-called nodes. Larger and complicated programs can also make use of the fault tolerance, scalability, and overall better performance. Longer durations of active operation were also facilitated by inter-station Wide Area Network (WAN) connections between individual stations and the main SCADA system. Advantages: The first-generation SCADA systems were expensive to procure, develop, and operate.

This Fig. 2.3 shows the DCN architecture to illustrate a dependable industrial communication infrastructure. All the control applications and the necessary support services are implemented on the servers. Stations give users an opportunity to make decisions in real time with an easy-to-use interface for monitoring and controlling the actions. Remote terminal units (RTUs) and other sophisticated endpoints allow for asset monitoring and control from afar. LAN-connected units offer industrial-specific data sharing, system coordination, and operational efficiency.

2.2.2.1 Algorithm: Distributed SCADA System

Input : Master Station Data: Temperature = 85 °F, Pressure = 55 PSI from substations. Substation Data: Local sensor readings (e.g., Pressure = 80 PSI, Temperature = 120 °F)

Output: Synchronized Data: Updated shared database across all stations. Alerts: High-temperature alert from the substation to the master station. Updated Stations: Periodic synchronization ensuring data consistency

```
// Initialization
```
Define `master_stations[]` as an array of control stations; Define `substations[]` as an array of local stations; Set `shared_database` as a distributed database;
```
// Main Process
```
foreach *master_station in master_stations* **do**
   | MonitorSubstations(); SynchronizeData(shared_database);
   | HandleLocalControls();
foreach *substation in substations* **do**
   | **while** *substation is active* **do**
   | | local_data ← CollectLocalData(); processed_data ←
   | | ProcessLocally(local_data); **if** *threshold_exceeded* **then**
   | | | AlertMasterStation();
   | | SyncWithMaster();
```
// Data Synchronization
```
`PeriodicSync()` LockDatabase(); MergeChanges(); ResolveConflicts(); UpdateStations(); UnlockDatabase();

Algorithm 1: Distributed SCADA System Algorithm

Input:

- **Master Station Data:** Temperature = 85 °F, Pressure = 55 PSI from substations
- **Substation Data:** Local sensor readings (e.g., Pressure = 80 PSI, Temperature = 120 °F)

Output:

- **Synchronized Data:** Updated shared database across all stations
- **Alerts:** High-temperature alert from substation to master station
- **Updated Stations:** Periodic synchronization ensuring data consistency

In the Distributed SCADA System Algorithm, masters and substation stations, as well as a shared distributed database, are initialized. In the major process, each master station polls its remote stations, synchronizes data in the shared database, and performs some local controls. The data of the local substation are constantly acquired and processed, with alarm messages sent to said master station when the threshold values are reached, and with the data lines of said substation synchronizing data with the master station. A periodic synchronizing function, which locks the database, reconciles changes, eliminates conflicts, and updates all stations with synchronized data. This is to have uniform data throughout the stations of the system. The input crops are sensor data of the substation and the master station, and the outputs are the synchronized data in the database, the alarms, and the updates.

2.2.3 Networked SCADA

The trend of distributed Supervisory Control and Data Acquisition systems increased significantly with the standardization of communication protocols. The development of these standards led to the incorporation of Commercial-Off-The-Shelf (COTS) technology and made SCADA systems simpler to install and operate. Their various components are structurally organized by means of a network structure. They offer remote operation and observation, very often from different places. An important progress in the networked SCADA was the use of the (Wide Area Network) WAN protocols like TCP/IP. This allowed continuous communication over wide distances between the master computer, Human-Machine Interfaces , Remote Terminal Units, and field devices.

Networked SCADA System Algorithm Initialize the Network of SCADA - Registration of Clients -Security Settings Searching for new data or data requests from each client is in the primary routine. If new data exists that it is to be transmitted is simply encrypted and sent out to the network. When a client needs data it makes a request the request is validated and data is returned to the client. Furthermore, the system in data transfer time frame not only authenticates the sender, ensures data integrity, and confirms the user's access rights, but it also logs information stay checking transactions. The input comprises data acquired from a client including, for example, a voltage or current reading, and the output data comprises encrypted or authenticated data transmitted via the network.

2.2.3.1 Pseudocode: Networked SCADA System

```
1. Initialize:
   - network = establish_WAN_connection()
   - clients = register_network_clients()
   - security = initialize_security_protocols()

2. Process:
   For client in clients:
       If client.has_new_data:
           broadcast(encrypt(client.data))
       If client.requests_data:
           send(validate_request(client))

3. Security:
   For data_transmission:
       authenticate()
```

```
16        verify_integrity()
17        check_access()
18        log_transaction()
```

Listing 2.2 Networked SCADA system

Code link is available in https://github.com/Sunzidasiddique1/ICS. There are three phases of the Networked SCADA System: initialization, main process, and security. During initialization, the Wide Area Network is configured, clients are registered, and security is established. The Foxhole-phase encrypts, broadcasts, and verifies data for real-time monitoring. The security level ensures authentication and access control, logs transactions, and intrusion detection to increase resilience.

2.2.4   Internet of Things SCADA

The modern generation of supervisory control and data acquisition (SCADA) systems makes use of cloud computing and the Internet of Things or Internet of Everything concept, to enhance its analytics to become more responsive and efficient. They also allow to connect a group of IoT devices with a SCADA, providing a higher level of automation and decision-making in the plant, due to the possibility to have sensors, actuators, and a SCADA system exchanging data in real time. Although the cloud-native SCADA has the advantages of scalability and cost saving, it is necessary to have significant contributions and security to compensate the risk in such a data-sharing network, which is real-time, cross-platform. IoT SCADA comes with transparent access to distributed devices and network resources and can even support cloud burst for borrowing storage. This type of architecture allows to have public or private, or hybrid topologies in the network due to the operational requirements.

1. **Public:** Public networks are those where networks are owned and managed by businesses that can also runs networks between an organization's SCADA systems via leased facilities. Since there is a similarity between the construction of cloud networks, the security measures to be applied to secure them may vary significantly. In addition to SCADA, other types of Internet-related services that are not related to SCADA could be offered on such networks. There are some computing tools on the market as well that don't require users to purchase SCADA features.
2. **Private:** And last but not least companies can either keep or out-source their SCADA networks. Customization: It can be customized to the specific needs of organizations and can even be linked to other business systems or business email system solutions [39]. It is expensive to maintain a private data center, and this is due to the space and hardware it occupies and the dedicated environmental control system that must maintain this hardware.

3. **Hybrid Network Services:** Network solutions that combine public and private ones. In order to design and implement such networks, it is required to standardize networks, protocols, and communication links. Combining various service providers to form a hybrid cloud system. That's why SCADA system operators can reach into think of as the cloud and provide cloud-like services, including "cloud bursting," otherwise known as buying more computers and adding them into systems.

In addition to these standard constructs, there are other cloud services modalities that can be observed, including community clouds, distributed clouds as well as multi-cloud environments in a broader sense. These are decisions that may help improve efficiency and client satisfaction.

2.2.4.1   Pseudocode: IoT SCADA System

```
1. Initialize:
   - cloud_platform = connect_cloud_service()
   - iot_devices[] = register_iot_sensors()
   - edge_processors[] = initialize_edge_computing()

2. Main Process:
   While system_operational:
     For each device in iot_devices:
       raw_data = collect_sensor_data(device)
       processed_data = edge_processing(raw_data)
       If requires_immediate_action:
         execute_local_response()

     Batch Upload to Cloud:
       compress_data()
       encrypt_data()
       send_to_cloud()

     Cloud Processing:
       analyze_big_data()
       generate_insights()
       update_dashboards()
       send_notifications()

3. Device Management:
   Periodic Maintenance:
     check_device_health()
     update_firmware()
     optimize_power_consumption()
     manage_security_certificates()
```

Listing 2.3 IoT SCADA system algorithm

Input:

- IoT Device Data: Temperature = 25 °C, Humidity = 70% from device 1
- Sensor Data: Temperature = 30 °C, Humidity = 65% from device 2

Output:

- Processed Data: Edge processing generates processed data:

  - Device 1: Temperature = 25 °C, Humidity = 70%
  - Device 2: Temperature = 30 °C, Humidity = 65%

- Cloud Upload: Processed data uploaded to the cloud and encrypted
- Insights Generated: "Temperature threshold exceeded in area 2," triggering a notification to the operator

The algorithm of IoT SCADA systems explains how to control IoT sensors placed in the network. At first it connects to the cloud, configures IoT devices, and boots for edge processors. First priority: retrieving sensor data and processing it, with local actuation (e.g., turning on a fan) as required. Processed data is transmitted in the cloud, where it is processed to transmit key insights and messages for operators. Additionally, and by way of example, such periodic maintenance can include device health monitoring, firmware upgrades, security maintenance, etc., for sustaining the operational integrity of the system.

2.3   SCADA Devices and Cloud System Connection

SCADA systems can connect to the cloud in two ways:

1. **On-site usage:** SCADA systems are installed on-site and directly send the real-time data to the cloud through a straight link. The controlling devices come with a cloud connection and can be monitored remotely. This architecture made it possible for people on a network to share and exchange information without considering where a given piece of it happened to be stored.
2. **Cloud-Based Application:** If the SCADA system is managed by a cloud-based application, it can therefore be managed from a remote distance. Data are stored and shared as in any common cloud computing environment. These are commonly referred to as "distributed configurations." In these settings, SCADA systems are run on-site, and data is communicated to the cloud. These configurations are often accommodated with the help of Personal Digital Assistants (PDAs) and/or Programmable Logic Controllers (PLCs).

Figure 2.4 illustrates a public cloud network system, which includes components such as:

- A laptop for monitoring and controlling the devices.
- A tablet/PDA to have wireless access to critical information and tasks.

Fig. 2.4 Public cloud
network (Reprinted from
[57])

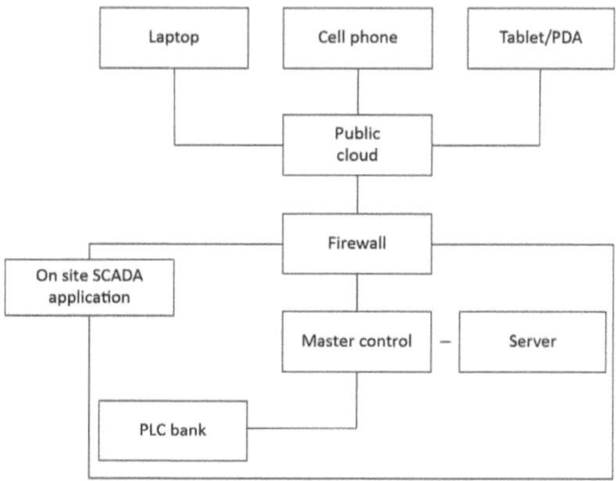

- Cell phone for instant messages, alerts, and remote viewing.
- An in-house SCADA program for real-time process surveillance, data gathering, and control.
- A fireball to ensure system and data elements are protected against unauthorized access and cyber attacks.
- A server that provides applications, database, and services for operation and data management.
- A master control station for controlling subsystems and devices, and a PLC station.

Hybrid SCADA features both private and public cloud capabilities and supports monitoring and remote accessibility via WAN connections. Cloud computing provides adaptable, configurable, and on-demand application services. And with those tools, automation decreases the price and eases the burden of managing resources.

Figure 2.5 illustrates a private or hybrid cloud network configuration. The network includes various devices, such as laptops, mobile phones, tablets/PDAs, and browser-enabled devices. Key elements of the network configuration are:

- Private or hybrid cloud infrastructure
- A distributed control network
- Communication devices, including radio modems, satellite receivers, and cellular antennas
- A PLC (Programmable Logic Controller) bank

2.3.1   Cloud Service Models

1. **Infrastructure as a Service (IaaS):**
   Infrastructure as a Service offers virtualized computing resources over the internet. IaaS enables users to interact with and control physical hardware devices–

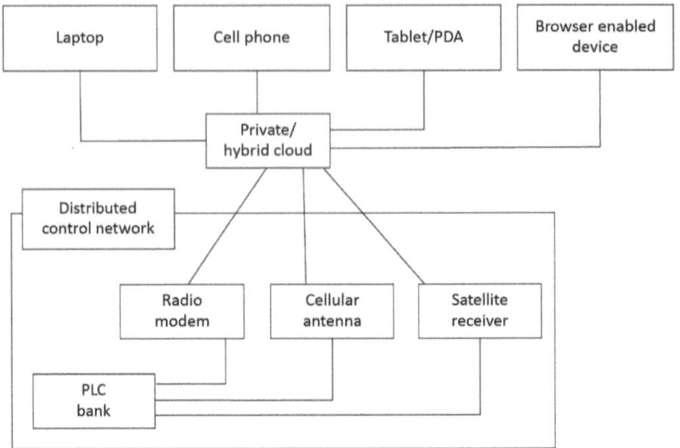

Fig. 2.5 Private or hybrid cloud network (Reprinted from [57])

such as servers, storage, and networking–remotely through a cloud provider. Typical IaaS services include virtual machines, containers, and networking components that users can scale and provision as needed. It is cost-efficient since there is no upfront investment required for hardware, and users are charged based on actual resource usage. Among the most well-known IaaS offerings are Amazon Web Services (AWS), Microsoft Azure, and Google Cloud Platform (GCP).

2. **Platform as a Service (PaaS):**
   Platform as a Service provides an environment for developers to build, deploy, and manage applications without the complexity of handling the underlying infrastructure. PaaS solutions include middleware services, runtime environments, frameworks, and development tools, which enable faster application development and deployment. The server, storage, and networking are fully managed by the cloud provider, allowing developers to focus solely on writing applications. This approach accelerates development, reduces time-to-market, and lowers costs. Popular PaaS providers include Google App Engine, Microsoft Azure App Service, and Heroku.

3. **Software as a Service (SaaS):**
   Software as a Service delivers software applications over the internet on a subscription or pay-per-use basis. Users can access and use software hosted and managed by a third-party provider via a web interface or application. Since SaaS eliminates the need for local installation, maintenance, and updates, management overhead is minimal. This model also provides flexibility and scalability, as applications can be accessed from any internet-enabled device. Examples of SaaS applications include Google Workspace (formerly G Suite), Microsoft Office 365, and Salesforce.

While cloud-based SCADA software is versatile and can be used on different devices from different locations, it has to be accompanied by strong security mechanisms to ward off potential attacks. Although traditional IT security principles apply, protective measures are differently implemented within SCADA systems.

2.4 Components of Supervisory Control and Data Acquisition Systems

The following are the components that make up a standard Supervisory Control and Data Acquisition (SCADA) system:

1. Human-Machine Interface (HMI)
2. Supervisory System
3. Remote Terminal Units (RTUs)
4. Programmable Logic Controllers (PLCs)
5. Communication Infrastructure
6. SCADA Programming

2.5 Human-Machine Interface

Human-Machine Interface (HMI) indexHuman-Machine Interfaces! HMIs magazine, the communication interface between industrial processes and human operators in SCADA systems. The HMI opens up the possibility for the real-time collection of industrial data, as it allows for control/decision-making. The SCADA databases are integrated into HMI as well, to offer a certain degree of intelligence, and reports can be accessed from the database. Instant troubleshooting and problem identification are enabled by the single point of view, which ultimately improves operational efficiency.

Figure 2.6 illustrates a Human-Machine Interface, highlighting key elements such as real-time data visualization, control panels, and alarm management. It provides operators with an intuitive view of system parameters, enabling efficient monitoring, adjustments, and response to alerts, ultimately enhancing decision-making and process control.

2.6 Supervisory System

The Supervisory System links the HMI software of the control room workstations to Programmable Logic Controllers, Remote Terminal Units, instrumentation, and other devices. The network structure of the system realizes the interaction of the system, real-time flow and information, and overall process control of the industry.

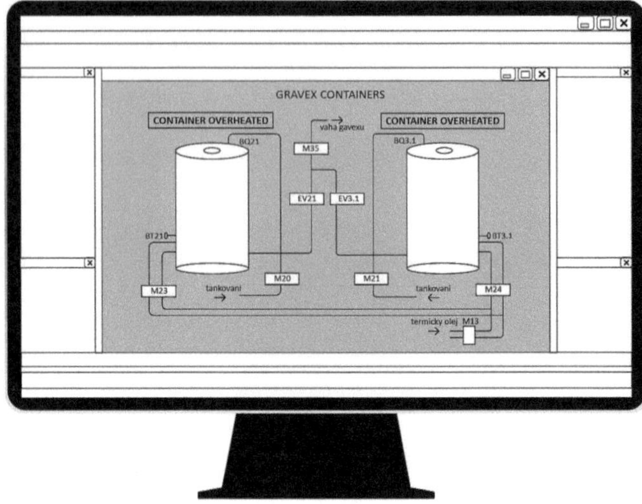

Fig. 2.6 Human-machine interface (Reprinted from [122])

In simpler SCADA systems, the supervisory/masters computer may be the only PC in the system. This smaller system makes communication and control simpler for smaller-sized businesses and will find greater SCADA systems in larger industrial complexes. They rely on the network of servers to support the supervisory backbone. And there's common software packages that take care of several websites, overseeing everything.

Redundancy is also a demand in big SCADA systems. Failover or hot backup server setups are extensively distributed to avoid a single point of server failure. This no-fail process increases system reliability and minimizes potential downtime by constantly monitoring for faults and switch-over. Disaster recovery is also provided by SCADA systems. The other sites are geographically distributed and can take over in case of a failure. In this way, continuity of the industrial process is guaranteed, thanks to the anticipation that allows the data and control flow to remain active even when unexpected shutdowns occur.

2.7   Remote Terminal Units

Remote Terminal Units (RTUs) play an essential role in modern industrial automation as the key contact point between the physical infrastructure and digital control processes. BSD The field-mountable, microprocessor-based control is designed to monitor, analyze, and report information associated with numerous sensors and actuators. They support transparent communication for on-site equipment with Control Room Supervisory Control systems (e.g., SCADA systems). RTUs enable operators to make decisions, identify issues, and drive efficiency unencumbered by the limitations of legacy, remote equipment monitoring solutions that only report static data, like temperature, pressure, voltage, and equipment states.

Remote Telemetry Units Posted on September 17, 2014, by admin RTU technology combines the latest technologies with solid field capabilities. They are considered smart agents that translate the raw physical signals into meaningful digital information, which can be processed in the master control systems. RTUs, possessing a built-in microprocessor, do more than simply collect and transfer to the host computer the data they receive. These commands could release or shut off valves, sound alarms, or initiate pumps. This two-way communication simplifies remote monitoring and control, reduces requirements for human intervention, and improves system latency.

Current RTUs are an embodiment of the union of electronic intelligence and old-school industrial hardware. However, they are no longer just relays of data but can work as independent agents that, when deployed, can perform some basic diagnostics, filter and manage data, and offer communication redundancy. The name "Remote Telemetry Units" is aptly descriptive; RTUs are not only observers, they are also transmitters that initiate an active response. Their ability to facilitate real-time interaction between physical processes and digital controllers is crucial in the era of smart manufacturing and the Industrial Internet of Things (IIoT), where connectivity, automation, and actionable data insights are essential for competitiveness and sustainability.

2.8 Programmable Logic Controllers

Programmable Logic Controllers (PLCs) have been extensively deployed in Supervisory Control and Data Acquisition (SCADA) systems for various kinds of automation requirements. PLCs connect to sensors and actuators, measuring process variables such as temperature, pressure, and positioning, and control process equipment such as valves, motors, and other devices. One of the primary applications is the transformation of analog values as read from a sensor into a digital format to make it available to the host microcontroller. PLCs are gaining popularity over Remote Terminal Units (RTUs) because of their more flexible configuration, lower power consumption, and a much larger range of capabilities and applications.

2.8.1 Programmable Logic Controller Work Process

1. **Input Monitoring:**
   The PLC observes input signals from a number of sources and transmits this data to a Central Processing Unit (CPU). Discrete (on/off) inputs are supported by some PLCs, whereas others are configured with analog modules to accommodate

variable analog inputs. The inputs can be from Human-Machine Interfaces (HMI), safety sensors, robots, and other inputs, including Internet of Things (IoT) devices.

2. **Logical Programming:**
   At the heart of each PLC is a 16-bit or 32-bit microprocessor-based CPU. The CPU is programmed by technicians and engineers so that it can respond to certain states and manipulate output signals according to predefined logic. The CPU checks the status of the input variables and acts according to the rules programmed. This logic programming provides unprecedented flexibility for industrial control systems of widely varying natures.

3. **Output Control:**
   The PLC controls outputs by standards of the logic that is used, applied to inputs under supervision. This allows for automated manipulation of the real world, e.g., machine control. PLCs can be cascaded to work in a series of operations by feeding the output signals of one PLC to the input terminus of another PLC, which enables the expansion and scaling of the complex control systems.

2.9 Advantages of Using Programmable Logic Controllers

Programmable Logic Controllers (PLCs) have played a major role in industrial automation for a long time. Here are some important benefits of PLCs:

1. **Ease of Programming:** PLCs are relatively simple to program, making them ideal for companies seeking to reduce costs and technical complexity. Their programming languages are often more accessible compared to those used in other industrial control systems.
2. **Mature and Well-Documented Technology:** PLCs are supported by decades of development, research, and testing. Extensive documentation, integration guides, and tutorials are available for most PLC types, which facilitates learning and implementation.
3. **Cost-Effectiveness:** PLCs are available across a wide range of price points. Budget-friendly versions make it feasible for small companies and startups to adopt industrial automation.
4. **Versatility:** Most PLCs can control a wide array of systems and processes due to their flexible programming and modular architecture.
5. **Reliability and Durability:** As solid-state electronic devices with no moving parts, PLCs are robust and can operate reliably in harsh industrial environments.
6. **Simplified Troubleshooting and Maintenance:** With fewer components and a streamlined structure, PLCs are easier to troubleshoot and typically require less maintenance, resulting in reduced system downtime.
7. **Energy Efficiency:** PLCs consume minimal electricity and offer high operational efficiency, helping to conserve energy and reduce wiring and power supply requirements.

2.9.1   Communication Infrastructure

Direct wire and radio connections are common in Supervisory Control and Data Acquisition (SCADA) systems. However, Synchronous Digital Hierarchy (SDH) and Synchronous Optical Networking (SONET) are often used in large-scale systems, such as railways and power generation. Only a few SCADA communication protocols are widely recognized or standardized; many deliver data to Remote Terminal Units (RTUs) only available upon request from the supervisory station. To facilitate data transmission and control commands, SCADA systems employ a variety of communication techniques and infrastructures. Some of the key methods are outlined below:

1. **Traditional Methods of Communication:** Early SCADA systems commonly relied on direct wiring and radio frequency communication. For large-scale applications like power plants and rail systems, technologies such as SONET and SDH remain prevalent [129].
2. **Telemetry Functionality:** Telemetry in SCADA refers to the system's capacity to be monitored or controlled remotely. Many users prefer SCADA data to flow through existing business networks or share resources with other networked applications [26].
3. **Legacy Protocols:** Older SCADA protocols were designed to be simple and cost-effective, transmitting data only when polled by the master station. Common legacy protocols include Profibus, Conitel, RP-570, and Modbus RTU. Notably, Schneider Electric made the Modbus protocol open for public use [68].
4. **Standardized Protocols:** Major SCADA vendors support standard protocols such as DNP3, IEC 61850, and IEC 60870-5-101/104. These protocols frequently use TCP/IP extensions, allowing easier integration with existing network infrastructures [116].
5. **Security and Reliability:** Modern SCADA systems prioritize secure and reliable communication. Protocols like DNP3, IEC 61850, and IEC 60870-5 are favored for their support of encryption, authentication, and resilience, especially over TCP/IP networks [164].
6. **Interoperability Challenges:** The lack of universal standards has led to the development of multiple proprietary control protocols by different vendors. This creates challenges in automation interoperability, as vendors often design their systems to lock customers into their ecosystem [128].
7. **Standardization Efforts:** Industry-wide standardization efforts are underway to promote interoperability. One such initiative is OPC-UA (Open Platform Communications Unified Architecture), originally known as "OLE for Process Control," which aims to unify communication protocols across platforms [150].

From traditional cable connections to advanced satellite-based communication, SCADA communication infrastructure continues to evolve with a strong emphasis on security, efficiency, and interoperability.

2.10   SCADA Programming

SCADA programming at the Human-Machine Interface (HMI) or master station level involves creating diagrams and maps that deliver critical information during process operations or in the event of system failures. Most commercial Supervisory Control and Data Acquisition (SCADA) systems utilize standardized programming interfaces, and tasks are typically implemented using the C programming language or one of its derivatives.

The programming process generally begins during the system design phase, once the software requirements of the SCADA system have been defined. At this stage, engineers determine which field devices will be integrated into the system, what data will be collected and displayed, and which control operations need to be executed.

Based on these decisions, appropriate programming languages and tools are selected. The development phase includes designing the user interface, writing routines for data acquisition and control, and performing rigorous testing to ensure proper functionality. Once the application is complete, it is installed and configured on the SCADA system, enabling real-time monitoring and control of industrial processes.

2.10.1   SCADA Software Overview

SCADA software is the backbone of industrial control and monitoring systems. Below are some popular SCADA software solutions:

(a) **RSView—Allen Bradley:** Offers a wide range of functions for SCADA and industrial automation applications. Scalable to systems of any size. Easily integrates with Rockwell Automation devices and applications [121].

(b) **Simatic WinCC—Siemens:** A robust and reliable SCADA system designed for complex industrial processes. Provides a graphical user interface for ease of monitoring and configuration, along with advanced trending, reporting, and alarm management features [110].

(c) **Flexible and Versatile SCADA Platform:** An adaptive SCADA solution supporting a wide range of industrial use cases. Compatible with many devices and protocols. Includes features like online visualization and mobile SCADA access [9].

(d) **FactoryStudio—Tatsoft:** A modular, object-oriented SCADA system ideal for small to medium-sized applications. Simple to use, scalable, and easily configurable for specific operational needs [15].

(e) **Cimplicity—GE Fanuc:** A comprehensive SCADA platform is widely used across diverse industries. Offers complete data acquisition, control, and monitoring functionality. Highly reliable with a large user base [67].

(f) **InduSoft Web Studio:** A scalable industrial automation and control solution. Its modular design and feature-rich environment support both small-scale and enterprise-level applications, improving usability and efficiency [114].

(g) **Litmus Edge:** An edge computing platform that complements existing SCADA systems. Enables real-time data capture, analysis, and decision-making at the edge, enhancing IIoT and business performance [69].

(h) **GENESIS64:** A powerful SCADA/HMI software with advanced visualization, analytics, and connectivity. Helps businesses optimize operations and increase productivity [146].

(i) **Ignition SCADA—Inductive Automation:** A flexible and complete industrial automation platform for building and maintaining SCADA systems. Supports rapid development and seamless deployment [158].

(j) **SIMATIC SCADA—Siemens:** A fully integrated SCADA system offering real-time monitoring, control, and optimization. Works seamlessly with Siemens hardware and software [113].

(k) **Action.NET:** A general-purpose SCADA software that simplifies complex automation processes. Offers a user-friendly interface and powerful features to improve production efficiency [19].

(l) **DAQFactory:** A user-friendly SCADA solution with essential data acquisition and control features. Suitable for manufacturing, research labs, and other industrial environments [51].

(m) **EisBaer SCADA:** A flexible and advanced SCADA platform designed for industrial automation. Offers a customizable interface and real-time monitoring capabilities to improve safety, efficiency, and reliability [28].

2.11 SCADA Architecture

Supervisory Control and Data Acquisition (SCADA) collects and analyzes data from machinery in the industrial sector [1]. The control system architecture employs computers, networked data transmission, and graphical user interfaces to manage machines and processes. SCADA is also an analytical process that controls industrial automation in manufacturing, production, development, and fabrication. The architecture of SCADA is based on four levels, which allow it to control and monitor large regions [164]. In addition to the hardware, SCADA includes software that collects and organizes process data. Sometimes, SCADA is referred to as a Remote Terminal Unit (RTU), and it performs real-time data processing to support effective decision-making and control.

The four levels in the SCADA architecture are as follows:

1. Process (Level 0)
2. Basic Control (Level 1)
3. Supervision (Level 2)
4. Operations Management (Level 3)

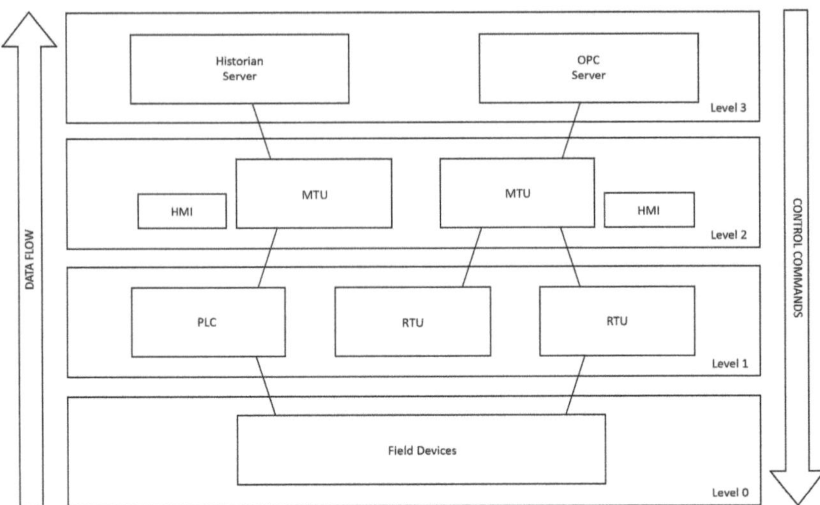

Fig. 2.7 SCADA system architecture

Data flows from Level 0 upward to Level 3 in SCADA. At the Operations Management level, data analysis leads to decision-making, which is then transmitted as control instructions to Level 0 devices. SCADA serves as a critical hub for industrial automation, organizing information and instructions hierarchically. Figure 2.7, taken from [8], illustrates this architecture.

The system architecture shown in Fig. 2.7 consists of several tiers of control and data transmission. These tiers include components such as the Historian, OPC Server, Control Commands, MTU, HMI, PLC, RTU, and other devices at different levels from 0 to 3.

SCADA systems are centralized and provide complete monitoring and control for specific areas. They receive information from processes and deliver control commands as a software layer interacting with hardware. The supervisory system receives process information and sends the necessary control commands to govern activities. The RTU, as the supervisory unit, powers SCADA. RTUs and PLCs automate many control actions, and they offer programmable logic converters to customize operations. For example, RTUs can regulate or dynamically modify water flow in a thermal power station. Figure 2.8, borrowed from [116], illustrates the SCADA system architecture.

Figure 2.8 displays a SCADA system architecture configuration that includes key components such as Intranet, Data Analytics, Data Storage, Internet, Operator, Remote Operator, HMI, Master Terminal, RTU, Field Devices, and more. The figure illustrates the communication and relationships between these various components in a structured manner.

SCADA allows operators to adjust flow rate set points and manage alarm conditions. When extreme conditions occur–such as flow loss or excessively high

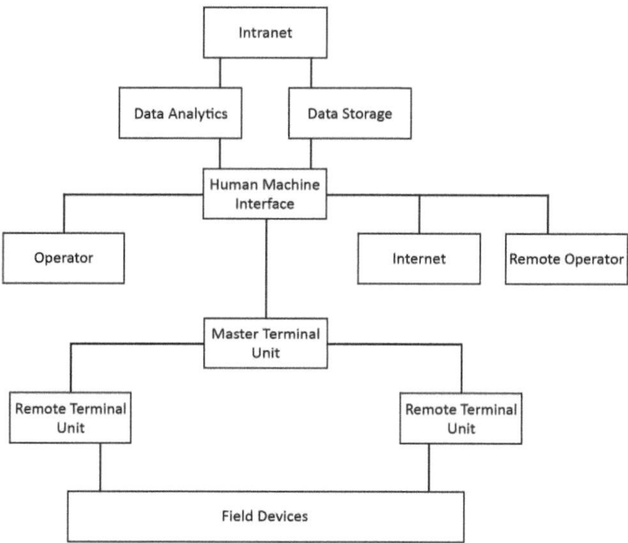

Fig. 2.8 SCADA system architecture configuration

temperatures–the system triggers alarms to alert operators. These conditions are displayed on the system interface and stored for historical analysis. SCADA also tracks controlled processes to ensure smooth and efficient operations.

The centralized nature of SCADA enables organizations to monitor and control multiple processes. Its PLCs and RTUs automate control actions, allow operators to modify parameters, and process alarms. The system's ability to interface with various technologies makes it ideal for managing and optimizing industrial processes, such as controlling gas pipeline pressure buildup.

Hardware Architecture
The design, development, and implementation of various hardware systems–including control systems–are central to the domain of hardware architecture. Control systems are vital to numerous industries such as industrial automation, automotive, aerospace, and energy. The hardware architecture of a SCADA system consists of two primary layers:

1. Client Layer
2. Data Server Layer

These layers prioritize human interaction and data processing, respectively. A typical SCADA station is composed of a single PC server that communicates with field devices via Programmable Logic Controllers (PLCs) or Remote Terminal Units (RTUs), utilizing both Wide Area Networks (WANs) and Local Area Networks (LANs). Effective monitoring and control are achieved through the interaction

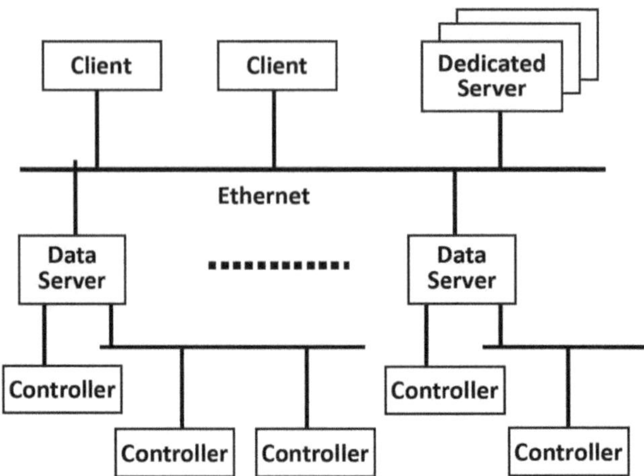

Fig. 2.9 Hardware architecture

between sensors and RTUs, which then interface with a central master system. These operations are predominantly executed by RTUs or PLCs.

Figure 2.9 illustrates a typical hardware architecture comprising clients, a dedicated server, Ethernet, data communication, and controllers. The client acts as a user interface, sending requests to and receiving responses from the server. The dedicated server functions as a central hub, managing data and services. Ethernet facilitates data transfer between devices, enabling seamless communication. Controllers are responsible for regulating system operations, and the term "data" here refers to the shared information within the system. Together, these components ensure the efficient transfer of data and control signals for optimal network performance.

In this architecture, the SCADA station is generally a single PC server. PLCs or RTUs establish connections between data servers and field equipment. PLCs connect using direct, networked, or bus-based methods, leveraging WAN and LAN protocols. Sensors feed data to PLCs or RTUs, which digitize the sensor inputs and transmit them to the master unit. Based on feedback from the master, RTUs activate relays to perform control operations. Monitoring and control are primarily handled by these RTUs or PLCs.

Software Architecture

SCADA software applications provide functionalities such as real-time trending, diagnostics, and information management [8]. These applications handle logistical information, maintenance scheduling, detailed schematics for machines or sensors, and expert-system troubleshooting. Figure 2.10 was adapted from [122].

Figure 2.10 shows the SCADA software architecture, highlighting components and operations such as the SCADA Client, Human-Machine Interface (HMI), Alarm Management, Data Management, Data Export, Recipe Management, and

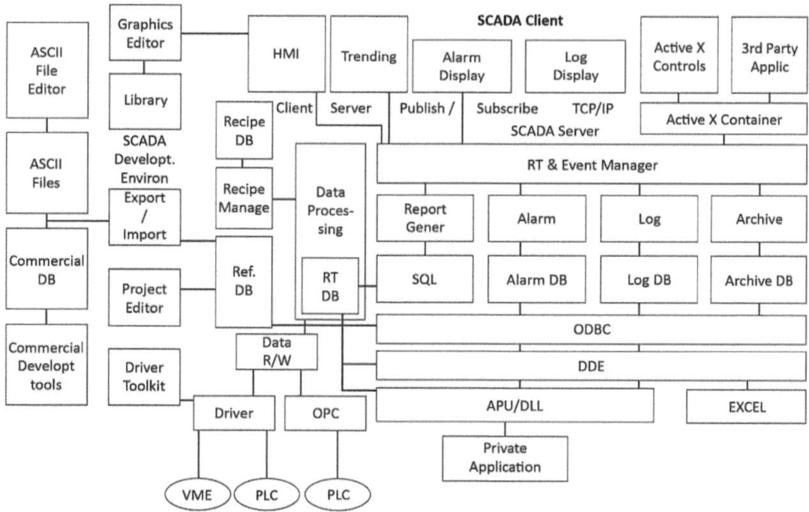

Fig. 2.10 Software architecture

Database Connectivity. The software suite also includes modules such as Graphics Editors, ActiveX controls, IP networking, and several databases used for logging and archiving.

Key features of SCADA software architecture include:

1. **Efficient Design:** SCADA software features a robust and intelligent design based on a real-time database (RTDB) and a server cluster that provides a solid foundation for data-driven control.
2. **Multitasking Capability:** SCADA platforms support multitasking, enabling the simultaneous execution of various industrial processes efficiently.
3. **Real-Time Database (RTDB):** The RTDB acts as a dynamic data repository, serving as the central hub for continuous data flow and supporting informed decision-making.
4. **Distributed Server Network:** SCADA architecture incorporates multiple servers, each assigned specific tasks to gather, analyze, and regulate data in real-time.
5. **Dynamic Data Management:** Servers update data dynamically, manage logs and archives, handle alarm systems, and resolve complex computational tasks.
6. **Parameter-Driven Control:** SCADA operations depend on precise parameter settings to direct servers to control the correct devices and components.
7. **Specialized Servers:** Modern SCADA systems may include dedicated servers for specific tasks, such as meticulous data logging or process optimization [8].
8. **Adaptability:** The SCADA design is versatile and can be tailored to meet the unique requirements of different industrial applications.

9. **Generalized Application:** SCADA software supports a wide range of products and processes, ensuring broad applicability across various industrial sectors [8].

2.12 SCADA Functionality

SCADA systems collect machine data by using sensors and networked devices linked with a Programmable Logic Controller (PLC) or Remote Terminal Unit (RTU). Devices and sensors capture parameters such as temperature, speed, pressure, and weight as raw data. This data is transmitted to a PLC or RTU, where it is processed into meaningful and actionable information.

1. **Trending:**

   (a) Predefined or online trending criteria
   (b) Eight or more trended parameters per graph with support for unlimited graphs
   (c) Support for both historical and real-time trends (not on the same graph)
   (d) Interactive parameters include scrolling, zooming, and cursor positioning
   (e) Integration with a dedicated trending module or ActiveX-based synoptic display

2. **Alarm Handling:**

   (a) Restrictions and status checks via an alarm data server
   (b) Support for complex derived parameter expressions
   (c) Centrally managed alarms with multi-level priority handling
   (d) Features include alarm acknowledgment, suppression, filtering, and grouping
   (e) Reconfigured alarm responses with options for email notifications

3. **Logging and Archiving:**

   (a) Logging and archiving capabilities for medium- and long-term data storage
   (b) Cyclic, time-stamped logging with filterable data options
   (c) Automatic data transfer from logs to archives when logs reach capacity
   (d) User action logging, identified by user ID or station ID
   (e) VCR-style playback functionality for reviewing recorded data

4. **Report Generation:**

   (a) Report generation using SQL queries from the data store, RTDB, or logs
   (b) Partial support for embedding charts in Microsoft Excel
   (c) Automated workflows for report generation, printing, and archiving

5. **Automation:**

   (a) Event-based automation using a scripting language
   (b) Triggerable actions such as loading displays, sending emails, or executing scripts

(c) Support for action sequencing and system configuration using "recipes"

(d) Capability to respond to external events for enhanced automation

2.13 Attack Types in SCADA Systems

Unauthorized access to a SCADA system can be carried out for harmful purposes [8]. To bring the machine out of control in a predetermined sequence, attackers may force pumps and motors to run at higher speeds than normal, turn equipment on and off irregularly, and operate valves and controls inappropriately or in reverse [116]. A dataset, as shown in Fig. 2.9, has been collected from [10]. Table 2.1 presents different types of cybersecurity attacks commonly encountered in SCADA systems.

2.14 SCADA Communication Protocols

SCADA systems require a protocol for data transmission to the operation. Conitel, Profibus, RP-570, and Modbus RTU are a few of the SCADA protocols [8]. These communication protocols are used extensively, although they are vendor-specific for SCADA vendors. Table 2.2 is given below for information:

2.15 Comparison of Control and Automation Systems

Automation systems are specifically designed to perform tasks or operations with minimal or no human intervention. These systems mechanize repetitive tasks, thereby enhancing productivity and efficiency. Automation systems utilize various technologies such as actuators, sensors, and Programmable Logic Controllers (PLCs) to carry out specific operations. Actuators physically move components of

Table 2.1 Types of cybersecurity attacks

Attack type	Targeted/ untargeted	Violated objectives
Denial of service	Targeted	Availability
Eavesdropping	Targeted	Confidentiality, Authorization
Man-in-the-middle	Targeted	Authentication, Confidentiality, Integrity
System break-in	Targeted	Authentication, Authorization
Virus	Untargeted	Availability, Integrity
Trojan	Untargeted	Confidentiality, Authentication
Worm	Untargeted	Confidentiality

Table 2.2 Communication protocols in industrial networks

Protocol	Network infrastructure	Topologies	Data rates	Maximum distance
BITBUS	Fieldbus	Bus	62.5 Kbps, 375 Kbps, 1.5 Mbps	1200 m
DC-BUS	2-wire cable	Line	115.2 Kbps up to 1.3 Mbps	100 km
Distributed network protocol	Ethernet	Line, Peer-to-Peer	100 Mbps, 1 Gbps	100 m
EtherCAT	Ethernet	Ring, Line, Daisy-chain	100 Mbps	100 m
Foundation HSE	Ethernet	Tree, Line, Star, Peer-to-Peer	100 Mbps	100 m
HART	2-wire cable	Point-to-point, Multi-drop	1.2 Kbps	3 km
IEC 60870	Serial, Ethernet	Ring, Tree, Line, Star	N/A	N/A
IEC 61850	Ethernet	Ring, Tree, Line, Star	N/A	100 m
Modbus	Serial, Ethernet	Line, Star, Ring, Mesh (MB+)	100 Mbps, 1 Gbps	N/A
PROFIBUS Fieldbus	Fieldbus	Point-to-point, Bus with spur, Daisy-chain, Tree	9.6 Kbps to 12 Mbps	100 to 1200 m, 15 km for optical channel
PROFINET [58]	Ethernet	Ring, Tree, Line, Star	100 Mbps, 1 Gbps	100 m
RAPIEnet	Ethernet	Line, Ring	100 Mbps	100 m

a machine or system, while sensors measure and monitor parameters like pressure, temperature, or velocity. PLCs control the operations of processes and machines by executing preprogrammed logic routines.

Control systems, on the other hand, are designed to achieve specific goals by manipulating and regulating the behavior of dynamic systems. These systems use feedback loops to observe outputs and adjust inputs in order to maintain the desired performance. Control systems are critical in applications where real-time regulation and precision are essential. They are widely used in engineering, robotics, aviation, automotive, and other domains requiring continuous observation and fine-tuned control of processes.

The differences between control systems and automation systems are presented in Table 2.3.

Table 2.3 Automation and control system comparison

System	Purpose	Control level	Key components	Functionality	Applications
SCADA	Monitor, control, data acquisition	Supervisory level	HMI, RTUs, PLCs, communication infrastructure	Real-time monitoring, simple control actions	Manufacturing, energy, water treatment, transportation
PLC	Automate individual processes	Control level	I/O modules, processor, programming interface	Discrete control tasks	Manufacturing, robotics, machinery automation
DCS	Control complex processes	Supervisory and control levels	Central controllers, field devices, communication networks	Continuous process control and coordination	Petrochemicals, oil refining, power generation
MES	Optimize production processes	Production and operational levels	Work order management, production scheduling, quality control	Improve and monitor production performance	Discrete manufacturing, automotive, electronics
IoT	Enable devices for data sharing	Multiple levels	Devices, sensors, connectivity protocols, analytics	Remote monitoring, data gathering and analysis	Industrial automation, smart homes, healthcare

2.16 Industrial Plant Case Studies

Industrial plants require robust management systems for handling multiple processes simultaneously. Supervisory Control and Data Acquisition (SCADA) systems are an integral component of industrial automation as they provide real-time monitoring and control of machines, resulting in efficient, safe, and reliable machine performance. The systems integrate various sensors, controllers, and communication networks to simplify the provision of data and automation [129].

Project Overview The primary objective of this case study is to develop a robust system that processes real-time data and keeps remote industrial sites under surveillance. Industries can attain maximum performance, prevent equipment breakdown, and maintain safety standards by implementing a SCADA-based temperature monitoring system.

(a) **Objective:** Process real-time data and keep remote industrial sites under surveillance
(b) **Scenario:** Temperature logging system for a remote plant

System Components and Functionality The industrial temperature monitoring system consists of a number of components that work in unison to ensure smooth operation. The key functions include the following:

1. Microcontroller-linked temperature sensors, linked with PC interface.
2. Data processing software installed on PC.
3. Continuous temperature data collection, values displayed by microcontroller.
4. Computer display temperature controls facilitated by software.
5. Microcontroller switches relay on in case of sensor temperature going beyond limit.
6. Relay contacts switch heaters on and off.

Eight multiplexed temperature sensors connect to a microcontroller via an ADC0808. The microcontroller serializes sensor readings through a MAX32 to the PC's COM port, storing them in "daq.mdb" and displaying them on the "DAQ System" interface. Figure 2.11 was from [1].

Figure 2.11 demonstrates the Block Diagram of the Industrial Plant's Temperature Controlsystem with Power supply, A to D converter, Temperature sensor, Relay driver, Microcontroller, 4 Relays, RS232 Interface, and connection to PC. These seem to demonstrate the components and interconnects of a system.

An interactive definition of lower and upper limits and set points is supported by the computer interface. The microcontroller drives a relay driver IC if the temperature from a sensor is approaching a defined threshold, controlling heaters. Temperature over a high/low limit triggers alarms. A Fig. 2.12 was borrowed from [1].

Figure 2.12 shows a GUI for a Data Acquisition and Logging System. The interface displays various recorded temperatures, limits, and related data in a tabular format. It includes temperature readings, limits, alarms, and other system parameters. This example demonstrates the SCADA system's ability to collect real-time data, automate responses to enhance performance, and monitor and control remote industrial activities.

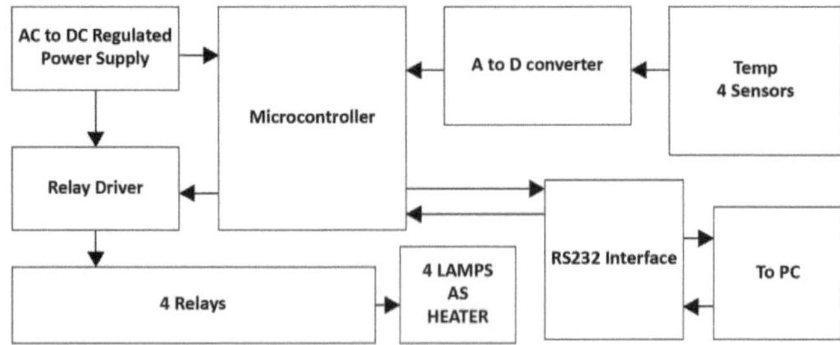

Fig. 2.11 Block diagram of temperature control industrial plant

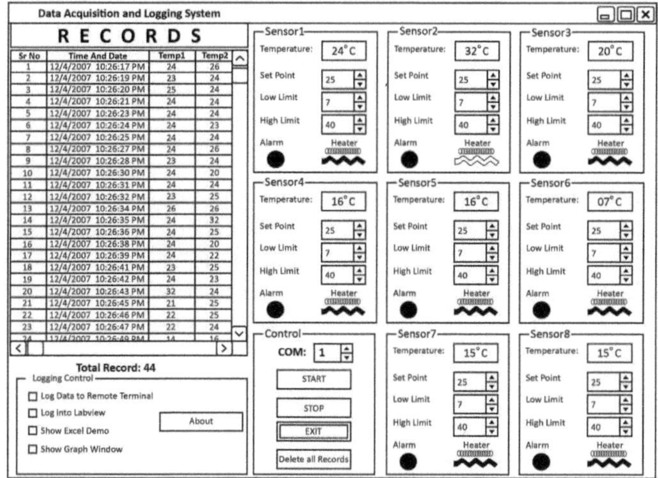

Fig. 2.12 SCADA for remote industrial plant

2.17   Security and Cybersecurity in SCADA Systems

SCADA systems play a significant role in monitoring and controlling industrial processes and critical infrastructure. Security and cybersecurity controls of SCADA systems are necessary to avert the risks of eventual disruptions, unauthorized use, loss of data, and physical harm. SCADA cybersecurity involves the establishment and safeguarding of SCADA networks. Millions of individuals depend on such networks for automatic control and remote human monitoring of critical goods and services, such as water, natural gas, electricity, and transportation.

2.18   Security and Cybersecurity in SCADA Systems

SCADA systems are a vital component of industrial process control and critical infrastructure monitoring. Ensuring robust security and cybersecurity measures is essential to prevent potential disruptions, misuse, data loss, and physical damage. SCADA cybersecurity involves both establishing and protecting SCADA networks. Millions of people rely on these systems for automated control and remote human oversight of critical utilities such as water supply, natural gas, electricity, and transportation services.

2.19   Security Challenges in SCADA Systems

Because they manage infrastructure that is often life-critical, SCADA systems are tempting targets for security exploits. Below are some key security challenges of SCADA systems:

1. **Cyberattack Susceptibility:** SCADA systems are more prone to various cyber threats such as ransomware, malware, and phishing. Exploitation of software or hardware weaknesses often involves external network connectivity. The unauthorized access could lead to substantial infrastructure takeover, operational disruption, and sensitive information theft, which are serious threats to industrial systems. This problem is aggravated by the fact that many SCADA systems are running on legacy technology that has poor security solutions, upgrades, and patches in place. These legacy devices are easy to compromise and, therefore, an attractive target for attackers who want to disrupt industrial systems.
2. **Weak Encryption:** Lack or misconfiguration of robust encryption is a common problem in SCADA environments, which facilitates eavesdropping on the communication between SCADA units and control centers. Lack of strong encryption may allow attackers to intercept communications, modify data, or launch sophisticated cyberattacks. These weaknesses compromise the integrity as well as the confidentiality of industrial control data and can lead to system compromise.
3. **Physical Risks:** Access policies play a critical role in security, as unauthorized individuals gaining access to key facilities (e.g., substations, control rooms) can disrupt electricity system operations and potentially lead to sabotage. Physical security measures such as barriers, access controls, guards, and video surveillance are essential to prevent unauthorized entry and protect SCADA hardware from intrusion. Lack of adequate physical protection leaves systems vulnerable to vandalism and unauthorized modifications, which can endanger the entire operation.
4. **Insider Threats:** SCADA systems are at a higher risk to insider threats, which can be intended or unintended. Trusted insiders, including employees, contractors, and third parties given broad access, can inadvertently or maliciously put the system at risk. Disgruntled employees may sabotage essential services, steal sensitive data, or cause service outages. Human oversight or carelessness can also have sensational consequences. To mitigate the risks, strict access control should be enforced, and security awareness training should be delivered regularly.

2.20   Strategies to Enhance SCADA Security

It is critical that Supervisory Control and Data Acquisition (SCADA) systems be secured as much as possible to prevent industrial processes from falling victim to cyber attacks and operational sabotage. Strong security is about using multiple levels of security to adequately reduce exposure. The following types of security ploys are mandatory for making SCADA systems safer:

1. **Network Segmentation:** SCADA networks should be isolated from external and enterprise networks to minimize the attack surface. Accordingly, firewalls and access controls should be implemented to achieve proper segmentation and prevent unauthorized access.
2. **Updating and Patching:** SCADA systems and devices must be regularly updated and patched to apply the latest security fixes. This helps close known vulnerabilities and strengthens defenses against emerging threats.
3. **Secured Communication Protocols:** Secure communication protocols, including encryption algorithms, should be employed to ensure data integrity and confidentiality during transmission between SCADA elements.
4. **Access Control:** Role-based mandatory access control should be enforced to prevent unauthorized access to the system. Only authorized personnel should be allowed to access or modify SCADA systems to avoid tampering.
5. **IDS or IPS:** Intrusion Detection Systems (IDS) and Intrusion Prevention Systems (IPS) are two distinct technologies that help secure organizational networks. IDS monitors network traffic for suspicious activity, while IPS not only detects threats but also takes automatic action to prevent them. Both systems work by analyzing network traffic to identify signs of malicious behavior [64].
6. **Security Audit and Penetration Testing:** Security audits and penetration testing should be conducted on SCADA systems at regular intervals to systematically identify vulnerabilities and weaknesses. These practices enable continuous improvements in the system's security posture.
7. **Employee Training:** Extensive training should be provided to employees to raise awareness about cybersecurity, including recognizing phishing attempts and adopting strong password management practices.

2.21 Evolution of SCADA Systems

Some good progress has been made toward more efficient, secure, and scalable industrial automation through the remarkable strides made in hardware, software, and networking technologies the development of SCADA systems. Both types of systems are updated incrementally and incrementally to bring them in line with market demands and technological advancements. These updates are designed to improve system performance, add new features, and address security issues in order to improve the operation.

SCADA architecture evolution. One of the major advances in SCADA architecture concerns the move away from atomic parameter range space design to an entity-based approach. Previous SCADA systems represented the process in separate variables as a list of single "tags," and the amount of these tags could be very large as the number of data to be handled grew, for example one could have the tags networked between two SCADAs for communications purposes but if two SCADA's need to access a tag on another SCADA slave, this tag could not been fully accessed to its fullest capabilities. The entity-based model, on the other hand, treats devices and processes as instances of (structured) objects or classes. This effectively simplifies system settings and management, and provides better data organization, scalability, and maintainability.

Another innovation is the support of multi-team development, which is necessary for large-scale industrial development with numerous engineering teams. Most contemporary SCADA programs allow collaboration and division of labor; teams aim to design and configure separate parts of the entire system, which they can all share. This leads to a drastic increase in development efficiency and a reduction in deployment time).

SCADA systems are also adopting new technologies to enhance their functionality and usability. Web technology adoption has made it possible to develop user-friendly user interfaces (UI) and provide remote control capabilities, so that operators can remotely monitor and control industrial processes on geographically distributed sites. Furthermore, improved operational process control is able to create standardized communication protocols between SCADA client and server modules, to ensure the secure and reliable transmission of data. Integration with third-party systems has also been strengthened, building a higher compatibility with various devices, software, and tools outside the Graphen product range.

One particularly transformative, well-known innovation is the introduction of OLE for Process Control (OPC), which is a collection of standards defining a reliable way for servers to interact with clients in a SCADA context. OPC, as an open interface standard, can make different system components communicate with each other transparently and can be easily integrated with third-party modules to improve SCADA systems' adaptability and interoperability. This normalization is particularly advantageous in sectors, for example, in which SCADA has o interface with a variety of industrial automation.

Also, modern SCADA systems focus on third-party integration in order to provide more extensibility and customization. Beyond this, their connections to third-party electronics, applications, or industrial equipment allow SCADA functionality to be adapted to individual processes. This flexibility allows for improved resource prioritization, optimized process chains, and deeper insights.

In conclusion, the development of SCADA systems is a never-ending attempt to further increase the efficiency, security, and flexibility of industrial automation. Thanks to frequent updates, the rise of an entity-centric architecture, ease of collaborative development, a web-based tech stack, and third-party integration options, the current generation of SCADA systems is up to the task of what today's complex industrial orgs need. All of this novelty is providing system flexibility and system agility, not to mention a system agility, but also ensures SCADA remains an industrial control building block for our times.

2.22 Benefits of SCADA Systems

There are many advantages of SCAD (Supervisory Control and Data Acquisition) systems, such as safety and reliability improvements, the decrease of downtime, and improving productivity and efficiency. Key benefits are explained as follows:

1. **Rich Functionality and Development:**
   By interoperating with Internet of Things (IoT) devices, SCADA systems support a wider range of connectivity and real-time connectivity to different data sources from heterogeneous domains. This makes the data-getting operation easier, and real monitoring, control, and decision-making are enhanced.

2. **Reduced End-User Development Effort:**
   Contemporary SCADA systems now use edge computing to perform data processing at the source. This cuts down on latency, improves the response action that's possible, and allows for real-time action in remote and bandwidth-constrained locations.

3. **Enhanced Reliability and Resilience:**
   The introduction of artificial intelligence (AI) and machine learning (ML) facilitates predictive analytics. SCADA monitoring systems can detect abnormalities, predict equipment failure, and identify usage patterns, leading to more informed maintenance and resource planning.

4. **Collaborative Development Framework:**
   Cloud SCADA solutions: With a cloud-based SCADA, this is not a problem and can scale as much as like and, of course, can access operations from anywhere. This fosters teamwork and accelerates the process and the rapid development and deployment of new features.

5. **Advanced Monitoring and Control:**
   To mitigate the growing cyber threats, contemporary SCADA systems utilize strong security functions including user authentication, data encryption, and intrusion detection systems to guarantee the integrity of data and its availability.

6. **Enhanced Efficiency and Productivity:**
   Novel technologies, such as AR (augmented reality) and VR (virtual reality), enhance interaction between the operator and the systems. AR displays real-time information in relation to the physical world to support situational awareness, and VR can facilitate assisted immersive training in complex tasks.

7. **Accelerated Decision-Making:**
   Next-generation SCADA systems will provide more user-friendly HMIs (Human-Machine Interfaces) using voice recognition, gesture interfaces, and touch screens that enhance control and minimize human errors.

8. **Data Visualization:**
   Intelligent features of SCADA systems organize intricate data into charts, graphs, and dashboards, enabling operators to quickly make sense of trends, thus optimizing operational choices.

9. **Remote Access:**
   Remote monitoring and control: SCADA's key capability is to provide operators the ability to monitor systems remotely 24/7 from multiple sites and to respond quickly to operational problems.
10. **Predictive Maintenance:**
    By examining historical and current data, SCADA techniques also facilitate predictive maintenance models, which can help in preventing equipment or asset failures and in lowering operational costs.
11. **Data Logging and Analysis:**
    Operating data is logged permanently by SCADA systems, through long-term monitoring, the inefficiencies can be discovered, performance can be assessed, and improvement opportunities can be found.
12. **Regulatory Compliance:**
    Compliance SCADA systems enable organisations to embed regulatory requirements in their daily operations to help them operate in compliance with industry standards and safety regulations, reducing regulatory risk.
13. **Scalability:**
    Scalability SCADA systems are inherently scalable, enabling organizations to grow or shrink the capacity of the system based on changing operational requirements.
14. **Improved Security:**
    Current SCADA systems, employing layered security with encrypted communication, access control, and real-time intrusion tracking protection, protect such critical infrastructure from cyber attacks.
15. **Centralized Control:**
    SCADA provides centralized control for various pieces of equipment and systems, as well as for many applications and processes. This degree of control serves to facilitate improved process control, better decision-making, and increased support of the administrative control of a legacy system.
16. **System Integration:**
    SCADA systems are compatible with integrating third-party hardware or software to facilitate smooth data transmission, process synchronization, and system-wide optimization.
17. **Operational Insights:**
    SCADA systems turn aggregated data from operations into actionable insight, and with it, organizations can increase efficiency and competitive edge.
18. **Real-Time Alerts:**
    SCADA systems provide immediate notification of abnormal or disastrous events, enabling the operator to react quickly enough to prevent a shutdown or at least lessen the damage.
19. **Support for Long-Term Planning:**
    SCADA platforms support trend analysis and forecasting–they can provide valuable strategic analytics that enable making better decisions about infrastructure planning and operation, and future-proofing.

2.23   Future Trends of SCADA Systems

Technological advancements such as increased automation, advanced analytics, and the incorporation of non-renewable sources will determine the future of Supervisory Control and Data Acquisition Systems (SCADA) in the utilities industry. Some primary trends that have been found would be the less requirement for human intervention with automation, real-time decision-making, or the usage of artificial intelligence (AI) and machine learning (ML) for predictive maintenance and advanced asset performance, and less downtime.

(a) **IoT Integration**

SCADA systems will depend more and more on real-time data provided by IoT-enabled sensors integrated into machines and devices. This will enable full monitoring and active control of the process, so that performances of the process can be improved and operational changes can be made immediately.

(b) **Edge Computing**

Edge computing trends will move data be processed closer to the edge, rather than in turning to traditional cloud infrastructure. The latency will be substantially decreased, and the speed of operational decision-making will be increased even faster.

(c) **Cloud Integration**

SCADA in the cloud system. The advantage that the cloud brings is that large amounts of scada data can be processed by such cloud platforms with the application of predictive capabilities by means of AI/ML. This will lead to more intelligent, data based operations and proactive maintenance, with product managers for operations.

(d) **Artificial Intelligence (AI)**

Discover how AI will leverage temporal data analysis to detect trends, anticipate machine failure, and execute inferences to improve overall system performance. SCADA systems will be using AI to optimize settings for efficiency.

(e) **Enhanced Cybersecurity**

Stronger security will also be critical as more networked items are incorporated. Multi-factor authentication, encryption, intrusion detection, and real-time threat monitoring will be norms for securing SCADA environments.

(f) **Data Analytics and Visualization**

Sophisticated data analytics and visualization capabilities will deliver greater understanding into operations, trends, and outliers. Real-time interactive dashboards will enable informed, data-driven decisions.

(g) **AR and VR** AR/VR in SCADA Operation and Maintenance In Mission Critical Use-Cases CVAR and VRA is going to disrupt the online mode of operating the SCADA system. Augmented reality and virtual reality can be used for the real time monitoring of the substations; these technologies will help in providing online monitoring to operators. These technologies will enhance context awareness, training, and user experience.

(h) **Energy Efficiency and Sustainability**

SCADA systems will help in the energy-saving initiatives through optimal operation and the incorporation of renewable energy. Sustainable development and eco-friendly energy management will be boosted.

(i) **Remote Access and Interoperability**

Better interoperability and a wider use of common communication protocols to encourage integration of heterogeneous systems. This would enhance remote monitoring and operation and flexibility, and scalability.

(j) **Predictive Analytics**

Predictive analytics will enable SCADA systems to predict component failures and pre-schedule maintenance. This will minimize downtime, maximize asset usage, and lengthen the life of highly stressed parts.

2.24   Summary

SCADA systems play a critical role in today's industrial automation, allowing real-time monitoring, control, and management to improve production, safety, and efficiency. SCADA for the Fourth Industrial Revolution backboned by IoT, AI, and Strong Cybersecurity, SCADA is propelling the Fourth Industrial Revolution. Advantages, challenges of SCADA systemsSCADA systems provide benefits, but also are difficult to integrate with legacy systems and secure against cyber threats. With the proliferation of digital connectivity, these are the kinds of issues that we need to overcome to keep SCADA secure and reliable.

Chapter 3
Understanding Communication and Protocols in ICS: Securing Network Infrastructure and Data Exchange

Abstract Communications and protocols in ICS are critical to ensuring the reliability and safety of data communication across various industries, including manufacturing. This chapter provides an in-depth exploration of the key components of data communication, focusing on the different protocols and their layers. It also examines various communication methods, industrial protocols, and network analysis techniques, alongside strategies for enhancing network security. By understanding these communications and protocols, readers will gain valuable insights into how they support and secure safe industrial operations.

Keywords Industrial Control Systems (ICSs) · Data communication · Communication protocols · Network security · Industrial protocols · Communication methods · Protocol layers · Manufacturing safety · Network analysis

3.1 Introduction

Industrial Control Systems (ICS) are crucial in modern industries and are responsible for controlling critical processes in areas such as production, energy, and transportation. The robust communication protocols that these systems rely on enable real-time data exchange between control centers and field equipment, resulting in accurate control and monitoring. Automation and data-driven decision-making are transforming ICS, as Artificial Intelligence (AI) and the Industrial Internet of Things (IIoT) continue to be integrated into industrial operations.

According to Nadir, communication protocols differ significantly across corporate, operational, and control networks, each tailored to meet distinct security, reliability, and performance requirements. LARGriff notes that networks used in corporate environments may employ protocols similar to traditional IT networks, whereas operational and control networks rely on custom protocols designed for resilience and efficiency. A key concept in control room and network architecture is that sensors, RTUs, and PLCs can communicate seamlessly.

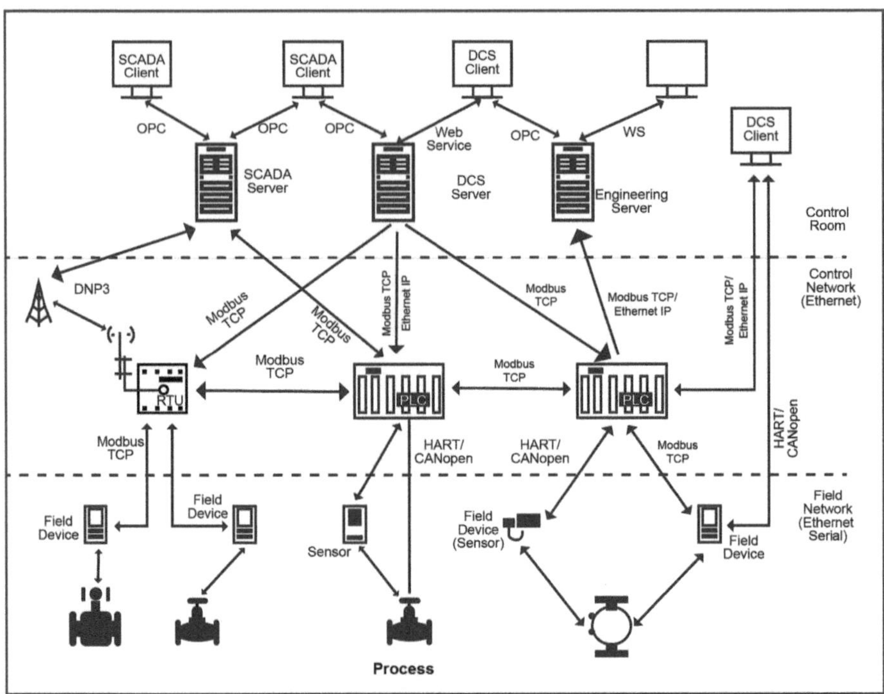

Fig. 3.1 Communication protocols in ICS architecture

Standards such as Modbus and Distributed Network Protocol 3 (DNP3) are commonly used to facilitate interoperability and standardization. An example from [54] is illustrated in Fig. 3.1.

Figure 3.1 illustrates communication protocols in an ICS architecture, including SCADA, DCS, clients, OPC, web services, servers, control rooms, and protocols such as Modbus, Ethernet/IP, DNP3, HART, and CANopen. Each component is essential for industrial communication, data sharing, and control. Central systems like SCADA and DCS monitor and manage industrial operations, while clients provide user interfaces.

3.2 Data Communication

Data communication is the exchange of information between sources, nodes, or devices. It enables the transmission and reception of information across networks, whether wireless or wired. In ICS, data transfer is critical for communication between sensors, actuators, controllers, and other elements, facilitating real-time monitoring, control, and synchronization of industrial processes and machinery.

3.2.1 Data Communication Components

A communication system includes, but is not limited to:

1. **Message:** Data or information transmitted from a source device to a destination device is called a message. Messages may take any suitable form, including text, audio, video, images, or combinations thereof. Messages are the essence of any communication network.
2. **Sender and Recipient:** For a message to be conveyed, there must be an identifiable source. The sender is a component of the data transmission system that generates and transmits messages. This device may be a computer, cell phone, telephone, laptop, video camera, or workstation.
3. **Receiver:** The receiver is the device where the message from the sender is ultimately received. It may be a personal computer, mobile phone, workstation, or other compatible device.
4. **Transmission Medium/Channel:** The physical medium through which data is transmitted from sender to receiver is called the transmission medium. Examples include fiber-optic, coaxial, and twisted-pair cables.
5. **Set of Rules (Protocol):** A protocol is a set of rules that govern the data communication process, specifying transmission, routing, and verification of data. Both communicating devices must support the same protocol to ensure interoperability.

3.3 Protocol

A protocol is a framework of guidelines and rules that govern communication between network devices. It ensures consistent, reliable communication between machines from different manufacturers, guaranteeing compatibility across diverse network settings.

Types of Protocols
There are various types of data communication protocols, each designed for specific functions at different layers of the Open Systems Interconnection (OSI) model. Major categories include:

1. **Communication Protocols**

   (a) **TCP/IP:** The Transmission Control Protocol/Internet Protocol (TCP/IP) suite forms the foundation of modern networking, including the Internet. TCP ensures reliable, connection-oriented data transfer, while IP handles routing and addressing of packets.
   (b) **HTTP/HTTPS:** Hypertext Transfer Protocol (HTTP) and its secure variant (HTTPS) are used to transfer data between web servers and browsers. These

protocols operate over TCP/IP, requesting and delivering web content such as HTML pages and multimedia files.

(c) **File Transfer Protocol (FTP):** FTP enables file upload and download interactions with remote systems over TCP/IP, facilitating easy access to file systems on servers.

(d) **SMTP/IMAP/POP3:** Simple Mail Transfer Protocol (SMTP) is the standard for exchanging email between servers. Clients retrieve and manage email using Post Office Protocol 3 (POP3) or Internet Message Access Protocol (IMAP).

(e) **Domain Name System (DNS):** DNS translates human-readable domain names (e.g., www.example.com) into IP addresses. It operates over UDP/TCP and maintains a hierarchical, distributed database of domain-to-IP mappings.

2. **Network Protocols**

(a) **Address Resolution Protocol (ARP):** ARP maps IP addresses to MAC addresses within a local network, enabling communication within the same subnet.

(b) **Internet Control Message Protocol (ICMP):** ICMP provides diagnostic and error-reporting capabilities for network devices. Utilities like `ping` and `traceroute` rely on ICMP.

(c) **Internet Group Management Protocol (IGMP):** IGMP manages multicast group memberships, allowing devices to receive multicast transmissions, useful for streaming and conferencing.

3. **Application Layer Protocols**

(a) **DNS:** Resolves domain names into IP addresses at the application layer.

(b) **Simple Network Management Protocol (SNMP):** SNMP monitors and manages devices on IP networks, supporting remote configuration, fault monitoring, and performance analysis.

(c) **HTTP/2:** HTTP/2 improves HTTP efficiency by multiplexing multiple streams over a single TCP connection and reducing latency through header compression and prioritization.

4. **Security Protocols**

(a) **SSL/TLS:** Secure Sockets Layer and Transport Layer Security (TLS) encrypt communications between devices to ensure confidentiality and data integrity.

(b) **IPsec:** Provides secure communication at the IP layer using encryption and authentication, enabling private data transfer over public networks.

(c) **Secure Shell (SSH):** SSH enables secure remote login and file transfer by encrypting data exchanged between two devices over unsecured networks.

5. **Data Link Layer Protocols**

   (a) **Ethernet:** The dominant wired networking technology for LANs and WANs, defining standards for cabling, signaling, and data framing.

   (b) **Point-to-Point Protocol (PPP):** PPP establishes direct connections over serial links, supporting multiple network layer protocols.

6. **Routing Protocols**

   (a) **Open Shortest Path First (OSPF):** A link-state interior gateway protocol that determines the shortest path between routers within an autonomous system.

   (b) **Border Gateway Protocol (BGP):** A path-vector protocol that manages routing between autonomous systems across the Internet based on policy and path attributes.

7. **Transport Layer Protocols**

   (a) **TCP:** Ensures reliable, ordered delivery of data packets between devices using error checking, acknowledgment, and retransmission.

   (b) **UDP:** A connectionless protocol providing low-latency communication without guaranteed delivery, suitable for real-time applications like streaming or gaming.

3.3.1 Layers of Operation Protocols

Protocols in the application layer facilitate message exchange between application processes (clients and servers) operating on various end systems. Specifically, the application layer represents an abstract layer responsible for providing communication services in both the OSI (Open Systems Interconnection) and TCP/IP (Transmission Control Protocol/Internet Protocol) models. While the OSI model consists of seven layers, the TCP/IP model adopts a simplified four-layer architecture. These four layers are:

(a) **Application Layer:**
   This layer provides a high-level interface for communication between networked hosts and is responsible for supporting application-level functions such as file transfers, email, and web browsing. It defines the protocols and data formats used in these interactions. In the OSI model, Layers 5 (Session), 6 (Presentation), and 7 (Application) collectively correspond to the application layer. In the TCP/IP model, all these functionalities are combined into a single Application Layer.

(b) **Transport Layer:**
   The transport layer provides reliable end-to-end communication services for applications across a network. It ensures that data is transferred accurately

and in sequence between host systems. Protocols such as TCP and UDP operate at this layer, offering mechanisms for flow control, error detection, and retransmission. In the OSI model, this is represented as Layer 4.

(c) **Network Layer:**

The network layer is responsible for the delivery of packets from the source host to the destination host across multiple networks. It manages logical addressing (such as IP addresses), routing, fragmentation, reassembly, and error reporting. Routers operate at this layer to forward packets across networks. Key protocols include IPv4, IPv6, ICMP, and routing protocols like OSPF and BGP. This layer is known as Layer 3 in the OSI model.

The network layer may also include the following sublayers:

- Subnetwork Access: Interfaces with data link protocols (e.g., X.25)
- Subnetwork Dependent Convergence: Ensures compatibility between dissimilar networks
- Subnetwork Independent Convergence: Manages data forwarding across multiple networks

The network layer supports connectionless packet forwarding, error detection (not correction), and uses interface-level addressing rather than node-specific identifiers.

(d) **Access Layer:**

Also known as the Network Access Layer in the TCP/IP model, this layer encompasses functionalities equivalent to Layers 1 (Physical) and 2 (Data Link) in the OSI model. It facilitates the physical connection between end devices and the network infrastructure. Devices such as computers, printers, and mobile phones rely on this layer for addressing and data link establishment. Protocols like Ethernet, Wi-Fi, and PPP operate here. This layer also provides basic mechanisms for authentication and access control, such as IEEE 802.1X.

The physical sublayer within the access layer is responsible for transmitting raw bits over a physical medium. Although security features are minimal at this level, authentication mechanisms like 802.1X are implemented to control access to the network. Advanced security features are typically provided by upper-layer protocols and will be discussed in the context of specific protocol implementations in later sections.

3.4 Overview of Industrial Control Protocols

Communication protocols play a pivotal role in the infrastructure of Industrial Control Systems, encompassing corporate, operational, and control networks. While corporate networks generally utilize standard Information Technology (IT) communication protocols, operational and control networks rely on specialized protocols tailored for enhanced performance, reliability, and precision. Among the most widely used communication protocols in industrial environments are Modbus and Distributed Network Protocol version 3 (DNP3).

Below is a discussion of some commonly employed industrial communication protocols:

(I) **Modbus Protocol**

Modbus is an open standard protocol originally developed by Modicon in 1979 for use with programmable logic controllers (PLCs) [31]. It is widely adopted across various industries due to its royalty-free nature, allowing easy integration by different device manufacturers. Modbus facilitates the transfer of signals from instrumentation and control devices to a supervisory controller or data acquisition system, such as a temperature and humidity sensor transmitting data to a central monitoring computer. In Supervisory Control and Data Acquisition (SCADA) systems, Modbus is commonly used to connect Remote Terminal Units (RTUs) with supervisory computers [8]. It supports both serial communication (RTU and ASCII formats) and Ethernet-based communication (TCP/IP).

(a) Modbus was introduced by Modicon an application layer protocol for industrial control systems [31].
(b) It is primarily used for real-time monitoring and control applications, particularly with Modbus TCP.
(c) Modbus TCP utilizes TCP packets containing Modbus Application Protocol (MBAP) headers and Protocol Data Units (PDUs).
(d) The protocol allows flexibility in transaction handling through user-defined function codes.
(e) While Modbus initially supported only serial communication, TCP/IP support was later added; however, security considerations were minimal in the original design.

(II) **Distributed Network Protocol (DNP3)**

The Distributed Network Protocol version 3 (DNP3) is primarily used in electric and water utility automation systems. It was designed to facilitate reliable data acquisition and control communications between SCADA Master Stations, Remote Terminal Units (RTUs), and Intelligent Electronic Devices (IEDs) [55]. Although DNP3 is mostly employed in utility sectors, it provides a robust solution for master-to-RTU or IED communications. For communication between master stations, ICCP (IEC 60870-6) is used. DNP3 competes with protocols such as Modbus and IEC 61850 [155].

(a) Developed by GE Harris, DNP3 enables seamless communication between SCADA components [127].
(b) It supports reliable data transmission over both serial and IP networks using Cyclic Redundancy Check (CRC) validation.
(c) Bidirectional communication allows interaction with remote field devices.
(d) Data exchange is structured through classification and instructions to transmit alarms and measurements.

(e) Messages consist of an application header, function code field, and data payload.

(f) Security is enhanced through certificate-based and shared key authentication mechanisms.

(III) **IEC 60870-5-104**

IEC 60870-5-104, commonly referred to as IEC 104, is a protocol standard for telecontrol equipment and systems. It facilitates serial data transmission with coded bits over TCP/IP networks, enabling real-time monitoring and control over geographically distributed systems.

(a) It is a globally recognized standard for communication in SCADA systems and electrical substations.

(b) The application layer encapsulates and transmits the Application Service Data Unit (ASDU).

(IV) **IEC 61850**

IEC 61850 is an international standard developed by the International Electrotechnical Commission (IEC) Technical Committee 57. It defines communication protocols specifically for Intelligent Electronic Devices (IEDs) used in electrical substations. It is a critical component of modern substation automation systems.

(a) It serves as a comprehensive standard for substation automation and communication.

(b) Communication architecture is divided into three layers: process layer (primary equipment), bay/interval layer (secondary equipment), and station layer (monitoring and control).

(V) **IEC 61400-25**

IEC 61400-25 is a multi-part standard that defines a communication framework for monitoring and controlling wind power plants. It ensures interoperability and data consistency across devices from different manufacturers.

(a) It extends the IEC 61850 standard for application in wind power plant systems.

(b) It facilitates communication between central supervision systems and local subsystems.

(c) It ensures interoperability among diverse equipment vendors.

(VI) **IEEE C37.118**

IEEE C37.118 is a standard that supports the exchange of synchronized phasor measurements, also known as synchrophasors, between power system components. It defines message types, structures, and synchronization requirements for real-time data communication.

(a) It is used to synchronize data communication between substations.

(b) It specifies data formats and time synchronization constraints for Phasor Measurement Units (PMUs) and Phasor Data Concentrators (PDCs).

   (c) It defines four message types: data, header, configuration, and command, supporting real-time system analysis.

3.4.1   Common Industrial Protocol

In industrial automation, the **Common Industrial Protocol (CIP)** is a widely used communication protocol designed to facilitate seamless data exchange between heterogeneous equipment. CIP, developed and maintained by the Open DeviceNet Vendors Association (ODVA), is implemented in various forms tailored to specific networking technologies.

Figure 3.2 is adapted from [35] to illustrate the communication structure and application of CIP in industrial systems.

The Association for Organics Recycling, in collaboration with the Environment Agency, has also introduced a standardized methodology for bioaerosol monitoring at open composting facilities, highlighting the relevance of reliable communication protocols in diverse industrial applications.

In ICS, common Ethernet-based industrial communication protocols include **EtherNet/IP**, **PROFINET**, **EtherCAT**, and **Modbus TCP**. These protocols differ in terms of data formats, transmission mechanisms, real-time performance, and embedded security features.

Among these, **Modbus** remains the most extensively adopted protocol in industrial environments due to its simplicity, open standard, and widespread compatibility with various devices and control systems.

Figure 3.2 illustrates the architecture of the Common Industrial Protocol (CIP) integrated with the OSI Model. It highlights various components including motor control systems, transducers, as well as associated profiles and protocols such as

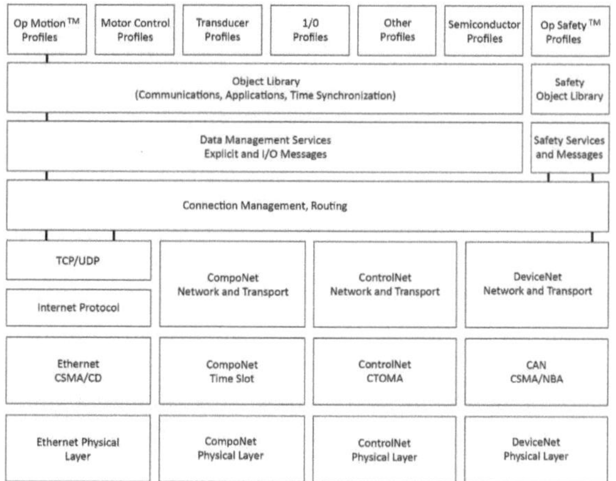

Fig. 3.2 CIP architecture with OSI model networks

CIP Safety, TCP/UDP, ControlNet, DeviceNet, and EtherNet/IP. These protocols collectively enable robust, efficient, and reliable industrial control and communication.

1. **EtherNet/IP:** EtherNet/IP (Ethernet Industrial Protocol) is an industrial network protocol that applies the Common Industrial Protocol (CIP) over standard Ethernet. It is one of the most widely adopted industrial communication technologies in North America and is extensively used across manufacturing, hybrid, and process industries. Managed by the Open DeviceNet Vendors Association (ODVA), EtherNet/IP utilizes TCP/IP and UDP/IP protocols for transport and provides real-time communication and control capabilities.
2. **ControlNet:** ControlNet is a high-speed fieldbus protocol that implements CIP over a Concurrent Time Division Multiple Access (CTDMA) data link layer. Designed for deterministic and repeatable communication, it supports both scheduled and unscheduled messaging. However, the adoption of ControlNet was relatively limited due to its fixed data rate of 5 Mbps, which restricted scalability and flexibility in evolving industrial environments.
3. **DeviceNet:** DeviceNet is a communication protocol used to interconnect industrial control devices such as sensors, actuators, and controllers. It operates by implementing the Common Industrial Protocol over the Controller Area Network (CAN) physical layer and defines application layers for device profiles, diagnostics, and configuration. DeviceNet supports distributed I/O systems and offers features such as data sharing, device interoperability, and real-time control.
4. **CompoNet:** CompoNet is a specialized industrial networking protocol based on Time Division Multiple Access (TDMA) communication techniques. TDMA enables multiple devices to share a common frequency channel by assigning distinct time slots for data transmission. CompoNet leverages this method to ensure high-speed and deterministic data exchange, making it suitable for applications requiring frequent updates with low latency and minimal communication overhead.

3.4.1.1 Comparative Table of Industrial Control Protocols

Industrial Control Systems (ICS) rely on diverse communication protocols for efficient data exchange, interoperability, and real-time process control. These protocols vary in speed, security, scalability, and application suitability, balancing high-speed transmission with reliability, security, and legacy system compatibility.

Selecting the optimal protocol for a given industrial environment requires understanding these differences. The following table compares common ICS protocols, highlighting transmission modes, security features, latency, and typical applications. Network topology, hardware support, and cybersecurity integration also influence protocol choice. With increasing interconnectedness, demand is growing for advanced, secure, and resilient protocols, particularly in critical infras-

Protocol		Characteristics	Security		Protocol layers implemented				Security measures
	Variant		Cipher	Authentication	Media Access	IP	Transport	Application	
Common Industrial Protoco (CIP)	DeviceNET	Adaptation for CAN	No	No	X	X	X	X	CIP have CIP Safety TM technology at application level Complement with well-known measures about control network segmentation and isolation
	ControlNET	Adaptation for CTDMA	No	No	X	X	X	X	
	CompoNET(CIP)	Adaptation for TDMA	No	No	X	X	X	X	
	Ethernet/IP	Adaptation for TCP/IP	No	No				X	
Modbus	Serial	Open standard Widely used in Industry	No	No	X			X	Adopt, if it's possible, cypher (SSL, VPN) or traffic inspection measures such as IDS (Snort), IPS (Tofino), etc.
	TCP		No	No				X	
DNP3		Electric Sector Low presence in Europe	No	Yes (DNP3 secure)		X	X	X	Implement DNP3 Secure
ProfiBus		Open standard ProfiBus DP & PA variants	No	No	X	X		X	Network segmentation, data cipher and traffic inspection
ProfiNET		Profibus evolution for Ethernet	No	No				X	Follow "Profinet Security Guide"
Powerlink Ethernet		Open Standard	No	No				X	Network segmentation and apply normal Ethernet security measures
OPC		Communications operation framework Originally Windows-based	No	Yes (OPC UA)				X	Implement OPC UA
EtherCAT		Open standard For very low update cycles	No	No				X	Network segmentation and apply perimeter security

Fig. 3.3 Comparative table of protocol

tructure sectors like energy, water treatment, and transportation, where encryption, authentication, and intrusion detection are crucial.

Figure 3.3 presents a detailed comparative analysis of the aforementioned protocols, as referenced from [35].

Figure 3.3 Illustrate ICS protocols. It includes information about security measures, protocol layers implemented, variants, ciphers, authentication methods, media types, access control, and specific protocols like DeviceNET, CompoNET, Ethernet/IP, Modbus, DNP3, ProfiBus, ProfiNET, Powerlink Ethernet, OPC, and EtherCAT.

3.4.2 Types of Communication Methods

Industrial Control Systems use several communication mechanisms to facilitate the transmission of data and directives across distinct components. These communication methods are crucial for maintaining seamless operation, ensuring data integrity, and enabling real-time control in industrial environments. The choice of communication method depends on factors such as speed, reliability, distance, security, and environmental conditions. The two primary communication techniques used in Industrial Control Systems are wired and wireless communication, each offering distinct advantages and trade-offs. These are the primary communication techniques used in ICS:

(I) **Wired Communication:** Data transfer using wire-based communication technologies, such as telecommunication cables, is referred to as wired communication. Another name for wired communication is wireline communication. Fiber-optic communication, cable television, and Internet connection are a few examples.

(a) **Ethernet:** Broadly utilized for quick, reliable wired data transfer.

```
Function InitializeEthernetConnection(ipAddress, port):
    Create TCP socket;
    Set socket parameters (buffer size, timeout);
    Connect to ipAddress:port;
    If connection successful:
        Return connection handle;
    Else:
        Return error status;
End
```

Listing 3.1 Pseudo Code for Ethernet Connection Initialization

The pseudocode for the `InitializeEthernetConnection` function is designed to establish a TCP connection to a specified IP address and port. Below is a summary of the function's steps:

(a) **Create TCP Socket:** A TCP socket is created to initiate communication.

(b) **Set Socket Parameters:** The function sets socket parameters such as buffer size and timeout, ensuring proper configuration for communication.

(c) **Connect to Target:** The function attempts to establish a connection to the given IP address and port.

(d) **Check Connection Success:** If the connection is successful, the function returns a connection handle, which is used for further communication.

(e) **Error Handling:** If the connection fails, an error status is returned to indicate the failure.

This function is used to initialize a TCP connection and handle the success or failure of the connection attempt.

(b) **Fieldbus:** Fieldbus is a standardized industrial network protocol used for real-time communication between field devices (e.g., sensors, actuators) and control systems in industrial automation [159]. It replaces traditional point-to-point wiring with a single bus, enabling efficient and reliable communication. Key characteristics of Fieldbus include:

- **Deterministic Communication:** Ensures timely delivery of data by assigning time slots or priorities to devices.
- **Scalability:** Supports many devices on the same network.
- **Error Handling and Diagnostics:** Includes mechanisms to detect and manage network or device issues.

The communication master initializes the system and assigns time slots to each field device for deterministic communication. It polls devices for data and transmits commands if the device is active. If an error occurs, it handles the issue and retries communication. This process ensures reliable and efficient real-time operations.

```
Begin
    Initialize communication master;
    Assign time slots to field devices;
    Poll each field device for data;
    if (device status is active) {
        Transmit control commands based on device status;
    } else {
        Handle device errors and retry;
    }
End
```

Listing 3.2 Pseudo Code for Fieldbus Communication

(c) **Serial Communication:** Serial communication is a method of transmitting data sequentially over a communication channel. It is commonly used for communication between devices like microcontrollers, sensors, and computers through interfaces such as RS232, UART, or USB. Unlike parallel communication, serial communication transfers one bit at a time, making it ideal for long-distance data transfer with minimal wiring.
Key Features:

- **Data Transmission:** Data is sent sequentially, bit-by-bit, over a single line, reducing the number of physical connections compared to parallel communication.
- **Standards:** Popular standards include RS232, RS485, UART (Universal Asynchronous Receiver Transmitter), SPI (Serial Peripheral Interface), and I2C (Inter-Integrated Circuit).
- **Efficiency:** The single-bit transfer makes serial communication suitable for long-distance transmission, as it is less prone to signal degradation compared to parallel communication.
- **Applications:** Commonly used in embedded systems, sensor interfacing, communication between computers and peripherals, and real-time data logging systems.

```
Begin
    Set baud rate, data bits, parity, and stop bits;
    Open serial port;
    while (data to send) {
        Transmit data byte-by-byte;
        Wait for acknowledgment from the receiver;
    }
    Close port after data transfer is complete;
End
```

Listing 3.3 Pseudo Code for Serial Data Transfer

Explanation:

The pseudocode demonstrates how serial communication works:

- **Initialize Communication Settings:** The baud rate, data bits, parity, and stop bits are configured to establish a standard communication protocol between the sender and receiver.
- **Open Serial Port:** The serial port is opened to enable data transfer.
- **Transmit Data:** Data is sent byte-by-byte to ensure accuracy and synchronization. After sending each byte, the sender waits for acknowledgment from the receiver to confirm successful transmission.
- **Close Port:** Once the data transfer is complete, the serial port is closed to release resources.

(II) **Wireless Communication:** Wireless communication is the transmission of information between two or more sites without the need for physical cables or other continuous guided mediums. Common wireless technologies mostly use radio waves.

(a) **Wi-Fi:** Uses WLANs for short-range communication. Wi-Fi is a wireless communication technology that enables devices to connect and exchange data over a Wireless Local Area Network (WLAN). It uses radio frequency signals to facilitate short-to-medium range communication without the need for physical cables. Wi-Fi operates on standardized protocols, such as IEEE 802.11, and supports multiple frequency bands, including 2.4 GHz and 5 GHz, to balance range and data transfer speeds. It is widely used in homes, offices, and public spaces for internet connectivity, file sharing, and IoT device integration, offering high flexibility and convenience for various applications.

(b) **Bluetooth:** Bluetooth is a short-range wireless communication technology designed for connecting and exchanging data between devices over a secure, low-power connection. It operates on the 2.4 GHz ISM (Industrial, Scientific, and Medical) band and uses frequency-hopping spread spectrum (FHSS) to minimize interference and ensure reliable communication. Bluetooth supports a range of up to approximately 10 m for most applications, making it ideal for personal area networks (PANs).

Explanation:

The pseudocode outlines the Bluetooth device pairing process:

- **Search for Devices:** The system scans for available Bluetooth devices that are discoverable.
- **Select Device:** The user or system selects a target device to pair with.
- **Pairing Check:** If the pairing is successful, the devices exchange pairing keys and authenticate the connection.
- **Establish Communication:** Once authenticated, a secure communication channel is established.
- **Retry:** If pairing fails, the process is retried until successful.

```
Begin
    Get CSI from the channel prediction algorithm;
    Get network traffic information;
    Generate one empty slot (B = 0) during each unit of
time;
    if (channel_state == GOOD_BER) {
        Buffer_length = 3; // optimum buffer unit length
        slot_status = busy;
        Transmit by combining up to three MPDUs;
    } else if (channel_state == BAD_BER) {
        buffer_length = 1; // same as 802.11b DCF
        slot_status = busy;
        Transmit a single MPDU;
    } else {
        Pause packet transmissions;
        Wait for the next empty slot;
    }
End
```

Listing 3.4 Pseudo Code for the Transmission Control Algorithm in Wi-Fi Networks

```
Begin
    Search for discoverable devices;
    Select target device from the list;
    if (pairing successful) {
        Exchange pairing keys;
        Authenticate connection;
        Establish secure communication channel;
    } else {
        Retry pairing process;
    }
End
```

Listing 3.5 Pseudo Code for Bluetooth Device Pairing

(c) **Cellular Networks:** Cellular networks provide long-range communication by dividing a region into smaller cells, each with a base station. These networks are widely used for mobile communication, allowing devices like smartphones to maintain connectivity while moving across different areas. The network is composed of multiple base stations that manage traffic within their respective cells.

```
Begin
    Monitor signal strength of the current cell;
    Identify neighboring cells with stronger signals;
    if (neighboring_cell_signal > current_signal) {
        Initiate handoff to the target cell;
        Update routing information at the network core;
        Resume data transmission through the new cell;
    } else {
```

```
 9              Continue transmitting via the current cell;
10          }
11      End
12
```

Listing 3.6 Pseudo Code for Cellular Data Handoff

Explanation:
The pseudocode explains the cellular data handoff process:

- **Monitor Signal Strength:** The system continuously monitors the signal strength of the current cell to assess connectivity.
- **Identify Stronger Signals:** It identifies neighboring cells that may offer a stronger signal.
- **Handoff Condition:** If the neighboring cell's signal is stronger than the current cell's, the system initiates the handoff process.
- **Update Routing Information:** The network core is updated with the new routing information for continued data transmission.
- **Resume Transmission:** Data transmission is resumed through the new cell after handoff.
- **Continue If No Handoff:** If no stronger signal is found, the system continues transmitting through the current cell.

(d) **WirelessHART:** WirelessHART is a wireless communication protocol designed specifically for process automation and control in industrial environments. It extends the HART (Highway Addressable Remote Transducer) protocol, which is widely used for connecting smart devices in industrial control systems. WirelessHART adds the capability of wireless communication to HART-enabled devices, allowing for remote monitoring and control in environments where wired connections are impractical or costly to implement.

WirelessHART operates on the IEEE 802.15.4 standard and is designed to provide reliable, secure, and low-power communication between field devices in industrial settings such as factories, refineries, and power plants. It supports a mesh network topology, where each device acts as both a transmitter and a repeater, ensuring that signals can travel long distances by hopping between devices.

```
1      Begin
2          Establish communication graph between devices;
3          Allocate time slots for each device communication;
4          if (link_quality < threshold) {
5              Adjust communication schedule dynamically;
6          }
7          Transmit and receive data within assigned slots;
8      End
9
```

Listing 3.7 Pseudo Code for WirelessHART Data Scheduling

Explanation:
The pseudocode for WirelessHART communication might look like this:

- **Initialize Network:** Configure WirelessHART devices with unique IDs and communication settings [115].
- **Transmit Data:** Send data from sensors to the network coordinator via a mesh network.
- **Mesh Network:** Devices forward data to ensure continuous communication, even if some devices are out of range.
- **Data Acknowledgment:** Each data transmission is confirmed by the receiver to ensure reliability.
- **Monitor Signal Integrity:** Devices check signal quality and reroute data as needed.
- **Security:** WirelessHART uses encryption and authentication for secure communication.

```
Begin
    Discover network neighbors;
    Build routing table using AODV protocol;
    Select optimal path to the destination;
    Transmit data packets through the mesh network;
    if (route changes detected) {
        Update routing table;
    }
End
```

Listing 3.8 Pseudo Code for Zigbee Mesh Network Routing

(e) **Zigbee:** Industrial and IoT low-power wireless communication standard.

(III) **Industrial IIoT:** Industrial IoT is a network of devices, sensors, apps, and networking equipment that collaborate to gather, supervise, and scrutinize data from industrial activities. Examining this data improves visibility and strengthens troubleshooting and maintenance skills.

(a) MQTT: A lightweight IoT messaging protocol for devices and apps.
(b) CoAP: Resource-constrained IoT device application protocol.
(c) LoRaWAN: Long Range Wide Area Network for low-power, long-range IoT communication.
(d) 5G: 5G mobile networks offer high data rates and minimal latency for IoT devices.

(IV) **Supervisory Control and Data Acquisition:** SCADA systems are used to monitor, control, and analyze industrial equipment and processes [8]. The system comprises software and hardware components that enable the collection of data from industrial equipment both remotely and on-site.

(a) DNP3: Reliable SCADA protocol for distant devices [127]

(b) Modbus: Popular SCADA-field device protocol [133]
(c) OPC UA: Secure, platform-independent communication with OPC UA [48]

(V) **Cloud-Based Communication:** Cloud communications use the Internet to access telecommunications apps, switching, and storage hosted by a third party. Cloud services refer to data-center-hosted, Internet-accessible services. VoIP has made voice part of the cloud, although these services were data-centric until recently. Hosted or cloud telephony replaces commercial telephone equipment like PBXs with third-party VoIP services. Cloud communications companies host voice and data communications apps and services on their servers, allowing consumers access to the cloud.

(a) RESTful API: Cloud-to-device connectivity
(b) Websockets: Full-duplex TCP channels
(c) Message queues: RabbitMQ and Apache Kafka provide asynchronous application communication

(VI) **HMI (Human-Machine Interface) Communication:** A Human-Machine Interface (HMI) is a user interface that links a person to a machine, system, or device. HMI may be used to describe any screen enabling user-device interaction; however, it is mostly associated with industrial processes.HMIs have similarities with Graphical User Interfaces (GUI), but they are not the same; GUIs are often used in HMIs to enhance visualization features.

(a) OPC (OLE for Process Control): Helps HMIs contact process control systems
(b) Web-based Interfaces: Interactive control and monitoring visual interfaces

(VII) **Data Historians and Analytics:** A data historian is a software application that logs the data from processes operating inside a computer system. Organizations use data historians to collect information on program operations for diagnosing errors in situations where dependability and uptime are crucial.

(a) Real-Time Analytics: Rapid data analysis for quick decisions
(b) Data Historians: Store and retrieve historical data for study and reporting.

(VIII) **Remote Access and VPN (Virtual Private Network):** A remote access virtual private network (VPN) allows users to connect to a private network from a distant location via a VPN. This kind of VPN is often used by employees requiring off-site access to their company's network or individuals seeking a secure connection to a private network from a public place.

(a) VPN: Allows secure Internet control system access.
(b) Remote Desktop: Remote desktop software controls and monitors systems.

3.5 Industrial Protocols with Corresponding Port Numbers

Industrial protocols are the essential sets of standards and rules that manage and control how data is communicated and exchanged between various devices within an industrial network. These protocols are crucial for ensuring reliable and predictable communication. They achieve this by establishing meticulously defined rules, a specific syntax for structuring the data, and operational constraints that must be adhered to during transmission. Furthermore, they incorporate robust error-handling mechanisms to detect and correct any errors that might occur during data transfer, maintaining data integrity. Synchronization methods are also a key component, ensuring that data is sent and received in the correct order and at the appropriate times, preventing conflicts and data loss.

Just as programming languages provide a structured framework for instructing computers, industrial protocols operate within networks to govern the communication between devices. These protocols define the specific rules and procedures that determine how various pieces of equipment interact with each other. Crucially, they also manage the exchange of information, ensuring that data is transmitted and received in a standardized and understandable format. In essence, industrial protocols are the network's equivalent of programming languages, providing the necessary instructions for seamless device coordination and effective data sharing.

This section will provide a detailed overview of several commonly used industrial automation protocols, highlighting their key features and functions. Crucially, we will also outline their corresponding port numbers. Port numbers are numerical designations that serve as entry points for network connections. They are essential for directing network traffic to the correct applications and services running on a device. Understanding these protocols and their associated port numbers is vital for ensuring seamless and secure data transmission between various industrial devices and systems, allowing for efficient automation and control processes.

Figure 3.4, adapted from [17], illustrates an overview of commonly used industrial protocols along with their respective network port assignments.

Figure 3.4 displays a collection of network protocols and their respective port ranges. The protocols offered include TCP, UDP, AMQP, BACnet/IP, CoAP, Modbus/TCP, MQTT, and others. Every protocol is linked to certain port numbers or ranges.

3.5.1 Top-5 Non-industrial Protocols

While the term "industrial protocols" often evokes images of automation and manufacturing, it's important to recognize that numerous communication standards exist outside of this specific industrial automation domain. These protocols play crucial roles in various sectors and applications. We can identify several key

Fig. 3.4 Industrial protocols with ports

Protocol	Port Ranges		Wireshark
	TCP	UDP	
AMQP	5671-5672	-	✓
ANSI C12.22	1153	1153	✓
ATG	10001	-	
BACnet/IP	-	47808	✓
COAP	5683	5683	✓
Codesys	2455	-	
Crimson v3	789	-	
DNP3	20000	20000	✓
EtherCAT	34980	34980	✓
Ethernet/IP	44818	2222	✓
FL-net	-	55000-55003	
FF HSE	1089-1091	1089-1091	✓
GE-SRTP	18245-18246	-	
HART IP	5094	5094	✓
ICCP	102	-	✓
IEC60870-5-104	2404	-	✓
IEC61850	102	-	✓
Modbus/TCP	502	-	✓
MELSEC-Q	5007	5006	
MQTT	1883,8883	-	✓
Niagara Fox	1911,4911	-	
OMRON FINS	-	9600	✓
OPC UA	4840	-	✓
PCWorx	1962	-	
ProConOS	20547	20547	
PROFINET	34962-34964	34962-34964	✓
S7comm	102	-	✓
Zigbee IP	17754-17756	17754-17756	✓

	Packet-based				Flow-based			
	Overall		ICS-to-ICS		Overall		ICS-to-ICS	
	Protocol	%	Protocol	%	Protocol	%	Protocol	%
Before IANA mapping	TLS	50.7	UDP	47.5	TLS	45	TCP	17.8
	HTTP	41.3	TCP	40.1	HTTP	23.7	TLS	15.8
	UDP	4.8	OpenVPN	8.8	TCP	18.4	UDP	8.2
	TCP	2.7	TLS	1.8	UDP	6.7	ICMP	4.1
	DNS	0.1	SIP	1.2	ICMP	3	RTCP	3.1
After IANA mapping	TLS	50.7	TCP	36.4	TLS	42.9	TLS	1
	HTTP	41.3	UDP	16	HTTP	22.6	TCP	0.6
	STUN	3.5	Open VPN	8.8	XMPP	6.6	UDP	0.4
	XMPP	1.7	RSF-1 clustering	5.4	HP V.ROOM	5.5	Reserved	0.4
	UDP	0.8	ActiveSync	2.4	TCP	3.2	ICMP	0.3

Fig. 3.5 Top-5 non-industrial protocols

communication protocols that, while not typically classified as industrial, are nonetheless vital in the broader technological landscape.

In the following section, we will highlight and briefly describe five important non-industrial protocols. These protocols have been selected based on their prevalence, significance in data communication, and impact on various industries outside of traditional industrial automation settings. Here are five important non-industrial protocols: Figure 3.5 was from [10].

Figure 3.5 displays protocol percentages across several scenarios: Packet-based, Flow-based, ICS-to-ICS, Before IANA, and After IANA. The protocols included are TLS, UDP, TCP, HTTP, OpenVPN, ICMP, DNS, SIP, STUN, XMPP, RSF-1 clustering, HP V.ROOM, ActiveSync, and Reserved. The percentages vary among the various protocol categories.

3.5.2 Virtual Private Network

A Virtual Private Network (VPN) employs encryption to secure data that is transmitted over public networks. VPNs create a secure tunnel between the user's device and the target server, ensuring confidentiality, integrity, and authenticity of the transmitted information. Several widely used VPN technologies implement encryption and tunneling in different ways to meet specific security and performance requirements:

1. **IPsec:** Internet Protocol Security (IPsec) is a suite of protocols that provides encryption and authentication at the network layer. It operates in two modes: transport mode, which encrypts only the payload of the packet, and tunnel mode, which encrypts the entire packet. IPsec supports features such as address authentication and Network Address Translation (NAT) traversal, allowing secure communication even across NAT-enabled devices.
2. **SSL/TLS:** Secure Sockets Layer (SSL) and its successor, Transport Layer Security (TLS), provide secure end-to-end encryption of data at the transport layer. VPNs based on SSL/TLS are commonly used in secure web browsing (HTTPS) and utilize standard web browsers as clients, making them highly accessible and user-friendly. These VPNs are particularly suited for remote access scenarios.
3. **SSH:** Secure Shell (SSH) is a protocol that provides secure remote access to networked devices through a command-line interface. The SSH-2 protocol, standardized by the Internet Engineering Task Force (IETF), is widely adopted by system and network administrators for secure remote server management, file transfer, and command execution.

Each VPN protocol offers unique advantages in terms of security, flexibility, and ease of deployment. The choice of protocol depends on the specific requirements of the application, including the level of encryption needed, compatibility with existing infrastructure, and user accessibility.

3.6 Types of Protocols

Communication protocols are essential for enabling effective data exchange between devices and systems. They define a standardized set of rules, formats, and procedures for transmitting data over various media, ensuring secure, reliable, and efficient communication. These protocols can be broadly categorized into serial communication protocols, network protocols, and wireless communication protocols.

- **Serial Communication Protocols:** Serial communication involves the sequential transmission of data bits over a single channel or wire. It is a cost-effective solution, particularly in systems that require minimal wiring and short-distance communication. Common serial communication protocols include:

  - **UART (Universal Asynchronous Receiver-Transmitter):** A commonly used protocol for asynchronous, low-speed, short-distance communication. It is widely employed to interface microcontrollers with peripheral devices such as sensors and displays.
  - **SPI (Serial Peripheral Interface):** A high-speed synchronous communication protocol that allows data exchange between a master device and one or more slave devices. SPI is extensively used in embedded systems for interfacing sensors, memory cards, and LCD displays.
  - **I^2C (Inter-Integrated Circuit):** A synchronous, multi-master, two-wire communication bus designed for short-distance communication between multiple integrated circuits. It is particularly suited for low-speed applications with minimal pin requirements, such as embedded control systems.

- **Network Protocols:** Network protocols manage the transmission of data across interconnected devices within a network. They are responsible for data routing, addressing, error checking, and ensuring the overall integrity of communication over local and wide area networks.

  - **TCP/IP (Transmission Control Protocol/Internet Protocol):** The foundational protocol suite of the Internet and most modern networks. TCP ensures reliable data delivery by establishing a connection and confirming receipt, while IP handles the logical addressing and routing of packets.
  - **MQTT (Message Queuing Telemetry Transport):** A lightweight, publish-subscribe messaging protocol designed for low-bandwidth, high-latency, or unreliable networks. MQTT is widely used in Internet of Things (IoT) applications for efficient communication between connected devices.

- **Wireless Communication Protocols:** Wireless protocols enable data transmission without the need for physical cables. These protocols utilize electromagnetic waves, such as radio frequency, for communication and are essential for mobile, IoT, and remote sensor network applications.

  - **Bluetooth:** A short-range wireless communication standard designed for connecting devices within a few meters. It is widely used in personal area networks (PANs), such as linking mobile phones, wireless headphones, and computers.
  - **Zigbee:** A low-power, low-data-rate wireless communication protocol tailored for industrial control and home automation systems. It is optimized

for long-term battery operation and supports mesh networking for improved reliability.
- **LoRa (Long Range):** A Low-Power Wide Area Network (LPWAN) protocol designed for long-range communication with minimal energy consumption. LoRa is ideal for applications in smart cities, agriculture, and remote monitoring systems where wide coverage is required.

3.7 Selecting a Microcontroller

Selecting an appropriate microcontroller is a critical step in the design and development of embedded systems. The choice depends on several factors, including processing power, memory capacity, supported communication protocols, power consumption, and ease of programming. As the microcontroller serves as the central processing unit in embedded applications, choosing the right one ensures optimal performance, system reliability, and project scalability.

3.7.1 Popular Microcontroller Options

- **Arduino:** Arduino is a widely adopted platform, particularly suitable for beginners and educational purposes due to its simple and user-friendly programming interface. It supports a variety of communication protocols, such as UART (Universal Asynchronous Receiver-Transmitter), I^2C (Inter-Integrated Circuit), and SPI (Serial Peripheral Interface). Its affordability and rich library ecosystem make it ideal for hobbyist projects and rapid prototyping.
- **Raspberry Pi:** Although technically a single-board computer rather than a traditional microcontroller, Raspberry Pi offers significantly higher computational capabilities. It supports a full Linux operating system, has more memory and processing power, and includes General Purpose Input/Output (GPIO) pins for protocol interfacing. It is well-suited for applications involving advanced computing requirements, such as image processing, data analytics, or complex network communication.
- **STM32 and PIC Microcontrollers:** STM32 microcontrollers, developed by STMicroelectronics, are based on ARM Cortex-M cores and are designed for high-performance embedded systems. They provide advanced features such as low-power consumption, high-speed interfaces, and extensive peripheral support. PIC microcontrollers, developed by Microchip Technology, are known for their robustness and broad compatibility with industrial and automotive standards. Both are extensively used in commercial and industrial environments due to their reliability and performance.

3.7.2   Key Considerations for Microcontroller Selection

- **Supported Communication Protocols:** Ensure that the microcontroller supports all required protocols for application, such as UART for serial communication, I^2C for sensor interfacing, or SPI for high-speed peripheral data exchange.
- **Library Availability:** Availability of well-documented libraries significantly simplifies development. Platforms like Arduino provide a wide range of libraries for communication protocols including I^2C, SPI, Bluetooth, and Wi-Fi, enabling faster and easier integration.
- **Processing Capability and Memory Size:** Choose a microcontroller with sufficient CPU speed and memory to meet application's processing demands. Simple microcontrollers may be insufficient for applications requiring complex calculations or large data handling, where higher-end options like STM32 or Raspberry Pi may be more appropriate.
- **Cost and Availability:** Budget constraints play a major role, particularly for large-scale deployments. Arduino and Raspberry Pi are cost-effective for prototyping and smaller projects, whereas STM32 and similar high-end microcontrollers may be more expensive but are justified for performance-critical systems.
- **Community and Technical Support:** A large and active developer community can be invaluable for troubleshooting and learning. Platforms such as Arduino, Raspberry Pi, and STM32 benefit from extensive online resources, including tutorials, forums, and open-source examples, which facilitate faster development and problem resolution.

3.7.3   Acquiring Required Hardware

Once the appropriate microcontroller has been selected, the next step is to acquire the necessary hardware components. The exact components required will depend on the specific microcontroller and the communication protocol used in project. However, there are several core components that are commonly utilized across most embedded systems and prototyping environments.

Essential Components
- **Microcontroller Board:** The micro-controller board serves as the central processing unit of the embedded system. Common options include the Arduino Uno, STM32 development boards, and the Raspberry Pi. Selection should be based on application's requirements for computational performance, input/output capabilities, and communication interface support.
- **Breadboard and Jumper Wires:** A breadboard allows for solderless prototyping and is essential for testing and modifying circuits during development. Jumper wires are used to establish electrical connections between the microcontroller and peripheral components, such as sensors and modules.

- **Peripheral Modules:** Depending on the communication protocols implemented in project, may require additional interface modules. These may include:
  - **Bluetooth Modules (e.g., HC-05, HC-06):** Used for short-range wireless serial communication.
  - **Wi-Fi Modules (e.g., ESP8266, ESP32):** Employed in Internet of Things (IoT) applications to enable wireless network connectivity.
  - **Radio Frequency (RF) Transceivers (e.g., NRF24L01, RFM69):** Suitable for wireless communication over medium to long distances or in mesh network configurations.
  - **Sensors and Actuators:** Specific sensors (e.g., temperature, humidity, accelerometers) and actuators (e.g., motors, LEDs) are selected based on the functional requirements of the application. These components allow the system to interact with the physical environment.

3.7.4 Set Up Development Environment

Establishing the appropriate development environment is essential for efficiently writing, compiling, and uploading code to micro-controller. Various development environments are available, each optimized for specific platforms and micro-controllers.

- **Integrated Development Environment (IDE)/Software Tools:**
  - **Arduino IDE:** The Arduino Integrated Development Environment is the most widely used for programming Arduino boards. It provides a user-friendly interface and built-in support for a variety of Arduino boards and communication protocols.
  - **PlatformIO:** A versatile and advanced IDE that supports multiple platforms, including Arduino, STM32, and Raspberry Pi. It offers features such as project management, debugging capabilities, and an extensive library integration.
  - **MPLAB X IDE:** This is the development environment specifically designed for programming PIC microcontrollers. It is suitable for more advanced embedded systems development and offers comprehensive debugging and simulation features.

- **Libraries:** Most communication protocols require specific libraries that simplify coding and enable seamless integration of the protocol into project. Some commonly used libraries include:
  - `Wire` library for I2C communication on Arduino boards
  - `SPI` library for handling SPI communication
  - `SoftwareSerial` library for serial communication, often used for Bluetooth and other serial peripherals on Arduino
  - `WiFi` library for connecting Arduino to Wi-Fi networks (e.g., using ESP8266 or ESP32 modules)

- **Drivers and Tools:** For certain microcontrollers, such as the STM32, additional drivers and utilities may be required to flash the microcontroller or to establish a USB interface. It is essential to refer to the official documentation of the specific microcontroller to obtain the correct drivers and tools.

3.7.5 Programming the Microcontroller

Microcontroller programming involves configuring the device to support communication through the specified protocols and subsequently enabling it to send and receive data. Below are the basic steps typically followed when programming a microcontroller to facilitate communication.

Basic Steps
- **Initialize the Communication Protocol:** Set up parameters such as the baud rate for UART, and configure the pins for SPI or I2C communication.
- **Send and Receive Data:** Use appropriate library functions to transmit and receive data over the selected protocol.
- **Error Handling:** Implement checks to verify successful communication and handle any errors appropriately.

3.7.5.1 Example Code for I2C (Arduino)

Here is a simple code snippet to use I2C communication in Arduino: referred from: https://github.com/Sunzidasiddique1/ICS.

```
#include <Wire.h>

void setup() {
    Wire.begin(); // Join I2C bus
    Serial.begin(9600); // Start serial communication
}

void loop() {
    Wire.requestFrom(8, 1); // Request 1 byte from device with
    address 8
    while (Wire.available()) {
        char c = Wire.read(); // Receive a byte
        Serial.println(c); // Print the received byte
    }
    delay(1000); // Wait before the next request
}
```

Listing 3.9 I2C Communication with Arduino

Explanation:
The provided Arduino code demonstrates how to use I2C communication to request and receive data from a slave device.

- **Wire.begin():** Initializes the I2C communication on the Arduino, joining the I2C bus.
- **Serial.begin(9600):** Starts serial communication at a baud rate of 9600, which is used to display the received data on the Serial Monitor.
- **Wire.requestFrom(8, 1):** Requests 1 byte of data from the I2C device with address 8.
- **Wire.read():** Reads the received byte from the I2C bus.
- **Serial.println(c):** Prints the received byte to the Serial Monitor.
- **delay(1000):** Pauses for 1 second before the next data request.

3.7.5.2   Testing and Debugging

- **Use Serial Monitor:** Utilize the serial monitor in the IDE for debugging purposes and to view output messages. This allows examination of real-time data from the microcontroller, helping identify problems early. You can print variable values, error messages, and system states, which is useful for troubleshooting and verification.
- **Logic Analyzer:** Use a logic analyzer, if available, to visually display the communication signals. It records digital waveforms and can help detect errors such as incorrect signal timing or data corruption. A logic analyzer provides a detailed view of the communication bus, making it easier to diagnose and troubleshoot issues that may not be evident through basic debugging methods.

By following these steps and utilizing the suggested resources, one can effectively learn and implement a communication protocol with a microcontroller.

3.7.6   *Importance of I2C in Industrial Control Systems*

I2C communication, demonstrated through Arduino example code, plays a significant role in industrial control systems (ICS) in several ways.

Communication with Sensors and Actuators is one of the most important applications, as industrial systems heavily rely on sensors (e.g., pressure, temperature) and actuators (e.g., valves, motors) for automation. The code illustrates how to initiate and exchange data with I2C devices, making it easier to integrate various modules into ICS.

Multi-device Communication is also critical in scenarios where multiple devices, such as numerous sensors, must communicate over the same bus. The use of unique I2C addresses is demonstrated to efficiently manage such communications.

To support **Prototyping and Development**, the code serves as a simple example for testing device communication, saving development time and effort.

Error Handling and Debugging is facilitated through serial communication , which enables data logging and debugging, thereby improving the robustness of industrial implementations.

The **Scalability and Integration** of I2C is another advantage, as it allows multiple devices to share a common bus, solving connectivity issues in expanding industrial systems.

So, the code serves as a valuable resource for **Training and Development**, allowing engineers and technicians to gain hands-on experience with I2C communication—an essential skill for maintaining and developing ICS.

3.8 Communication Protocol Messaging Patterns

There are standard message patterns that most communication protocols will follow. In industrial applications, the following two patterns are frequently considered:

1. **Request-Response:** In this pattern, a client requests data or a service, and the server replies with the required information. It is a standard industry technique that ensures smooth and predictable message exchange.
2. **Publish-Subscribe:** In this pattern, an intermediary node acts as a distributor, routing messages to multiple subscribers based on predefined topics or classifications. Publishers send messages, and subscribers receive messages according to their specified interests or filters.
3. **Other Patterns:** In addition to these, communication protocols may employ patterns such as asynchronous messaging, multicast, message queues, and broker-based communication. The choice of pattern depends on the specific requirements of the industrial communication system.

Additionally, communication protocols are commonly positioned in suites, which:

1. **Protocol Suites:** A set of protocols cooperate for industrial communications. These suites are layered in a stack, where the lower layers are responsible for hardware-level transmission, and the upper layers are in charge of higher-level communication services.
2. **Layered Design:** The layers in the protocol stack should be designed in accordance to network components and applications. This modular design offers the scalability and flexibility which are required to provide solutions for a wide range of industrial communication situations.

When it is aimed at designing robust and efficient communication systems in an industrial environment, these types and structures of messages are important to know.

3.9   Ensuring Network Security

Securing the integrity and reliability of industrial control systems (ICS) is the primary objective of ICS security. This encompasses both the hardware and software components used by system users and operators to ensure safe and continuous operation.

Figure 3.6, adapted from [101], illustrates a framework related to ICS security.

Figure 3.6 presents information on security layers, application Figure 3.6 provides a comprehensive overview of security layers, application security, threats, service security, and cybersecurity. It also offers wide views on the various forms of security such as interruption, vulnerabilities, interception, modification, privacy, fabrication, integrity, availability, authentication, non-repudiation, infrastructure security, attacks, data confidentiality, access control, and communication security as well as security planes like the end-user plane, control plane, and management plane. Moreover, it reveals not just one but several facets of security concerns.

1. **Authentication:** Authentication of the communicating parties (tables) and protection against unauthorized access and unauthorized replication of data.
2. **Availability:** Against denial-of-service attack, or in a case of failure, it ensures continued operation within a reasonable time, with no or minimal disruption.
3. **Access Control**: Responsible for authorization management and operation accessibility, which can decrease the cases of illegal operation and stolen data.
4. **Data Privacy:** Encryption-based technique is used, and sensitive information are protected from unauthorized access.
5. **Non-non-repudiation:** Ensures the origin and integrity of transferred data so that those involved cannot deny having sent or received data.
6. **Data Integrity:** Data integrity is the property that data will not be modified in an unauthorized way, meaning that data is valid and accurate and exists as it was intended—if somebody tries to an unauthorized modification will be detected and the data tampered margin of error would be eliminated.

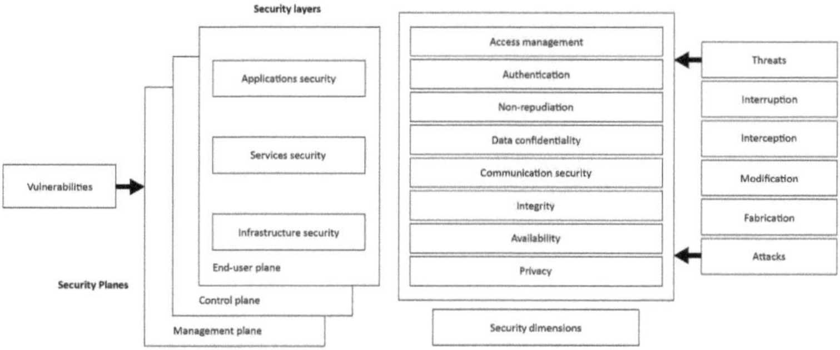

Fig. 3.6 The network security framework

7. **Packing Privacy:** Protecting the content of Internet operations, ensuring that third parties do not eavesdrop or steal signals, and providing a mechanism to authenticate the communication system to secure the data transmission.
8. **Privacy:** Shields users, critical infrastructure, and the data that it monitors from unauthorized inspection or revelation, promoting trust and usage.

3.10   Techniques of Redundancy for Reliability in SCADA Systems

The reliability of Supervisory Control and Data Acquisition (SCADA) systems—which play vital roles in the supervision and management of industrial processes and infrastructure—is of paramount importance. Redundancy techniques are employed to enhance system resilience and ensure stable performance. Below are key redundancy mechanisms commonly used in SCADA systems:

Path Redundancy:

Path redundancy involves establishing multiple communication paths between devices or control centers within the SCADA system. These redundant paths ensure continuous data transmission by providing alternative channels in the event of network congestion, device failure, or other disruptions.

Device Redundancy:

SCADA systems incorporate redundant key components such as sensors, actuators, or controllers. Backup units are configured to automatically take over if a primary device fails, thereby maintaining uninterrupted system operations.

Server Redundancy:

Redundant servers are used in SCADA systems to ensure continuous data processing. Typically, one server operates in active mode while the other remains on standby. In case the main server fails, the backup server takes over immediately, minimizing service interruption.

Redundancy in Protocols:

SCADA systems often use multiple communication protocols to transfer data [8]. Redundant protocols allow the system to switch to an alternative protocol if the primary one fails or encounters compatibility issues, ensuring uninterrupted communication.

Network Redundancy:

Redundant networking involves using multiple communication networks, channels, or equipment. In the event of a failure, traffic is rerouted through alternate paths to preserve connectivity and avoid communication loss.

Power Redundancy:

SCADA systems are equipped with redundant power sources such as generators or Uninterruptible Power Supplies (UPS). These backup power systems prevent downtime by supplying continuous power during outages or voltage fluctuations.

Data Redundancy:
Critical data in SCADA systems is replicated across multiple storage devices or locations. This redundancy protects against data loss due to device failure or corruption and ensures data integrity and availability.

Hot Standby:
In a hot standby configuration, a backup system runs in parallel with the main system and remains fully synchronized. If the main system fails, the hot standby seamlessly takes over with no interruption in operations.

Cold Standby:
Cold standby systems are inactive backups that require manual intervention to activate. Although less costly than hot standby systems, they involve longer recovery times due to the manual steps needed to bring the backup online.

Load Balancing:
Load balancing in SCADA systems distributes processing and network traffic across multiple servers or communication channels [8]. This prevents overloading, improves performance, and increases reliability by efficiently managing system resources.

SCADA systems can maintain critical industrial processes and infrastructure with high availability, fault tolerance, and robustness by employing these redundancy techniques.

3.11 Integration of Emerging Technologies

Advanced technology-driven communication and control protocols are transforming industrial and business systems more dramatically than ever. Artificial Intelligence (AI) enables predictive analysis, the Internet of Things (IoT) allows enhanced real-time data sharing, blockchain ensures secure and immutable data exchange, and 5G technology provides ultrafast and reliable communication systems [91].

However, adopting these technologies is not straightforward, due to challenges such as the need to coexist with legacy systems and address cybersecurity concerns. Successful implementation requires thoughtful integration of existing systems, collaboration among cross-functional teams, and scalable solutions to accommodate future expansion.

These advancements improve operational efficiency, reduce costs, and enhance security across industries ranging from healthcare to government. The impact of these innovations is redefining communication in a highly connected world. Integrating new technologies into communication protocols represents a significant shift in the way businesses function and interact today.

AI helps businesses by providing predictive insights that help in taking proactive actions and processes improvement. The IoT enables devices to share data with one another on a continuous basis so that operational monitoring, control, and management become possible in real time. Blockchain makes data secure and immutable, which in turn establishes trust in some communication system [123].

Furthermore, 5G technology supports the applications and services in high speed and large capacity, which could be realistically developed.

Solutions to compatibility with older systems and best practices for security adoption are necessary for these technologies to realize their full potential. Functioning seamlessly takes careful planning, coordinated teamwork, and an adaptable infrastructure, capable of adjusting to changing needs.

The implementation of these emerging technologies represents the commencement of communication systems into a new generational phase—one that is more efficient, more integrated, and more secure—for industrial and digital transformation.

3.12 Challenges in ICS Communication

Several Industrial Control System (ICS) architectures were developed during a period when cybersecurity was not considered a major concern. Consequently, many of these systems lack fundamental security mechanisms such as encryption, authentication, and access control. This omission makes ICSes vulnerable to a variety of cyber threats, including data breaches and denial-of-service (DoS) attacks.

1. **Real-Time Requirements:** In ICS environments, real-time communication is essential for maintaining control and safety. Network delays or interruptions can hinder prompt decision-making and negatively affect operational efficiency and safety outcomes.
2. **Security Vulnerabilities:** The use of outdated communication protocols and inadequate security practices increases the risk of cyberattacks. Cyberattacks targeting critical infrastructure can result in severe disruptions and have far-reaching consequences.
3. **Legacy Systems:** Many ICS components are based on legacy hardware and software that lack support for modern security standards and protocols. This poses significant challenges in system integration and modernization efforts.
4. **Regulatory Compliance:** Meeting industry-specific standards and regulatory requirements, such as NIST or IEC 62443, adds complexity to network design and protocol selection. Compliance often requires significant planning, documentation, and resource allocation.
5. **Interoperability Issues:** The coexistence of multiple proprietary and open communication protocols in ICS environments can hinder seamless data exchange between devices from different vendors, reducing overall system efficiency.
6. **Scalability Constraints:** As the number of devices within ICS networks increases, managing scalability becomes a challenge. Without appropriate infrastructure, the system's performance and reliability may degrade over time.

7. **Remote Access Challenges:** Enabling secure remote access to ICS components is complex and requires robust authentication and encryption mechanisms to prevent unauthorized access and maintain system integrity.
8. **Communication Reliability:** Communication within ICS environments may be compromised by physical interference, hardware failures, or network congestion. These disruptions can lead to operational instability and process failures.
9. **Human Error:** Errors in configuration, operation, or maintenance by human operators can introduce vulnerabilities or lead to system downtime. This highlights the need for continuous training and strict procedural controls.
10. **Maintenance and Software Updates:** Implementing software patches and updates in continuous 24/7 operational environments is challenging. Interrupting industrial processes for maintenance may not be feasible, making non-disruptive update strategies essential.

These challenges underscore the critical need to enhance the security, interoperability, and resilience of communication mechanisms within Industrial Control Systems (ICS) to ensure reliable and safe industrial operations [142].

3.13 Summary

ICS communication protocols ensure seamless data exchange, coordination, and control in industries. Robust communication mechanisms are vital as industries adopt advanced technologies. Protocols like TCP/IP and specialized solutions offer tailored communication for industrial needs. Security is paramount against cyber threats due to increased connectivity. Human collaboration and training enhance protocol utilization. In conclusion, ICS communication aligns tradition, innovation, security, and efficiency for a data-driven future.

Chapter 4
Exploring Industrial Automation Systems: Security Strategies, Optimization of Control Mechanisms, and AI Integration

Abstract Industrial Automation Systems are fundamental to modern manufacturing and industrial processes, driving cost reduction and enhancing flexibility, efficiency, and quality. This chapter provides a comprehensive examination of various types of automation, including fixed automation, programmable automation, and flexible automation. It also covers essential equipment used in industrial automation systems, detailing their operational processes, application areas, and associated benefits. Additionally, the chapter evaluates security measures for industrial automation systems, addressing cyber threats, data integrity, and system reliability. The discussion extends to sensory technologies for real-time data collection and monitoring, actuator control optimization, and mechanical systems within industrial automation. This analysis offers valuable insights into how industries are evolving through innovation, enhancing operational efficiency and safety.

Keywords Industrial automation systems · Fixed automation · Programmable automation · Flexible automation · Automation equipment · Sensory technologies · Actuator control

4.1 Introduction

Industrial automation and control (IACS) systems are the foundation of essential infrastructure, acting as a pivot for manufacturers as well as the entire country. The smooth functioning of numerous vital industries, including chemical plants and key supply facilities like power generators, greatly depends on these systems. IACS is important, but it goes beyond individual factories; it includes resilience and national security.

OT networks, with IACS at their core, orchestrate and optimize activities within the complex web of contemporary industrial processes. The stakes are high at chemical facilities, where any disruption can have disastrous consequences in addition to compromising production efficiency. Similarly, to prevent significant

M. A. Rahman et al., *Securing Industrial Control Systems*,
https://doi.org/10.1007/978-3-032-03018-4_4

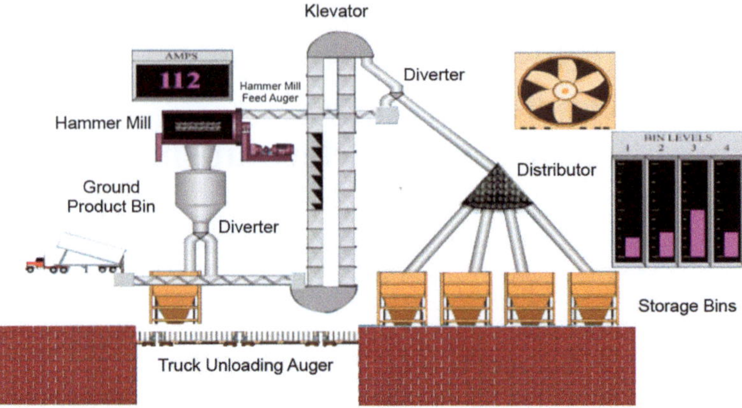

Fig. 4.1 Use of automation

disruptions and disasters that could have a nationwide impact, power generators—essential nodes in the energy infrastructure—must run with consistent reliability. Figure 4.1 is shown below.

Figure 4.1 shows a detailed machine model of an automation system. It includes a hammer mill, a ground product bin, a diverter, an elevator, a distributor, storage bins, and a truck unloading auger, among others. It displays "BIN LEVELS" and "112 AMPS" with three bars representing levels 1, 2, and 3. The plant layout indicates that materials should move from the auger for unloading to the hammer mill, then to the bin for ground products, and finally to the storage bins following the elevator and distributor.

Essentially, the strong operation of IACS is a strategic issue for the welfare of the country, not just a problem for individual businesses. The stability of the country's industrial and infrastructure landscape, the provision of essential services, and the prevention of disasters all depend heavily on the resilience of these systems.

4.2   Types of Industrial Automation Systems

Industrial automation systems cover a broad array of technologies and methods for automating tasks and processes in manufacturing, production, and other industrial environments. These systems provide a variety of advantages, such as higher efficiency, better product quality, and increased safety for employees. Some main types of industrial automation systems are provided as follows:

1. **Fixed Automation:** Designed for high-volume and uniform production processes with limited variation. Suited for tasks where the production requirements remain constant.

2. **Programmable Automation:** Allows for the reprogramming of tasks, making it ideal for small-batch production where tasks need to be adjusted or modified frequently.
3. **Flexible Automation:** Adapts to variations in products and processes through the use of computer-controlled systems. Suited for industries with diverse and changing production requirements.
4. **Integrated Automation:** Connects various automated systems within an organization to achieve seamless and coordinated operations. Ensures efficient communication and collaboration between different components.
5. **Robotics and AI-Based Automation:** Utilizes advanced robotics and artificial intelligence for handling complex tasks and decision-making. Enables automation in areas requiring cognitive abilities.
6. **Process Automation:** Optimizes processes, particularly in industries such as chemicals and pharmaceuticals. Focuses on enhancing efficiency and reliability in the production of goods.
7. **Discrete Automation:** Automates specific tasks in manufacturing, such as assembly lines and quality control processes. Commonly used in industries with distinct and separate production stages.
8. **Batch Automation:** Tailored for industries involved in batch processing, such as food and beverage production. Optimizes the handling of tasks in batches rather than continuous flow.
9. **Continuous Automation:** Applied in sectors requiring uninterrupted and continuous production, such as energy and chemicals. Ensures a steady and continuous flow of production processes.

These types of industrial automation systems have diverse needs and production scenarios, providing flexibility and efficiency in various industrial settings.

4.3 Equipment for Industrial Automation Systems

In industrial automation, sensors and actuators are essential components that operate as the sensory inputs and outputs of the automated system. Control system components are essential in coordinating the functioning of equipment and processes in industrial automation systems. These components ensure the reliable and effective operation of the automation system.

1. **Sensors:** Sensors are devices that identify and quantify different physical attributes, including temperature, pressure, location, proximity, and others. They serve as the sensory inputs of the automation system, collecting live data from the surroundings or the machines.
2. **Actuators:** Actuators respond to sensor data by triggering physical actions or modifications in the system. They convert control inputs into mechanical motion to adjust processes or conditions as needed by the automation system.

3. **Control System Components:** Control system components are essential in coordinating the functioning of equipment and processes in industrial automation systems. These components ensure the reliable and effective operation of the automation system.
4. **Supervisory Control Elements:** Supervisory control elements are tools and systems that monitor and control industrial processes at a higher level to ensure coordination and optimization across numerous subsystems.

4.4   Industrial Automation Work Process

Industrial automation typically consists of three primary components. The first component is a controller, which may be a PLC or an industrial computer. It gathers data from the factory setting and carries out preset commands. The second component is an input device, such as a sensor, that gathers data from the surroundings. The data is then sent to the controller for processing. The third component is an output device, such as a motor or valve, that executes the action determined by the controller.

Automation in Production System

Automated production is essential for businesses seeking to achieve high levels of productivity and operational efficiency. Several key benefits and facilitations of automation are outlined as follows:

1. **Real-Time Data Analysis:** Leveraging real-time data analysis helps to promptly identify and address potential issues, thereby reducing downtime and ensuring the smooth operation of industrial processes.
2. **Automation of Data Collection and Analysis:** Automating data collection and analysis saves time, minimizes human errors, and allows personnel to focus on critical tasks, ultimately enhancing overall operational efficiency.
3. **Root Cause Analysis:** Automated processes facilitate detailed performance analysis, enabling companies to identify the root causes of inefficiencies and operational challenges. This leads to targeted interventions and improved outcomes.
4. **Resource Optimization:** Automation contributes to optimal utilization of time and human resources, resulting in significant cost savings and improved monitoring and control of production activities.
5. **Informed Decision-Making:** Access to accurate and timely information empowers organizations to make well-informed decisions, adapt effectively to changes, and enhance productivity and performance.
6. **Operational Excellence and Competitiveness:** Automation in manufacturing helps businesses achieve operational excellence, boost productivity and profitability, and maintain a competitive edge in the marketplace.

4.5   Exploring Key Aspects of Robust Security Measures

In the modern age of technology, robust security controls are essential to safeguard privacy, sensitive data, and critical infrastructure from various forms of cyberattacks. As society becomes increasingly dependent on interconnected technologies and networked systems, ensuring effective industrial security is more important than ever. The following points outline key factors that enhance the effectiveness of security mechanisms:

1. **Facilitation of Automated Operations:** Industrial security serves as the foundation for the autonomous operation and control of industrial processes. It ensures that these processes are executed reliably and safely, minimizing the risk of errors and potential hazards. Automation plays a vital role in maintaining productivity, reducing manual intervention, and adapting efficiently to evolving industrial demands.

2. **Assurance of Availability, Integrity, and Confidentiality:** A cornerstone of industrial security is the assurance of the three core principles—availability, integrity, and confidentiality. Well-designed security controls ensure system functionality even under potential threat conditions, while also maintaining accessibility when needed. Data confidentiality safeguards sensitive information from unauthorized access, and integrity measures protect data and processes from tampering or corruption.

3. **Creation of a Secure Operational Environment:** Comprehensive security practices not only maintain system availability but also establish a protected environment for industrial operations. These practices shield systems from both malicious cyberattacks and unintentional security breaches. Proactive implementation of security measures provides a safeguard that enables uninterrupted and secure operations.

4. **Integration of Sustainable Methodologies:** Once security concerns are addressed, integrating sustainable and adaptive methodologies becomes essential. This approach supports the secure connection between industrial systems, enterprise networks, and cloud platforms. It promotes interoperability and long-term functionality without compromising existing technologies, thereby ensuring a balanced and durable technological ecosystem.

5. **Flexibility to Evolving Technologies:** Industrial security is dynamic and must continuously evolve alongside technological advancements. Flexible and adaptive security strategies are critical to maintaining the resilience of security protocols against newly emerging threats. This flexibility is vital for future-proofing industrial systems in an era of rapidly changing cybersecurity challenges.

4.6 AI in Industrial Automation: Communication Aspects

Communication plays a vital role in industrial automation by streamlining Artificial Intelligence (AI) capabilities for efficient and adaptive operations. The following are some essential communication components that contribute to successful AI-driven automation:

1. **Data Collection and Analysis:** Effective communication allows AI systems to process large volumes of automation data in order to make informed decisions. Real-time data provides insights into current operations and equipment status, while historical data supports AI algorithms in identifying patterns and adapting to evolving conditions. Continuous data analysis enables AI to detect potential issues early and prevent them from developing into critical failures.
2. **Command and Control:** Reliable communication ensures the accurate execution of tasks in AI-driven automation systems. Instructions are securely delivered to robotic components, ensuring error-free task performance. Control systems guarantee that commands from AI are precisely followed, enabling seamless execution of automated functions and maintaining operational integrity.
3. **Feedback Loop:** Communication plays a key role in enabling AI to learn from processes and improve its performance over time. AI systems evaluate the outcomes of their actions through feedback and refine their decision-making accordingly. This iterative process enhances system efficiency and contributes to the development of self-improving automation frameworks.
4. **Synchronization:** Effective communication ensures the synchronization of different AI subsystems to achieve unified automation goals. It facilitates the exchange of information, coordination of tasks, and resource sharing among multiple AI-driven components. Such synchronization ensures all subsystems work collaboratively, improving overall system efficiency and reliability.
5. **Adaptive Responses:** Strong communication capabilities allow AI to adapt to dynamic operational environments. By continuously receiving real-time sensor inputs and feedback, AI systems can adjust their responses to meet changing conditions. This adaptability supports resilient and efficient operations even in uncertain or fluctuating scenarios.
6. **Anomaly Detection:** Communication enables AI to detect and respond to anomalies in automated systems. Through constant access to sensor readings and system parameters, AI can identify deviations from normal operating patterns. Communication channels support real-time alerts and rapid implementation of corrective actions, minimizing disruption and improving fault tolerance.
7. **Scalability:** Communication systems must scale effectively as the number of devices, data streams, and AI components increases. Scalable communication protocols are essential to accommodate the growing volume of data generated by expanding automation networks. Supporting scalability ensures the system remains efficient and adaptable as new AI technologies are integrated.

8. **Human Interaction:** Effective communication also supports meaningful interaction between human operators and AI systems. Human oversight is critical for setting goals, verifying decisions, and intervening when necessary. Clear communication channels allow operators to monitor, adjust, and collaborate with AI systems to optimize performance and safety.

9. **Remote Monitoring and Management:** Communication facilitates real-time remote monitoring and control of AI-driven automation, which is particularly valuable in industrial settings that require continuous supervision. Remote access allows operators to analyze both current and historical data, update system parameters, and make informed decisions from a distance. This capability is crucial for optimizing processes and ensuring system resilience.

These communication elements are fundamental in enabling the practical application of AI within industrial automation, ensuring efficient, scalable, and adaptive operations.

4.7 Optimizing Actuator Control Strategies for Automation Mechanisms

Actuators are among the most critical components within automation systems, as they translate control signals into mechanical motion. To ensure that actuators function with maximum accuracy and efficiency, different control methods and technologies are applied depending on the actuator type and its intended application.

The most common actuator used in automated systems is the **stepper motor**. Stepper motors are capable of moving in fixed, small increments and are ideal for applications requiring precise positioning. To understand stepper motor functionality, it is essential to be familiar with concepts such as *micro-stepping* and *half-stepping*. These techniques allow the motor to perform finer and smoother motions, thereby improving positional accuracy and minimizing mechanical wear.

However, stepper motors can sometimes exhibit *rotor oscillation*, where motion causes vibrations that affect system stability. Engineers address this issue by implementing **damping techniques** to reduce oscillations and achieve smoother motor performance. Additionally, stepper motors equipped with **permanent magnets** offer higher torque and efficiency, making them suitable for more demanding industrial applications.

To operate effectively, stepper motors rely on **stepper motor drives**, which generate the precise electrical signals required for controlled movement. These drives manage the motor's direction, speed, and position, playing a crucial role in overall system performance. In some automation scenarios, **linear stepper motors** are used to produce direct linear motion, thereby expanding the applicability of stepper motor technology.

Besides stepper motors, several other types of actuators are employed in automation systems, each with unique control requirements based on their operating principles. Examples include:

- **Brushless DC motors**: Known for their high efficiency and low noise levels, these motors are ideal for continuous motion applications.
- **Direct drive actuators**: These eliminate mechanical transmission components, enabling higher speed and more precise control.
- **Hydraulic actuators**: Characterized by their high force output, they are primarily used in heavy-duty industrial applications.
- **Pneumatic actuators**: Valued for their high speed and ease of operation, these actuators are suitable for lightweight or repetitive tasks.

Each type of actuator requires a specialized control strategy to achieve optimal performance. By applying the appropriate control methods, automation systems can significantly enhance productivity, reliability, and energy efficiency. Optimizing actuator control not only improves the performance of individual components but also contributes to the overall efficiency and effectiveness of industrial automation processes.

4.8 Sensory Technologies in Automation

Sensors are essential components in automation systems as they enable machines to perceive and respond to their environment. They exist in various forms, each designed for a specific purpose. Sensors are used to acquire critical information such as position, velocity, force, and distance, which is processed by control systems to make decisions and take appropriate actions.

Sensors can be classified based on their location, operational principles, and the type of information they transmit. Some sensors provide **absolute values**, delivering precise positional or status data. Others provide **incremental values**, indicating the amount of change or displacement from a reference point. Key performance characteristics of sensors include **linearity** (how accurately the sensor output follows a predictable pattern), **resolution** (the smallest change the sensor can detect), and **dynamic response** (how quickly the sensor responds to changes). These attributes are crucial to ensure the accuracy and reliability of the collected data.

For position detection, two main types of sensors are utilized: **angular position sensors** and **linear position sensors**. These sensors operate using various technologies, including resistive, capacitive, inductive, and optical methods, to detect rotational or linear movement. Some sensors employ **absolute encoding** to provide precise positional information, while others utilize **incremental encoding** to track relative movement over time.

To measure motion, **velocity** and **acceleration sensors** are employed. Common examples include **tachogenerators**, which generate voltage proportional to speed; **optical encoders**, which detect movement using light; and **micromechanical sen-**

sors (MEMS), which are compact and efficient for detecting motion and vibrations. These sensors help automation systems regulate speed and respond promptly to dynamic changes.

Contact sensors are used to detect physical interactions such as touch, pressure, or bending. Examples include **piezoresistive sensors**, which vary resistance upon pressure; **capacitive tactile sensors**, which detect changes in electric charge due to contact; and **strain gauge sensors**, which measure deformation due to stretching or bending. These sensors are commonly used in robotic hands, pressure-sensitive interfaces, and safety mechanisms.

Advanced sensors capable of highly accurate **distance** and **speed** measurements include **Time-of-Flight (ToF) sensors**, **Laser Rangefinders**, and **Laser-Doppler Velocimeters**. These technologies are vital for tasks such as object detection, navigation, and speed control in autonomous systems.

Sensors are fundamental components of any automated system. They provide the essential data required for machines to operate efficiently, intelligently, and safely. By integrating various types of sensors, automation systems can achieve reliable monitoring and control, leading to enhanced process performance and system effectiveness.

4.9 Dynamics Mechanical Systems in Industrial Automation

Mechanical systems are crucial in industrial automation systems, enabling everything from simple motions to complex operations. Understanding and optimizing mechanical systems require an in-depth knowledge of their dynamic behavior under the influence of forces and motion. The following are important dynamic factors in the context of mechanical systems in industrial automation:

1. **Kinematics and Dynamics Concepts:** Kinematics and dynamics principles are employed to analyze the motion and the forces acting on automated machinery and equipment. A thorough understanding of these concepts aids in designing accurate motion profiles, improving system performance, and ensuring operational safety.
2. **Vibration Analysis:** Vibration analysis is essential for identifying and mitigating unwanted vibrations in automated systems. Excessive vibration can lead to performance degradation, increased wear and tear of components, and potential structural damage. Effective vibration control contributes to longer equipment life and stable operation.
3. **System Response:** The dynamic response of mechanical systems to inputs and disturbances must be carefully assessed to develop effective control strategies. A well-characterized response helps improve stability, accuracy, and overall system performance in industrial automation systems.

4. **Modal Analysis:** Modal analysis is employed to determine the natural modes of vibration in mechanical systems. This includes identifying resonance frequencies and mode shapes, which are critical for vibration suppression and mechanical integrity. In industrial automation, modal analysis supports the design of robust and stable mechanical structures.

4.10 Control of Actuators in Automation

Stepper motors are widely used as drivers in automation systems due to their simplicity in control and ability to move in precise, small steps. Stepper motors are controlled by supplying specific electrical signals to their windings, causing the rotor to rotate incrementally. Below is an overview of how stepper motors are utilized to control devices in robotic and automation systems:

1. **Stepper Motor Operation Principles:** Stepper motors provide precise control of position and velocity by dividing a full revolution into discrete, equal steps. This capability makes them highly suitable for applications requiring accurate motion control.
2. **Half-Step Mode Operation:** In half-step mode, the motor alternates between full and intermediate step positions, resulting in smaller steps and smoother motion compared to full-step operation. This mode also enhances the resolution of the motor.
3. **Micro-Step Mode:** Micro-step mode divides each full step into multiple smaller steps, enabling even finer resolution and smoother motion. This mode is especially beneficial for applications requiring high precision in positioning.
4. **Methods of Damping Rotor Oscillations:** To minimize rotor oscillations and improve motor performance, various techniques are employed, such as vibration suppression, dynamic braking, and the incorporation of damping resistors. These strategies help enhance motor stability and operational smoothness.
5. **Permanent Magnet Stepper Motors:** Permanent magnet stepper motors utilize permanent magnets to generate magnetic fields, resulting in higher torque and efficiency compared to other types of stepper motors. They are often used in applications demanding reliable performance and efficiency.
6. **Stepper Motor Drives:** Stepper motor drives manage the operation of stepper motors by providing precise electrical pulses to energize the motor coils. This ensures accurate positioning, smooth motion, and efficient control of the motor.
7. **Linear Stepper Motors:** Linear stepper motors convert rotational motion into linear motion, offering precision positioning and actuation capabilities. These motors are commonly used in applications such as linear stages, conveyors, and precision machinery.

4.11   Application of Industrial Automation

Industrial automation is widely implemented across various sectors to enhance efficiency, reduce operational costs, and ensure safety. Key application areas of industrial automation are illustrated in Fig. 4.2, and briefly described below:

1. **Automated Surveillance by Video**

   (a) **Industrial Safety:** Automated video surveillance plays a vital role in ensuring safety in industrial environments. It enables continuous monitoring of hazardous areas, detection of suspicious activities, and prevention of accidents.

   (b) **Protection of Critical Infrastructure:** Video surveillance systems are essential for safeguarding critical infrastructure such as water treatment plants and power stations. They help detect potential threats and prevent unauthorized access.

   (c) **Emergency Response:** Real-time video analytics support timely and effective emergency responses by providing situational awareness during crises and emergencies.

2. **Automated Highway Systems**

   (a) **Traffic Optimization:** Automated highway systems improve traffic flow by regulating vehicle movements, reducing congestion, and minimizing travel time.

   (b) **Fleet Management:** These systems facilitate efficient fleet management through route optimization, vehicle condition monitoring, and predictive scheduling of maintenance in logistics and transportation industries.

Fig. 4.2 Some application areas of industrial automation

   (c) **Autonomous Vehicles:** Automation of highway systems is a foundational requirement for the development and operation of autonomous vehicles, which enhance road safety and offer innovative transportation solutions.

3. **Automated Production**

   (a) **Collaborative Robotics:** Collaborative robots (cobots) work alongside human operators to increase productivity and safety in modern manufacturing facilities.

   (b) **Predictive Maintenance:** Predictive maintenance systems use sensors and data analytics to anticipate equipment failures, thereby reducing unplanned downtime and minimizing maintenance costs.

   (c) **Smart Factories:** In the context of Industry 4.0, smart factories utilize automation, the Internet of Things (IoT), and data analytics to enable real-time decision-making and intelligent manufacturing processes[91].

4. **Home Automation**

   (a) **Security Systems:** Home automation enhances residential security through smart devices such as locks, door/window sensors, surveillance cameras, and alarm systems.

   (b) **Energy Efficiency:** Automated thermostats, smart lighting, and energy-efficient appliances help reduce energy consumption and utility costs in smart homes.

   (c) **Elderly Care:** Home automation systems support elderly care by integrating fall detection devices, health monitoring tools, and emergency response systems to improve quality of life and safety.

4.12 The Benefits of Industrial Automation

Industrial automation enhances the efficiency of operations for original equipment manufacturers (OEMs), manufacturers, and various industrial processes by offering several significant advantages.

1. **Lower Risk:** Industrial automation systems provide a safe working environment and protect workers from accidents or injuries that may occur during tasks that are otherwise performed manually. These tasks include material processing, machining, welding, and the operation of hazardous substance containment systems. Robots and automated-guided vehicles can also minimize the risks associated with manual handling. Machining and welding can be performed by highly automated machines in controlled environments, reducing workers' exposure to sparks, fumes, or heat. Additionally, automated systems can handle toxic or radioactive chemicals, preventing workers from being exposed to these hazardous substances.

2. **Increased Productivity:** Automatons and monitors can pinpoint areas of waste in the form of production slowdowns, equipment idle time, or workflow blockages. This allows for maximum processing and efficiency.
3. **Improved Quality:** With interconnected automation equipment, improved product quality, waste reduction, and higher profitability are realized. Such systems provide better process repeatability, real-time data acquisition, and advanced quality check.
4. **Improved Decision-Making:** The use of industrial automation devices allows easier access to much better quality and real-time data, which leads toward the creation of more accurate analytical models. This facilitates greater supply-chain coordination, enhances customer engagements, and enables new revenue streams to be discovered.

4.13 Summary

Industrial automation solution has revolutionized many industries and brought numerous benefits, advances, as well as new process creations. The potential applications are extensive, including optimizing traffic congestion and empowering drones to travel on autopilot highways, through to enhancing safety in factories by offering automated surveillance.

Beyond efficiency gains, integrating robots with predictive maintenance in manufacturing facilitates the "smart factories" envisioned by Industry 4.0. As a result of this paradigm shift, by bringing data analytics together with the Internet of Things (IoT) and automation in one place, we can now make decisions instantaneously. Industrial automation is increasingly influencing homes, offering benefits such as elderly care, energy saving, and enhanced security. As technology advances, the advantages of these systems become more apparent across various sectors. Industrial automation prioritizes convenience, efficiency, and safety, ushering in an era of intelligent, networked systems that are revolutionizing daily life. While still evolving, industrial automation promises further innovation and positive changes for numerous sectors.

Chapter 5
Mitigating the ICS Attack Surface: Identifying Attack Vectors, Reducing Vulnerabilities, and Security Mapping Techniques

Abstract One of these critical components is an understanding of the industrial attack surface that enables the detection of unauthorized or malicious access within industrial networks. This chapter delves into different attack surface categories and how to measure attack surface metrics and provides mapping and guidance on how to reduce the attack surface. It also deals with various attack vectors (and prevention systems for that) and access attacks, namely logical and physical access. The chapter also explores typical attack scenarios and presents best practices for mitigating the attack surface. This book will help to uncover the major security gaps in a given industrial automation system and show what attackers could potentially exploit to infiltrate other parts of the network.

Keywords Industrial attack surface · Attack surface metrics · Attack surface mapping · Attack vector prevention · Access attacks · Logical access attacks · Physical access attacks · Industrial automation systems · Network security · Attack techniques

5.1 Introduction

For Industrial Control Systems (ICSes), the attack surface refers to all channels, communication protocols, and weak points that malicious actors, hackers, or other malevolent entities may exploit to gain access to systems or network control applications. The attack surface is an aggregation of vulnerabilities and weaknesses within the ICS environment that can be targeted by cyber attackers.

Securing the cyber environment of an ICS requires sophisticated attack surface management. Organizations must ensure that their attack surface is continuously monitored and that threats are promptly identified and mitigated. Maintaining a minimal attack surface is also crucial to reducing the risk of cyber-attacks. To protect ICSes, organizations should enforce stringent access controls, segment networks, update software regularly, deploy intrusion detection systems (IDSes), and provide comprehensive end-user training.

M. A. Rahman et al., *Securing Industrial Control Systems*,
https://doi.org/10.1007/978-3-032-03018-4_5

">

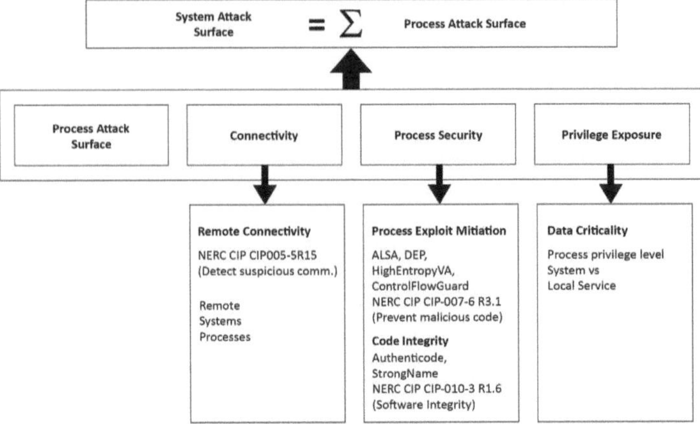

Fig. 5.1 Attack surface management

An attack surface specifies the number of possible attack vectors an intruder can use to enter a system and filter sensitive data. Therefore, a smaller attack surface is easier to protect. Figure 5.1 is taken from [76].

Figure 5.1 illustrates the correlation between process attack surfaces and system attack surfaces. The components consist of a network diagram, a process attack surface box, and three sub-boxes dedicated to remote connectivity, process exploit mitigation, and data criticality. The main point is to keep the system safe at both the system and process levels.

5.2   Attack Surface

The term *attack surface* refers to all potential vectors that attackers can exploit to breach an organization's network, applications, or data by taking advantage of vulnerabilities. The attack surface can be physical or digital—including servers, endpoints, cloud infrastructure, IoT devices, or even human factors such as employee access credentials. A larger attack surface provides attackers with more opportunities to identify and exploit vulnerabilities, making it essential for organizations to adopt a proactive security approach. Reducing the attack surface is a fundamental principle of cybersecurity because the fewer people and devices that have access, the lower the likelihood of unauthorized access or malicious activity, such as breaches.

The total vulnerable space of an organization or system is also referred to as the attack surface. Figure 5.2 (adapted from [34]) illustrates this concept. One of the primary objectives of attack surface analysis is to systematically identify, classify, and map all possible attack vectors within an organization. This process involves locating risk-critical zones, identifying vulnerable systems, and assessing

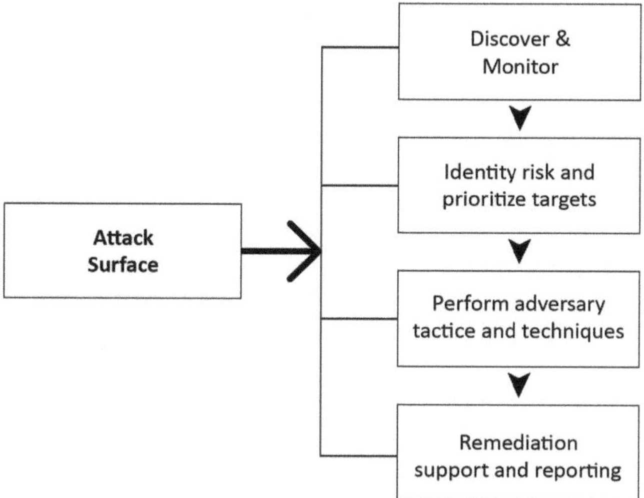

Fig. 5.2 Attack surface

their susceptibility to threats. By understanding these risk factors, organizations can deploy security controls such as network segmentation [160], access controls, multi-factor authentication (MFA), and real-time monitoring to reduce the number of exploitable attack vectors.

It is important to account for a wide range of both internal and external threats when evaluating the attack surface. This may include modeling the potential impact of a successful cyberattack, documenting all open ports, publicly available cloud services, exposed APIs, and legacy insecure software components, and determining whether existing security controls are sufficient. Emerging threats, including zero-day vulnerabilities and increasingly sophisticated social engineering tactics, further complicate the traditional IT infrastructure. Attack surface management—analogous to regularly trimming the garden of security—enables organizations to strengthen their security posture and minimize cyber risks.

Figure 5.2 shows a cybersecurity framework. The system attack surface displays all the ways an attacker may penetrate or abuse a system. This diagram illustrates several steps of the attack surface, including risk identification, discovery and monitoring, target prioritization, remediation, attack support, surface reporting, and adversary tactics and strategies.

5.2.1 Pseudocode for Attack Surface

```
BEGIN AttackSurfaceAnalysis

    // Define system components
    INITIALIZE components = ["WebServer", "DatabaseServer",
    "ApplicationServer", "NetworkDevices"]

    // Define potential entry points for each component
    FUNCTION IdentifyEntryPoints(component):
        IF component == "WebServer" THEN
            RETURN ["HTTP", "HTTPS"]
        ELSE IF component == "DatabaseServer" THEN
            RETURN ["SQLPorts", "SSH"]
        ELSE IF component == "ApplicationServer" THEN
            RETURN ["APIEndpoints", "WebSockets"]
        ELSE IF component == "NetworkDevices" THEN
            RETURN ["Telnet", "SNMP"]
        END IF
    END FUNCTION

    // Analyze attack surface
    FUNCTION AnalyzeAttackSurface(components):
        attackSurface = {}
        FOR EACH component IN components:
            entryPoints = IdentifyEntryPoints(component)
            attackSurface[component] = entryPoints
        END FOR
        RETURN attackSurface
    END FUNCTION

    // Main process
    attackSurface = AnalyzeAttackSurface(components)
    PRINT "Attack Surface: ", attackSurface

END AttackSurfaceAnalysis
```

Listing 5.1 Attack surface analysis algorithm

This pseudocode, referred from: https://github.com/Sunzidasiddique1/ICS, describes the entry points and components of a system. The function `Identify EntryPoints` returns possible entry points based on the component type, such as HTTP for web servers and SQL ports for database servers. The **Analyze Attack Surface** function collects all these entry points to form an overall context for the attack surface of that particular system.

5.3 Categories of Attack Surface

The **attack surface**: Then sum the number of possible access points, or attack vectors, an unauthorized user can exploit to enter and extract sensitive information from within [156]. This minimizes the attack surface and improves security and makes it easier to defend against potential attacks. We can break down the attack surface fairly easy into two unique groups:

1. **Digital Attack Surface:** All vulnerabilities and entry points available to cyber attackers for targeting an organization in its digital environment. This includes software, networks, protocols, user access points, and more.

   a. **Weak Passwords:** Easily guessable or poorly constructed passwords pose significant security risks. Cybercriminals can exploit such weaknesses to hijack user accounts, access sensitive data, spread malware, or penetrate internal networks.

   b. **Misconfiguration:** Attackers often target ports, communication channels, wireless networks, and firewalls that are improperly configured. Weak encryption standards can also make communication channels vulnerable to man-in-the-middle attacks.

   c. **Software and OS Bugs:** Vulnerabilities in third-party applications, operating systems, or firmware can be exploited by attackers. Poorly written code may allow unauthorized access to systems, data breaches, or malware installation.

   d. **Web-Facing Assets:** Weak or vulnerable code in web applications, edge servers, and exposed APIs can lead to data leaks or unauthorized access.

   e. **Obsolete Components:** Failing to regularly update systems leaves them exposed. Attackers often target outdated endpoints and legacy components with known vulnerabilities.

   f. **Shared Directories and Databases:** These may be breached to gain unauthorized access and launch ransomware or other forms of cyberattacks [149].

   g. **Shadow IT:** Unauthorized devices or software used without the knowledge of IT teams introduce unmonitored vulnerabilities, increasing the risk of successful attacks.

2. **Physical Attack Surface:** This refers to all corporate physical assets and devices that are vulnerable to physical attacks. Examples include servers, desktop computers, mobile devices, hardware components, and physical security systems.

   a. **Malicious Insiders:** Employees or contractors with authorized access may misuse their privileges to steal data, sabotage systems, or install malware.

   b. **Device Theft:** Adversaries may steal hardware or gain physical access to facilities to exfiltrate data and compromise operational procedures. Remote work environments, personal devices, and improperly discarded hardware are particularly vulnerable.

   c. **Baiting:** In baiting attacks, attackers place infected USB drives or other removable media in public locations, hoping potential victims will pick them up and connect them to their devices, which can lead to malware installation.

5.4 Measuring Attack Surface Metrics

Measuring the attack surface is essential for strengthening cybersecurity efforts [99, 143]. By identifying and counting vulnerabilities and entry points, organizations can turn abstract security concerns into measurable facts. This process not only shows how exposed a system is but also helps organizations make better decisions [143]. With a clear understanding of the security landscape, they can focus their attention, use resources wisely, and fix critical vulnerabilities quickly. Measuring the attack surface is not just a technical task—it is a key step toward proactive and complete cybersecurity [99].

1. **Number of Entry Points:** During the planning and development stages, identify all possible ways an attacker could interact with the system. These include user interfaces (such as web forms and APIs), network ports, and other points that could be used to gain unauthorized access to system features or private data. Fewer entry points usually mean a smaller attack surface and easier security control [37].
2. **Complexity of Interactions:** Review how different parts of the system interact with one another. Pay special attention to how the system responds to unusual behavior or harmful input. Keeping these interactions simple and clear helps reduce security risks, since complex setups can hide weak points.
3. **External Dependencies:** Check the security of external libraries or APIs used in the system. If these components are not secure, they can become weak spots. Keeping dependencies updated ensures that security patches from the vendors are applied.
4. **User Privileges:** Set up a clear and controlled access system that assigns the right level of permissions to different user roles. Consider the risk of users trying to take advantage of weaknesses to gain higher access than they are allowed.
5. **Data Exposure Points:** List all places where sensitive information is stored, processed, or shared. This includes network communications, database queries, and temporary storage in application memory. Studying these areas can uncover risks where attackers could steal, change, or damage data.

5.5 Steps to Reduce Attack Surface

Reducing the attack surface is crucial for mitigating cybersecurity risks and preventing hacking attempts. A smaller attack surface limits the number of exploitable vulnerabilities, making it more difficult for attackers to gain unauthorized access to critical systems and data. Understanding the security environment allows organizations to safeguard sensitive data by securing vulnerable attack vectors, eliminating unnecessary access points, and continuously monitoring for potential threats. One of the most effective methods for reducing the attack surface is managing access and user permissions, with a particular focus on revoking or modifying access

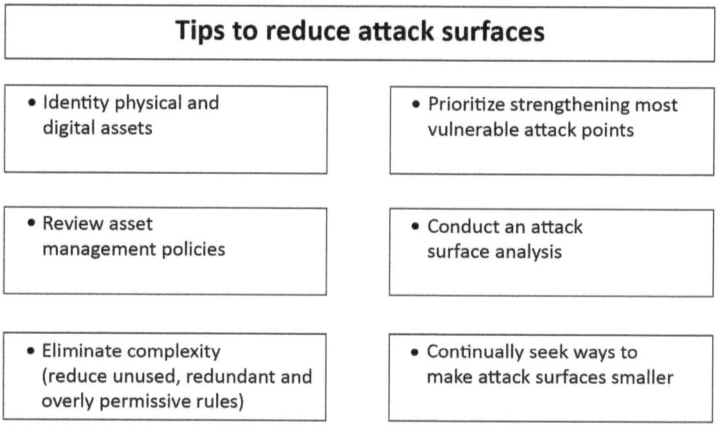

Tips to reduce attack surfaces

- Identity physical and digital assets

- Prioritize strengthening most vulnerable attack points

- Review asset management policies

- Conduct an attack surface analysis

- Eliminate complexity (reduce unused, redundant and overly permissive rules)

- Continually seek ways to make attack surfaces smaller

Fig. 5.3 Steps to reduce attack surface

levels based on user roles and responsibilities. Implementing the principle of least privilege ensures that users and applications have only the permissions necessary for their tasks, minimizing the risk of privilege escalation attacks.

Attack surface analysis helps identify both immediate and potential future risks by evaluating vulnerabilities across an organization's infrastructure. This includes analyzing misconfigurations, monitoring network activity, and assessing third-party integrations that could introduce security weaknesses. Organizations should conduct regular security audits and vulnerability assessments to ensure that emerging threats are promptly addressed. Below are some steps to reduce the attack surface. Figure 5.3 is adapted from [148].

Figure 5.3 illustrates the five recommendations for mitigating attack surfaces, including steps such as identifying physical and digital assets, prioritizing the strengthening of vulnerable attack points, reviewing asset management policies, eliminating complexity, and continually seeking ways to reduce unused and redundant elements to make attack surfaces smaller.

5.6 Attack Surface Mapping

Attack surface mapping is an effective technique that extends beyond the fundamentals of cybersecurity. It identifies, visualizes, and documents potential entry points and vulnerabilities within a digital ecosystem. This practice enhances the understanding of system security by systematically identifying exploitable vulnerabilities. It functions as a floodlight, illuminating potential security violations that might otherwise remain concealed. This visual audit enables organizations to allocate security resources more efficiently and strategically, prioritizing efforts to protect critical vulnerabilities. Attack surface mapping is more than just a diagram; it is a proactive stance against threats, a blueprint for securing the digital domain, and a testament to informed cybersecurity practices.

5.7 Components of Mapping

Attack surface mapping involves dissecting several key components:

a. **Interfaces:** This component focuses on identifying all user interfaces (UIs), application programming interfaces (APIs), and other communication channels used by users, applications, or external systems to interact with an organization's systems. Interfaces help in identifying potential attack routes such as injection attacks via user input forms, unauthorized API access, or vulnerabilities in communication protocols.

b. **Data Flow:** This component tracks the flow of data within the system, including its sources, destinations, transformations, and storage locations. Understanding data flow aids in recognizing potential vulnerabilities at different stages, such as unauthorized data access at rest, data manipulation during transit, or insecure storage methods.

c. **Dependencies:** This component identifies third-party integrations and external libraries used by the organization's systems. Examining dependencies helps in detecting vulnerabilities in external components that attackers might exploit to gain access to the organization's systems.

d. **Access Points:** This component identifies all possible vulnerabilities that an attacker might exploit to gain unauthorized access to an organization's systems. This includes conventional network ports, web applications, remote access points, physical access points, and other potential avenues for system entry.

5.7.1 Mapping Techniques

Attack surface mapping employs various techniques to identify potential entry points and vulnerabilities in a system. These techniques help assess the security posture by mapping out all possible ways an attacker could interact with the system. Some common mapping techniques include:

1. **Data Flow Diagrams :** These diagrams illustrate the data flow within a system, including sources, destinations, transformations, and storage locations. Security specialists can detect potential weaknesses in data flow stages, such as unauthorized access points, data manipulation opportunities, and insecure storage locations, through DFD analysis. DFDs help visualize the entire data ecosystem, facilitating the identification of vulnerabilities that attackers may exploit to gain unauthorized access to sensitive data.

2. **Network Diagrams:** Network diagrams depict the physical and logical arrangement of a network infrastructure, showcasing devices, connections, and communication protocols. Security experts can identify exposed services, misconfigured network devices, and segmentation issues by analyzing network diagrams. These illustrations highlight the network's attack surface, emphasizing potential entry points that attackers may exploit for unauthorized access or network disruption.
3. **Dependency Trees:** Dependency trees illustrate the interdependencies among various system components, including software programs, libraries, and external services. Security experts can identify weaknesses in dependent components that may affect the system's overall security. These illustrations highlight the interconnected nature of systems, emphasizing the cascading effects of vulnerabilities and facilitating targeted efforts to secure the most critical components.

5.8 Attack Surface Measurement Tools

Attack surface measurement tools are software programs or platforms designed to assess and evaluate the vulnerability of a system, application, or network to potential cyberattacks and threats. These tools analyze various aspects of a system's configuration, structure, and dependencies to identify possible entry points, weaknesses, and attack vectors [156]. By providing detailed insights, they enable security professionals and organizations to understand the scope and severity of their vulnerabilities and implement appropriate security strategies to mitigate risks. Common features of these tools include vulnerability detection, automated scanning, and integration capabilities. Notable examples include:

1. **OWASP ZAP (Zed Attack Proxy):** A robust web application security scanner that identifies vulnerabilities during development and testing phases. It offers automated scanning, manual testing tools, and a comprehensive API for integration with development pipelines. ZAP helps developers and security experts detect common threats such as injection attacks, cross-site scripting (XSS), and insecure configurations.
2. **Nessus:** One of the most widely used vulnerability scanners, Nessus features an extensive database of known vulnerabilities and misconfigurations. It supports both scheduled and on-demand scanning and generates customizable reports to aid in effective remediation.
3. **Burp Suite:** A comprehensive set of tools for performing security assessments of web applications. It includes a web vulnerability scanner, proxy, and intruder, supporting both automated and manual testing. Its modular architecture allows professionals to conduct in-depth analyses and exploit assessments.
4. **OpenVAS:** A free and open-source vulnerability scanner capable of detecting various security issues, such as software vulnerabilities, misconfigurations, and weak credentials. It features a user-friendly interface for scheduling scans and generating detailed reports to support vulnerability management workflows.

5. **Nexpose:** A vulnerability management solution designed to help organizations identify, prioritize, and remediate security risks. It combines scanning capabilities with asset discovery and risk assessment based on exploitability and business impact. Nexpose integrates with various security tools to streamline vulnerability management.
6. **OWASP Dependency-Check:** A specialized tool for scanning project dependencies to identify known vulnerabilities in third-party libraries and components [112]. Supporting multiple programming languages and package managers, it helps reduce risks associated with external dependencies.
7. **Qualys Vulnerability Management:** A cloud-based solution for continuous vulnerability detection and remediation. It includes features such as asset discovery, real-time monitoring, and customizable reporting. Qualys integrates with other security platforms to provide a holistic risk mitigation approach.
8. **SecurityScorecard:** A platform that assigns security ratings to organizations based on their external security posture. It continuously monitors metrics like network security, patch cadence, and DNS configurations. These ratings assist organizations in evaluating their performance, setting remediation priorities, and ensuring compliance.
9. **UpGuard:** Provides real-time scanning of an organization's external attack surface, identifying digital assets and vulnerabilities across web applications, cloud services, and infrastructure. It offers automated risk assessments, threat intelligence integration, and remediation tools to improve overall security posture.

5.8.1 Cyber Attack Surface

The term "cyber attack surface" refers to the complete set of potential points, paths, and vulnerabilities within an organization's digital environment that adversaries could exploit to launch cyberattacks. It includes a wide range of elements that may be targeted, compromised, or breached—placing sensitive data, systems, networks, applications, and other digital assets at risk of unauthorized access or control. Figure 5.4 is reproduced from [61].

Figure 5.4 illustrates the cyber attack surface. It demonstrates the interconnection of many networks, including personal, business, and public environments, and their vulnerability to hacking. This figure highlights how hackers target workers via home networks and business partners or clients.

5.8.2 Commonly Used Attack Methods

Any weak spot in a network that has the potential to result in a data leak falls under the term "surface vulnerabilities." Data may be leaked to hackers from various

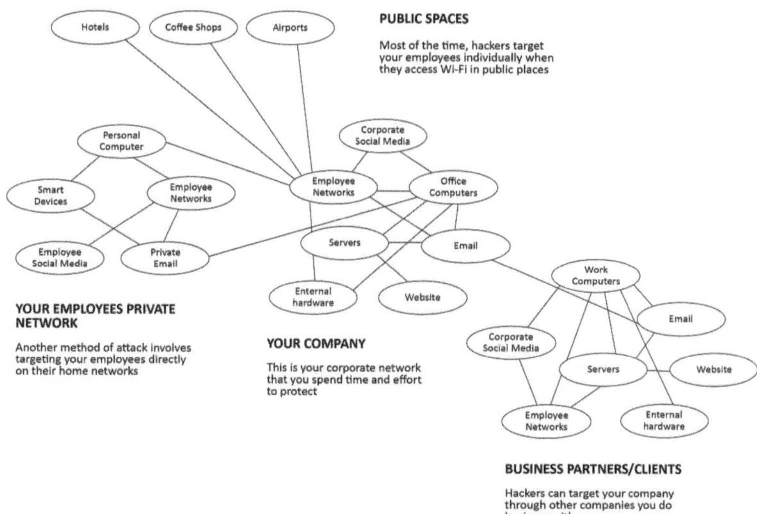

Fig. 5.4 Cyber attack surface

sources, including users themselves, as well as through devices such as computers, mobile phones, and hard drives.

The use of weak passwords, lack of email security, unsecured ports, and failure to patch software are all examples of vulnerabilities. These weaknesses create backdoors for attackers to exploit, targeting individuals and organizations to conduct cyberattacks. Man-in-the-middle (MITM) attacks are enabled by weak web-based protocols, which hackers may leverage to intercept and steal data.

Classifying the Cyber Attack Surface

The attack surface of a system encompasses all potential entry points that attackers might exploit to gain unauthorized access, steal data, or disrupt operations. By categorizing the attack surface, security experts can gain a comprehensive understanding of the various elements that require protection. The attack surface can be classified into three main categories:

1. **Network vulnerabilities:** Malicious actors may exploit weaknesses in network access points, unauthorized hardware, insecure protocols, and misconfigured interfaces.
2. **Software vulnerabilities:** Attackers may exploit flaws in applications, operating systems, and third-party software, including vulnerabilities in code, configurations, and access controls.
3. **Physical vulnerabilities:** Tangible assets such as servers, computers, and other hardware within an organization's premises must be secured to prevent unauthorized access and protect sensitive data.

5.9   Network Surface Components

Network surface components are distinct elements within a network that attackers may exploit as potential entry points. These components can be categorized based on their functionality and their exposure to risks. If not properly secured, they represent vulnerabilities that can be exploited. Below is an analysis of key network infrastructure components:

1. **Open Ports:**

   a. Unsecured open ports serve as entry points for attackers, allowing external traffic to access internal network services.
   b. Attackers can exploit these ports to gain unauthorized access, exfiltrate data, or launch attacks such as DoS and Remote Code Execution.
   c. Proper firewall configurations and port scanning tools help identify and close unnecessary open ports, reducing the attack surface.

2. **Insecure Protocols:**

   a. The use of insecure protocols, such as HTTP instead of HTTPS or outdated versions of FTP and Telnet, exposes data to interception through MITM attacks.
   b. Insecure protocols often transmit sensitive information, such as usernames, passwords, and credit card numbers, in plaintext, making it easier for attackers to compromise confidentiality.
   c. Secure alternatives like HTTPS, SSH, and SFTP should be used to encrypt data in transit and ensure data integrity.

3. **Multiple Users:**

   a. Multiple users with varying levels of access complicate access management and may lead to misconfigured permissions.
   b. The presence of numerous user accounts increases the risk of unauthorized access through social engineering, phishing, or brute-force attacks.
   c. Implementing a least-privilege access model and regularly reviewing user permissions can mitigate these risks.

4. **Multiple Administrative Accounts:**

   a. Multiple users with administrative privileges expand the attack surface by providing more potential targets for privilege escalation attacks.
   b. Attackers may attempt to compromise administrative accounts to gain full control over a system.
   c. Limiting the number of admin accounts, enforcing MFA, and monitoring admin activities can reduce risks.

5. **Low Bandwidth:**

   a. Limited bandwidth negatively impacts network performance, making it difficult to handle high volumes of traffic, especially during security incidents or attacks.
   b. During a security breach, low bandwidth can delay incident response efforts, hindering data backup, mitigation strategies, or network isolation.
   c. Optimizing network bandwidth, implementing Quality of Service protocols, and ensuring sufficient bandwidth during peak periods enhance security and incident response efficiency.

5.9.1   Software Surface Components

Microsoft Surface is a collection of personal computers, tablets, and interactive whiteboards created by Microsoft. These devices include touchscreens, and most of them use the Windows operating system. While offering various functionalities, they can present significant vulnerabilities if the software is not properly managed or secured.

1. **Improper Coding:**

   a. Flaws in software code create vulnerabilities that attackers can exploit to compromise the system.
   b. These weaknesses can lead to unauthorized access, data breaches, or system failures if not addressed during development.
   c. Ensuring secure coding practices and conducting thorough code reviews can help mitigate these risks.

2. **Privacy Settings:**

   a. Weak or improperly configured privacy settings expose sensitive data, making it accessible to unauthorized individuals or applications.
   b. This can result in data leakage, identity theft, or other privacy breaches that compromise user confidentiality.
   c. Regular audits and updates to privacy settings are essential for maintaining a secure environment for personal data.

3. **Open-Source Apps Without Active Community Support for Upgrades and Patch Management:**

   a. Open-source applications that lack an active community or formal support structure are more vulnerable to security risks.
   b. These apps may go unpatched for extended periods, increasing the likelihood of unaddressed security vulnerabilities that attackers can exploit.
   c. Encouraging community engagement and establishing robust patch management practices can help ensure timely updates and prevent exploitation.

5.9.2 Physical Surface Components

Physical surface components refer to the visible and directly interacted-with elements of a system or device. These components are essential for functionality, usability, and security. They include hardware, interfaces, and tangible devices, all of which can present security risks if not properly managed or protected. Physical security is often overlooked, yet it plays a crucial role in maintaining overall system integrity. Below is an analysis of key physical surface components vulnerable to security threats:

- **Internal Employees:**

  a. Insider threats from employees with direct access to critical systems and data pose significant security risks.
  b. Employees may intentionally or unintentionally misuse their access, leading to unauthorized access, data theft, or system manipulation.
  c. Regular training, monitoring, and enforcing strict access controls can mitigate risks associated with internal threats.

- **Rogue Devices:**

  a. Unauthorized devices connected to the network, such as personal laptops, smartphones, or external drives, can create vulnerabilities.
  b. These rogue devices may introduce malware, compromise network security, or facilitate data breaches by bypassing security measures.
  c. Strict device management policies, network segmentation, and endpoint protection software can help prevent unauthorized devices from gaining access.

- **Social Engineering:**

  a. Social engineering involves manipulating individuals into divulging confidential information by exploiting human psychology.
  b. Attackers often use tactics such as pretexting, baiting, or impersonating trusted individuals to gain unauthorized access to sensitive data or systems.
  c. Raising awareness, conducting regular security training, and implementing verification protocols can reduce the success rate of social engineering attacks.

- **Passwords on Sticky Notes:**

  a. Poor password practices, such as writing passwords on sticky notes, create significant security vulnerabilities.
  b. These notes can be easily accessed by unauthorized individuals, leading to potential breaches of sensitive data and systems.
  c. Encouraging the use of password managers, enforcing strong password policies, and educating employees on secure password practices can minimize this risk.

- **Phishing Emails:**

  a. Phishing emails are deceptive messages designed to trick users into divulging personal information or clicking on malicious links.
  b. These attacks can lead to system compromises, data breaches, and financial losses if users are exposed to them.
  c. Educating users aboutout recognizing phishing attempts, implementing email filtering systems, and regularly testing security protocols can help prevent phishing attacks.

5.10 Attack Surface Reduction in Cybersecurity

Attack surface reduction in cybersecurity is a proactive strategy aimed at minimizing the number of vulnerabilities and potential access points in an organization's digital infrastructure. By identifying and addressing weaknesses in the system, this approach reduces overall exposure to cyberattacks. A critical aspect of attack surface reduction is the continuous process of finding and mitigating security gaps, limiting unnecessary access, and removing obsolete or unused components that could serve as entry points for attackers.

Effective implementation of this strategy includes vulnerability management, where known weaknesses are patched, and threats are proactively identified. Access control mechanisms are strengthened by ensuring that only authorized users have access to critical systems and sensitive data. Network segmentation further limits the spread of potential attacks, ensuring that a compromise in one segment does not jeopardize the entire network. Additionally, safe development practices—such as secure coding, regular code reviews, and the use of secure frameworks—play an integral role in minimizing the introduction of vulnerabilities into systems.

Furthermore, reducing the attack surface involves utilizing advanced security technologies, including firewalls, IDS, and encryption techniques, to safeguard the perimeter and internal systems. Regular audits and penetration testing are crucial for assessing the security posture and identifying new potential risks that may emerge over time. By continuously evolving and refining security measures, organizations can stay ahead of cybercriminals and protect their critical assets from compromise.

Attack surface reduction is not a one-time effort but an ongoing process that demands continuous monitoring, updating, and adaptation to emerging threats. This comprehensive approach significantly decreases the likelihood of cyberattacks, enhances the overall cybersecurity framework, and strengthens the resilience of an organization's digital environment against potential exploitation.

5.11   Attack Vectors

Attack vectors are the pathways hackers use to exploit security weaknesses and gain access to systems and networks [156]. These vectors encompass both external threats like malware, phishing, and DDoS attacks, and internal threats, such as misconfigurations, insider threats, and privilege misuse. Common examples include social engineering, credential theft, vulnerability exploitation, and insufficient insider protection. Given the evolving nature of cybercrime, organizations must proactively identify and mitigate these risks. Information security aims to block attack routes through layered defenses, continuous monitoring, and incident response. Like a museum safeguarding artwork by securing entry points (doors, windows, etc.) and vetting employees, organizations deploy digital security layers like encryption, firewalls, intrusion detection systems, and MFA to reduce exploitation risks.

While completely blocking all attack vectors in complex digital ecosystems is nearly impossible, secure protocols, rigorous access controls, regular patching, and real-time threat detection significantly reduce attack surfaces, making it harder for attackers to exploit vulnerabilities. Cybersecurity awareness training is also crucial for educating employees and fostering a security-first culture. Figure 5.5 is adapted from [112].

Figure 5.5 demonstrates the outline of attack vectors, including phishing, vulnerability exploits, insider threats, and open ports. Phishing is connected to "Weakness," and endpoints are connected to the servers. This implies that attack vectors are directed at endpoints. Some of the most common ways attacks occur

Fig. 5.5 Attack vectors

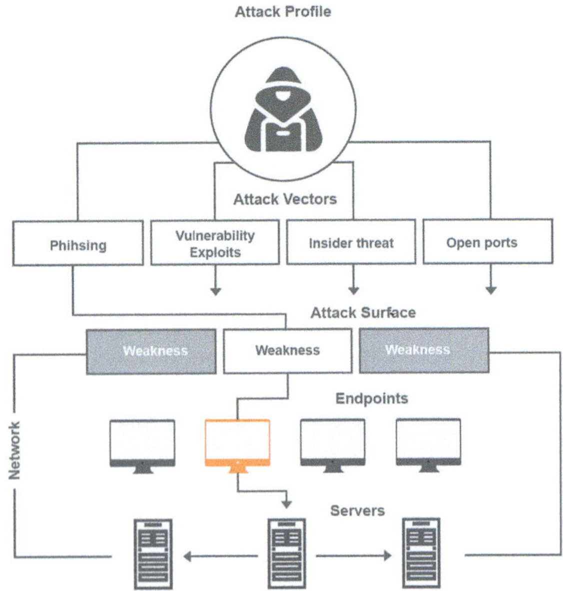

include phishing emails with malicious attachments, stolen passwords, unprotected data, insiders abusing their access, unpatched software vulnerabilities, malicious code on websites, compromised applications, and devices with open ports.

5.11.1   Classifying Attack Vectors

Attack vectors are the paths or methods used by attackers to gain unauthorized access to systems, networks, or data. These vectors can be categorized into several groups based on factors such as the target of the attack, the technique used, and the entry point exploited by the attacker. Classifying attack vectors helps organizations prioritize their defenses and understand the potential risks associated with each type of attack [156]. Below are several typical categorizations of attack vectors based on their risk levels:

Low Risk

Low-risk attack vectors typically involve methods or vulnerabilities that pose minimal threats to a target system or data. While these attacks might not immediately cause harm, they can serve as stepping stones for more sophisticated attacks. Examples of low-risk attack vectors include minor social engineering attempts or the exploitation of insignificant vulnerabilities. These vectors are less likely to cause significant damage but should still be addressed to maintain overall security.

1. **Minimal impact on security:** These attack vectors result in negligible effects on system integrity or data confidentiality [156].
2. **Limited threat to assets:** While a threat exists, the potential for damage to valuable assets, such as intellectual property or sensitive data, is low.
3. **Unlikely to lead to major breaches:** These attacks are typically non-disruptive and would not cause widespread damage or severe security breaches if exploited.

Medium Risk

Medium-risk attack vectors pose a more substantial threat to the security of systems and data. These attacks may exploit known vulnerabilities or weaknesses that could allow attackers to gain unauthorized access or disrupt services. While these attacks may not be catastrophic, they can still cause significant damage, compromise sensitive information, or disrupt critical operations if successful. Addressing medium-risk attack vectors requires proactive monitoring and risk mitigation strategies to prevent exploitation [156].

1. **Moderate impact on security:** These vectors can lead to potential data corruption, unauthorized access, or service disruptions.
2. **Some threat to assets:** Assets such as personal data, intellectual property, or proprietary systems may be at risk, though the attacker may need additional steps to fully exploit the vulnerability.

3. **Could result in breaches if exploited:** While not catastrophic, a medium-risk attack could lead to significant breaches, potentially resulting in data loss or the need for substantial incident response.

High Risk

High-risk attack vectors are the most dangerous and present an immediate and severe threat to an organization's security. These vectors often exploit critical vulnerabilities, provide direct access to sensitive data or systems, and can lead to catastrophic breaches if not addressed. Attacks exploiting high-risk vectors can result in major financial losses, data breaches, reputational damage, or full-scale system compromise. Timely identification and response to high-risk vectors are crucial for minimizing potential damage.

1. **Significant impact on security:** High-risk attack vectors have the potential to compromise entire systems, leak sensitive data, or destroy critical infrastructure.
2. **High threat to assets:** Attackers using these vectors could cause substantial damage to valuable assets, such as customer databases, financial records, or intellectual property.
3. **Severe breaches if exploited:** High-risk attacks, if successful, can lead to severe breaches, potentially resulting in the exposure of vast amounts of sensitive information, operational shutdowns, or legal and compliance repercussions.

5.11.2   Pseudocode for Attack Vectors

```
BEGIN AttackVectorsAnalysis

    // Define known attack vectors
    INITIALIZE attackVectors = ["Phishing", "Malware", "
    BruteForce",
    "SQLInjection", "CrossSiteScripting"]

    // Define entry points and associated attack vectors
    FUNCTION MapAttackVectorsToEntryPoints(entryPoints):
        vectorMap = {}
        FOR EACH entryPoint IN entryPoints:
            IF entryPoint == "HTTP" OR entryPoint == "HTTPS"
    THEN
                vectorMap[entryPoint] = ["Phishing", "
    CrossSiteScripting"]
            ELSE IF entryPoint == "SQLPorts" THEN
                vectorMap[entryPoint] = ["SQLInjection"]
            ELSE IF entryPoint == "SSH" THEN
                vectorMap[entryPoint] = ["BruteForce"]
            ELSE IF entryPoint == "APIEndpoints" THEN
                vectorMap[entryPoint] = ["Malware"]
            ELSE IF entryPoint == "Telnet" THEN
                vectorMap[entryPoint] = ["BruteForce"]
```

```
21          END IF
22        END FOR
23        RETURN vectorMap
24    END FUNCTION
25
26    // Main process
27    entryPoints = ["HTTP", "HTTPS", "SQLPorts", "SSH", "
      APIEndpoints", "Telnet"]
28    attackVectorMap = MapAttackVectorsToEntryPoints(entryPoints
      )
29    PRINT "Attack Vectors: ", attackVectorMap
30
31 END AttackVectorsAnalysis
```

Listing 5.2 Attack vectors analysis algorithm

This pseudocode, referred from: https://github.com/Sunzidasiddique1/ICS, analyzes attack vectors associated with specific entry points in the system. It maps attack vectors like phishing, malware, and SQL injection [4] to entry points such as HTTP and SQL ports. The function `MapAttackVectorsToEntryPoints` helps in associating these vectors with corresponding entry points.

5.12 Common Attack Vectors

Attack vectors are the specific methods or pathways that attackers use to gain unauthorized access to a system or network, steal data, or cause disruption. These vectors take advantage of vulnerabilities within the target system, application, or network and are often the means by which cyberattacks are executed. Understanding these attack vectors is essential for developing effective cybersecurity defenses. Below are some of the most common attack vectors frequently exploited in cybersecurity:

a. **Phishing:** Phishing is one of the most common attack vectors, where attackers send fraudulent emails that appear to come from legitimate sources. These emails often contain malicious links or attachments designed to deceive users into revealing sensitive information such as login credentials, credit card numbers, or personal data. Phishing attacks can be highly effective because they exploit human trust and often appear to be from well-known organizations or contacts.

   - Phishing emails may impersonate banks, e-commerce platforms, or even colleagues to trick users into taking action.
   - Once the victim interacts with the malicious link or attachment, their personal data or system access may be compromised.

b. **Malware:** Malware refers to any malicious software specifically designed to cause damage or gain unauthorized access to a system. Types of malware include viruses, Trojans, ransomware, worms, and spyware. These programs often infect devices by exploiting vulnerabilities in software or through malicious downloads from untrusted sources.

- **Trojans**IT often disguise themselves as legitimate software, enabling attackers to control systems or steal data without detection.
- **Ransomware** locks critical data or entire systems, demanding a ransom for release.
- **Viruses** can self-replicate and spread across systems, causing damage to data and system operations.

c. **Compromised Passwords:** Weak, stolen, or reused passwords are significant attack vectors. Many users rely on simple or repeated passwords across multiple services, making it easier for attackers to gain access to accounts. Compromised passwords often result from phishing attacks, keylogging, or brute-force attacks. Once a password is compromised, attackers can gain unauthorized access to sensitive systems, applications, or data.

- Passwords obtained through phishing attacks or data breaches are often sold or used for further exploits.
- Using password managers and multi-factor authentication (MFA) can help mitigate the risk associated with compromised passwords.

d. **Encryption Issues:** Encryption plays a crucial role in protecting sensitive data from unauthorized access. However, poor or improper encryption practices can lead to serious vulnerabilities. Inadequate encryption allows attackers to intercept and decrypt sensitive data, such as passwords, credit card numbers, and confidential documents. This can occur if weak encryption algorithms are used, if encryption keys are poorly managed, or if the encryption process itself is bypassed.

- Attacks such as man-in-the-middle attacks exploit weak encryption protocols to capture and decrypt data transmitted over insecure networks.
- It is essential to use strong, modern encryption standards and ensure that data in transit and at rest is properly encrypted.

e. **Unpatched Software:** Unpatched software refers to systems, applications, or operating systems that have not been updated with the latest security patches. Hackers often exploit known vulnerabilities in outdated software to gain unauthorized access to systems. These vulnerabilities may be publicly documented, making unpatched systems easy targets for attackers looking to exploit these weaknesses.

- Cybercriminals often use automated tools to scan for unpatched systems and exploit them to deploy malware or steal data.
- Regular software updates, automated patch management, and vulnerability scanning are key practices for minimizing the risk associated with unpatched software.

5.12.1 Prevention from Attack Vectors

To prevent attack, it is essential to adopt a comprehensive strategy that tackles vulnerabilities in all parts of an organization's architecture, technology, and human factors. This multi-layered approach helps minimize the risk of cyberattacks by addressing both technical and human vulnerabilities. Below are some overarching tactics for mitigating typical attack vectors:

1. **Honeypots:** Honeypots are decoy systems or networks set up to attract attackers. By mimicking real systems, they serve as a trap for cybercriminals. These systems allow organizations to gather intelligence on attack tactics, techniques, and procedures (TTPs) used by attackers. This information can then be used to improve the organization's security posture and anticipate future attacks. Honeypots also distract attackers from valuable targets, thus reducing the potential impact on critical systems.

   - Honeypots can be deployed within critical systems to monitor unauthorized access attempts and study attack patterns.
   - The data collected from honeypots can be analyzed to strengthen defensive strategies and patch vulnerabilities before they are exploited in real-world attacks.

2. **Load Balancers:** Load balancers help manage incoming network traffic by distributing the load evenly across multiple servers. This reduces the risk of server overload, especially during high-traffic periods, and ensures the availability of services even if some servers are compromised. Additionally, load balancers can play a key role in mitigating Distributed Denial-of-Service (DDoS) attacks, where the goal is to flood a server with excessive requests, overwhelming its capacity.

   - By using load balancers, an organization can prevent a single point of failure, which could be exploited by attackers to disable systems or take them offline.
   - Load balancers can also detect unusual traffic patterns that might indicate the presence of a DDoS attack, triggering mitigation measures.

3. **Advanced Firewall Protection and Updated Antivirus:** Firewalls act as a barrier between a trusted internal network and untrusted external networks, such as the Internet. Advanced firewalls use a variety of filtering techniques, including stateful inspection, deep packet inspection, and application-layer filtering, to block malicious traffic. Combined with up-to-date antivirus software, which helps detect and neutralize malware, this dual-layer approach strengthens an organization's defense against a wide range of cyber threats.

   - Firewalls can block suspicious traffic based on IP addresses, protocols, and port numbers, preventing unauthorized access to internal systems.
   - Antivirus software scans files, programs, and network traffic for known malware signatures, preventing viruses, Trojans, and other malicious software from infecting systems.

- Regular updates to both firewall rules and antivirus signatures are crucial to staying protected against evolving threats.

4. **Enforcing Security Policies:** A strong security policy sets clear rules for how systems and data should be protected within an organization. This includes guidelines for user access, password management, encryption, data handling, and incident response. By enforcing these policies consistently, organizations ensure that everyone adheres to security best practices, reducing the likelihood of vulnerabilities being introduced due to human error or negligence.

   - Security policies should be regularly reviewed and updated to address emerging threats and changes in the organization's environment.
   - Enforcement mechanisms, such as user training and periodic security audits, help ensure compliance with the policies.
   - Policies should also cover physical security, ensuring that sensitive equipment is protected from unauthorized access.

5. **Education:** Employee education is one of the most effective ways to defend against social engineering attacks, such as phishing, spear-phishing, and baiting. Educating staff members about recognizing suspicious activity, understanding the risks associated with weak passwords, and following proper security protocols can significantly reduce the likelihood of a successful attack.

   - Regular security awareness training should be provided to employees, including simulated phishing exercises to test their ability to identify malicious attempts.
   - Employees should be trained to report suspicious activities promptly, enabling a quicker response to potential threats.
   - Encouraging a security-first mindset throughout the organization helps foster a culture of vigilance against cyber threats.

6. **Proper Use of SIEM Tools:** Security Information and Event Management (SIEM) tools provide real-time analysis of security alerts and logs generated by applications, devices, and systems within an organization. These tools aggregate data from multiple sources to detect and respond to potential security incidents more effectively. By using SIEM tools, organizations can improve incident detection, response, and recovery, ensuring that security breaches are identified and addressed swiftly.

   - SIEM tools help identify anomalies in network traffic, failed login attempts, and other indicators of compromise, enabling quick identification of potential attacks.
   - The integration of SIEM with incident response systems can automate responses to certain security events, reducing the time taken to mitigate attacks.
   - SIEM tools also support compliance efforts by maintaining detailed logs of security events for audit purposes.

5.13 Access Attacks

Access attacks are also associated with unauthorized use of another person's equipment. Such attacks can sidestep security controls, leaving data deep breaches and potentially unaltered or even damaged. The attackers were taking advantage of different vulnerabilities, like software defects, weak passwords, phishing tactics, misconfigurations, or even insider privilege, to get unauthorized access. Once attackers gain access, they may exfiltrate sensitive data or manipulate systems to disrupt services. Unauthorized access is a serious security risk that demands strong defense mechanisms. They are using advanced tactics to obtain unauthorized access, but strong countermeasures include effective authentication practices employing meticulous security updates and training for users along with regular network monitoring.

Pseudocode for Access Attack

```python
# Define potential access methods
access_methods = [
    "PasswordCracking", "SocialEngineering",
    "ExploitingVulnerabilities", "ManInTheMiddle", "
    SessionHijacking"
]

# Simulate access attack
def simulate_access_attack(access_method, target):
    if access_method == "PasswordCracking":
        return f"Attempting to crack password for {target}"
    elif access_method == "SocialEngineering":
        return f"Attempting social engineering on {target}"
    elif access_method == "ExploitingVulnerabilities":
        return f"Exploiting known vulnerabilities in {target}"
    elif access_method == "ManInTheMiddle":
        return f"Performing Man-in-the-Middle attack on
    communication to {target}"
    elif access_method == "SessionHijacking":
        return f"Hijacking session for {target}"

# Main process
targets = ["WebServer", "DatabaseServer", "UserAccount"]
for target in targets:
    for method in access_methods:
        attack_result = simulate_access_attack(method, target)
        print(attack_result)
```

Listing 5.3 Access attack simulation algorithm

This pseudocode simulates different access attacks on various targets, such as a web server, database server, or user account. The `SimulateAccessAttack` function simulates different attack methods like password cracking, social engineering, or session hijacking, providing an output for each attempt. The main process runs these simulations for each target and attack method.

Categories of Access Attacks

Access attacks can be broadly categorized based on the techniques used and the objectives of the attackers. These categories provide insight into how unauthorized access is gained and the methods used to exploit vulnerabilities. Below are some of the most common types of access attacks:

5.13.1 Logical Access Attacks

Logical access attacks are carried out through digital means without the need for physical access to systems or devices. These attacks primarily focus on exploiting software vulnerabilities or using methods to bypass security measures designed to control access to sensitive information.

a. **Brute Force Attacks:** Brute force attacks involve attempting to guess passwords by trying numerous combinations until the correct one is found. Attackers can use automated tools to quickly test a vast number of potential passwords. To speed up this process, attackers may use techniques like rainbow tables, which store precomputed hash values of passwords to reduce the time required for cracking.

   - Brute force attacks create noticeable traffic on the network, which can be detected by security systems, especially if the system is well-configured to detect excessive login attempts.
   - Organizations can mitigate this type of attack by implementing account lockout policies, password complexity requirements, and multi-factor authentication (MFA).

b. **Reconnaissance and Credential Theft:** Attackers often perform reconnaissance to gather information about a system or network before launching an attack. This phase may involve scanning for weak points, open ports, or publicly exposed services. Once they acquire valid credentials (via phishing or data breaches), attackers can leverage them to gain unauthorized access to systems.

   - Credential theft can occur through techniques like phishing or exploiting weak passwords. Once attackers gain access to valid credentials, they can easily bypass security measures that rely solely on passwords.
   - Strong user authentication systems and the regular monitoring of login patterns can help detect such attacks.

5.13.2 Physical Access Attacks

Physical access attacks involve directly gaining access to the hardware or devices used in a network or system. These attacks can be more difficult to defend against because they often target human weaknesses or physical vulnerabilities within an organization's infrastructure.

a. **Physical Access to Hardware:** Attackers may attempt to physically access an organization's servers, workstations, or storage devices to extract data or plant malicious software. This can include stealing or tampering with hard drives, USB drives, or other hardware that contains sensitive data.

   – Such attacks can be mitigated by enforcing strict physical security controls, such as restricted access to server rooms, surveillance cameras, and biometric authentication for access to sensitive areas.

b. **Social Engineering and Human Manipulation:** Social engineering is a key tactic in physical access attacks, where attackers manipulate individuals to gain unauthorized access. This can involve tactics such as impersonating authorized personnel or using psychological manipulation to convince employees to grant access to sensitive areas or information.

   – Social engineering attacks can also include impersonating IT support to convince employees to provide passwords or other access credentials.
   – Employee training on security awareness, recognizing suspicious behavior, and confirming identities before granting access are essential countermeasures.

c. **Phishing Emails and Keyloggers:** Phishing emails and keyloggers are commonly used tactics in physical access attacks. Phishing emails may trick users into clicking on malicious links or downloading harmful attachments, which can compromise their devices. Keyloggers, which are often installed through phishing or physical access, record every keystroke entered by a user, allowing attackers to capture sensitive data like passwords.

   – To defend against phishing, organizations should use email filtering tools, enforce strong password policies, and educate employees on how to recognize phishing attempts.
   – Regularly updating software to patch known vulnerabilities and using anti-malware tools can prevent the installation of keyloggers and other malicious software.

d. **Exploiting Human Fallibility:** Physical access attacks often exploit human fallibility, such as negligence, forgetfulness, or lack of awareness. Attackers can take advantage of situations where employees fail to properly secure devices or leave confidential information exposed in public or common areas.

   – Effective training, access control policies, and regular security audits help mitigate the risk posed by human errors. Secure disposal of confidential materials and the use of encryption for sensitive data are also essential safeguards.

5.14   Common Attack Techniques

Common attack techniques refer to the methodologies used by cybercriminals to exploit weaknesses in systems, networks, or applications. These strategies are constantly evolving as attackers devise new methods to circumvent security safeguards and gain unauthorized access. Below are some prevalent attack methods:

  I. Password Attacks: Threat actors utilize phishing, brute force, and Ophcrack to crack passwords.
 II. Pass-the-Hash: Attackers use malware to steal password hashes and gain unauthorized access.
III. Trust Exploitation: VPN-connected hosts are abused to access internal networks.
IV. Port Redirection: Compromised systems are used to attack other targets.
 V. Man-in-the-Middle: Attackers intercept, alter, or divert data.
VI. IP, MAC, and DHCP Spoofing: Attackers falsify addresses to impersonate devices for various attacks.

5.15   Access Attack Exploit Tools

Access attack exploit tools are in the form of specialized software created to perform unauthorized access to systems, networks, or applications. These tools leverage badly implemented authentication mechanisms such as weak passwords, or flawed encryption etc. Following are a few popular tools for performing access vulnerability exploitation:

  1. **Ncrack:** A high-speed network authentication cracking tool designed to detect and exploit weak passwords on network services. It supports multiple protocols such as SSH, RDP, and FTP.
  2. **Aircrack-ng:** A powerful suite of tools for assessing Wi-Fi network security, capable of cracking WEP and WPA-PSK keys through packet capture and brute-force methods.
  3. **CeWL:** A custom word list generator that scrapes websites to gather words for password cracking, increasing the chances of discovering weak passwords.
  4. **Crunch:** A command-line tool for creating custom wordlists for password cracking, often used with brute-force tools like Hydra.
  5. **Medusa:** A fast, parallelized login brute-forcer supporting multiple protocols, used for dictionary attacks on various login services.
  6. **Hydra:** A widely used password-cracking tool capable of performing brute-force and dictionary attacks on numerous network services.
  7. **John the Ripper:** John the Ripper is a fast password cracking tool, among many other uses that can be used to crack many different types of passwords using methods such as the dictionary and brute force attacks

8. **THC-Hydra:** A high-quality brute-force tool supporting multiple protocols and very fast login system jamming.
9. **RainbowCrack:** A tool using precomputed hash values (rainbow tables) to crack encrypted passwords faster than brute-force methods.
10. **Hashcat:** It is a password cracking tool that is available for both CPU and GPUs, it supports various algorithms.

5.16 Attack Surface Management

Organizations must continuously defend a perpetual catalogue of all internal and external Internet-connected assets that tracks the full attack surface and exposures over time. Attack surface management is the process through which it happens.

a. **Known Assets:** Systems, devices, and software that security personnel are aware of and have authorized for access, regularly assessed and monitored.
b. **Unknown Assets:** Programs, systems, or hardware that an organization is unaware of or has not authorized, posing security risks.
c. **Rogue Assets:** Unauthorized or compromised assets that can be exploited and are challenging to detect and control.
d. **Vendors:** External vendors, including software providers and third-party service suppliers, must be carefully managed to mitigate vulnerabilities.

5.17 Importance of Managing Attack Surface

Effective attack surface management [34] enables effective, continuous discovery across network to recover everything and monitor the risks from these assets and provide organizations the ability to act upon threats intentionally.

Strategies for Managing Attack Surface

Strategies to reduce the overall attack surface so as to minimize potential entry points through which an attacker could gain unauthorized access to systems, networks, or data. Below are some effective tactics:

a. **Patch Management:** Regularly update software, operating systems (OSes), and libraries to fix vulnerabilities.
b. **Access Control:** Enforce the principle of least privilege to limit unnecessary access.
c. **Network Segmentation:** Divide networks into isolated segments to prevent breaches and lateral movement.
d. **Secure Coding Practices:** Implement security measures early in the development process.
e. **Regular Audits:** Conduct frequent security audits to identify and remediate vulnerabilities.

5.18 Summary

Industrial Control Systems Cybersecurity from attack surfaces understanding and control that also includes vulnerabilities for penetration and weaknesses to exploit when breaking down critical infrastructure. The attack surface becomes more complex as technology evolves, which requires precaution to mitigate the risks. Techniques such as network segmentation, access controls, and then continuous monitoring are employed to reduce the attack surface and decrease risk of a cyberattack. However, they need to continue supporting critical operational safety and security as these organizations digitize, which means that reducing the attack surface is just as important.

Chapter 6
Network Segmentation in Industrial Operations: Enhancing ICS Security Through Threat Mitigation and DNS Leak Prevention

Abstract The industrial attack surface is a crucial component in gaining visibility of the various ways unauthorized or malicious access could be fostered in industrial networks with types of attack surfaces. This chapter goes over a few different types or categories of attack surfaces and then transitions into how to measure the size of an attack surface in terms of metrics, how to map out that surface, and finally concludes with strategies for reducing the size of the attack surface. It helps to counter various attack vectors and provides solutions to the prevention for these vectors along with logical and physical access-related attacks. This chapter will extend to address common attack techniques and ways to manipulate the attack surface. This will help readers to understand the weaknesses and entry points used by unauthorized or malicious actors to attack industrial automation systems.

Keywords Attack vector prevention · Access attacks · Logical access attacks · Industrial automation systems · Network security · Attack techniques · Vulnerability management

6.1 Introduction

Network segmentation prevents lateral propagation of attacker threats within data centers, cloud environments, or university networks. When networks are appropriately divided into segments, security measures can be implemented to stop malicious lateral movement across an organization, even if an adversary has managed to breach one part. Only a limited portion of a system is exposed to threats, which is enclosed by security controls, preventing attacks outside that area from affecting the broader network. Thus, this containment strategy decreases the impact of cyber incidents and supports overall compliance with regulatory requirements and data protection regulations.

This approach allows organizations to isolate minor security events, preventing intrusions from escalating and reducing the overall attack surface. It also enhances network throughput by preventing congestion and ensuring effective utilization of resources. In computer networking, it is common practice to break down a network

M. A. Rahman et al., *Securing Industrial Control Systems*,
https://doi.org/10.1007/978-3-032-03018-4_6

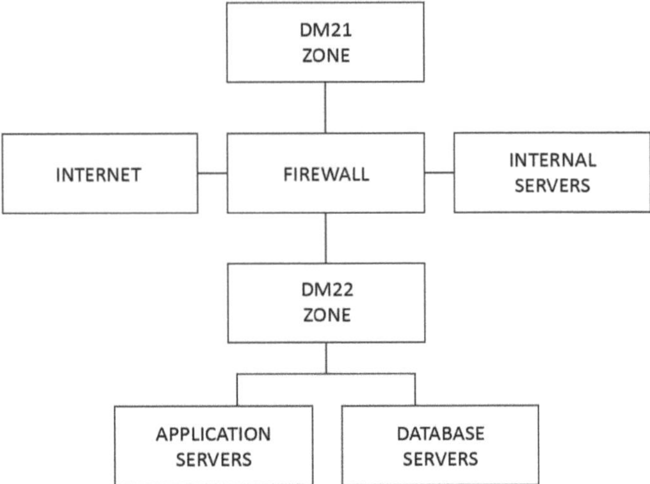

Fig. 6.1 Network segmentation

into separate subnets, making them more secure behind firewalls, VLANs, and micro-segments.

In the context of ICS, this means that elements within an ICS network are regulated and separated to prevent unauthorized access or to stop potential attacks in their tracks. The critical importance of ICS environments (e.g., power grids, manufacturing systems, and water treatment plants) necessitates enforcing network segmentation to defend these infrastructures from cyber threats. Network segmentation can prevent lateral movement by separating critical assets and control system components from less secure hardware and software. It helps isolate specific types of data traffic, protecting the integrity, availability, and confidentiality of industrial operations protocols, while mitigating outages that could lead to physical damage or operational downtime. Figure 6.1 is adapted from [160].

Figure 6.1 illustrates network segmentation, including two DMZs, with labeled components such as the Internet, firewall, internal servers, application servers, and database servers. This demonstrates how the network architecture is divided into internal and external connections.

6.2 Network Segmentation on ICS Threat Mitigation

Network segmentation in ICS environments isolates the system, limiting dangerous lateral movement. By controlling access within each segment, attackers are restricted in their ability to navigate the network. As a result, a compromise only affects the targeted segment, preventing critical system threats from spreading rapidly. This containment provides security professionals with more time to detect, respond to, and mitigate threats before they cause widespread damage. Segmentation reduces asset exposure, decreasing the attack surface and lowering the impact of potential breaches.

6.2.1   Pseudocode for Network Segmentation

```python
# Define network zones
zones = {
    "ControlZone": [],
    "DemilitarizedZone (DMZ)": [],
    "EnterpriseZone": [],
    "ExternalZone": []
}

# Assign assets and systems within each zone
zones["ControlZone"] = ["criticalSystems"]
zones["DemilitarizedZone (DMZ)"] = ["intermediarySystems"]
zones["EnterpriseZone"] = ["businessSystems"]
zones["ExternalZone"] = ["publicSystems"]

# Establish segmentation policies
def establish_segmentation_policies():
    for zone in zones:
        print(f"Configuring firewall rules for {zone}")
        print(f"Configuring access control lists (ACLs) for {
        zone}")
        print(f"Configuring intrusion detection system (IDS)
        for {zone}")

# Isolate communication between zones
def isolate_communication():
    for source_zone in zones:
        for target_zone in zones:
            if source_zone != target_zone:
                print(f"Isolating traffic from {source_zone} to
                {target_zone}")
                print(f"Allowing only authorized traffic
                between {source_zone} and {target_zone}")

# Monitor and log network activity
def monitor_network_activity():
    for zone in zones:
        print(f"Enabling logging for all network traffic in {
        zone}")
        print(f"Setting up alerts for suspicious activity in {
        zone}")

# Periodically review and update segmentation policies
def review_and_update_policies():
```

```
38      periodic_review_time = "Scheduled"
39      system_operational = True
40      while system_operational:
41          current_time = "Scheduled"
42          if current_time == periodic_review_time:
43              print("Reviewing segmentation policies")
44              print("Updating policies as needed")
45              break
46
47  # Initialize network segmentation
48  def initialize_network_segmentation():
49      establish_segmentation_policies()
50      isolate_communication()
51      monitor_network_activity()
52      review_and_update_policies()
53
54  # Main process
55  initialize_network_segmentation()
```

Listing 6.1 Network Segmentation Algorithm

The pseudocode, referred from: https://github.com/Sunzidasiddique1/ICS for network segmentation, outlines the process of dividing a network into distinct zones to enhance security. It defines four zones: Control Zone, DMZ, Enterprise Zone, and External Zone, each assigned specific systems based on their roles. Security policies for each zone include configuring firewall rules, access control lists, and IDS. Communication between zones is isolated, allowing only authorized traffic. Network activity is continuously monitored, with alerts set for suspicious activity. Segmentation policies are periodically reviewed and updated. This approach ensures secure, controlled communication and improves overall network security [101].

6.3 Types of Network Segmentation

Network segmentation is a technique used to divide a network into smaller, manageable, and more secure sub-networks or segments. This approach enhances performance, security, and management by isolating network traffic. Using logical or physical techniques, network segmentation divides an organization's systems into smaller parts to prevent the spread of malicious activities and optimize resource utilization.

1. **Physical Segmentation:** Physical segmentation involves dividing the network into multiple physically separated areas or subnets, each with its own set of hardware resources and security controls. This is typically achieved using firewalls, routers, and access points to regulate traffic flow between subnets. These devices filter, route, and enforce security policies for each segment. While this technique provides a high level of isolation and security, it often requires substantial investment in hardware infrastructure. Managing physical segmentation becomes

more complex as the network grows, posing challenges such as maintaining consistent configurations across segments and troubleshooting issues spanning multiple physical devices.

2. **Logical Segmentation:** Unlike physical segmentation, logical segmentation does not require additional hardware and utilizes the existing network infrastructure. This method is often more cost-effective and easier to implement. Logical segmentation works by using network addressing schemes, such as subnetting or Virtual Local Area Networks, to organize and segment traffic. While network addressing schemes define complex relationships between devices, VLANs automatically group devices into logical subnets based on specific criteria, such as function, location, or department. Logical segmentation enables more flexible and scalable network management, allowing administrators to reconfigure segments without making physical changes to the infrastructure. However, ensuring proper isolation and security between logical segments may require additional measures, such as access control lists or virtual firewalls.

6.4   Segmentation in Industrial Control System Environments

In the defense-in-depth security strategy for industrial control system environments, an important method is to segment networks and devices. It is considered a best practice in security, and we all know that it has provided better security benefits as compared to defenses that are positioned at the corners of the network. By segmenting an industrial control system network into logically or physically isolated segments, organizations can establish security zones that confine unauthorized communication between critical and non-critical systems. This method minimizes the threat of cyberattacks, effectively prevents the spread of malware, and ensures that disruptions in one area do not affect the rest of the network. Businesses can prevent unauthorized access to their networks and limit attacker lateral movement, reducing their attack surface area against both targeted attacks as well as accidental system failures, in order to enhance their security posture by implementing segmentation.

While many industrial firms isolate corporate information technology networks from industrial control systems or operational technology networks to minimize exposure to external threats, achieving further segmentation within industrial control system networks can be a difficult and expensive process. Industrial control system environments typically contain legacy systems, closed-source protocols, and real-time operational demands that make changes in the network difficult. Operators must understand their networks and devices, and IT/OT cooperation is vital to maintain plant operations through segmentation. Network segmentation also calls for more access control policies, firewall rules, and intrusion detection devices in order to be able to better administrate communication among separate networks.

Regardless, segmentation is consistently rated as necessary to securing critical infrastructure against cyber attacks, insider threats, and accidents. Through the adoption of frameworks such as the Purdue Enterprise Reference Architecture, organizations can enforce hierarchical segmentation, which isolates industrial control system control layers from enterprise IT environments for secure and reliable industrial operations.

6.5 Enhancing ICS Security through Network Segmentation

Network segmentation [160] is essential for ICS cybersecurity. Segmentation confines cyberattacks to isolated segments, preventing rapid propagation to critical systems. Access controls within segments restrict unauthorized movement, mitigating the risk of breaches. This containment strategy accelerates incident response and improves incident management. Smaller segments further reduce the attack surface [76, 148], limiting lateral movement. Additionally, segmentation isolates functions, ensuring that breaches within the ICS infrastructure do not spread.

The following measures enhance ICS security through network segmentation:

1. Assess risks to identify critical assets, threats, and vulnerabilities. The organizational segmentation plan should address operational requirements and compliance needs.
2. Implement firewalls, access control lists, and authentication mechanisms to restrict access to segments.
3. Develop security incident detection, containment, and mitigation strategies.
4. Utilize network monitoring tools and intrusion detection systems for continuous threat detection [5].
5. Establish and enforce rules for segment communication.
6. Conduct penetration testing, vulnerability assessments, and security audits regularly.
7. Provide ongoing training for employees on network segmentation and security best practices.
8. Integrate segmentation with other security measures as part of a comprehensive cybersecurity strategy.

To address emerging threats and evolving business needs, organizations should continuously review and update their cybersecurity strategies.

6.6 Implementing an ICS Segmentation Approach

Industrial organizations can enhance the security of their ICS networks by implementing a specialized segmentation strategy tailored to their specific needs. This approach utilizes customized industrial security appliances designed for operational teams, as opposed to conventional IT segmentation techniques.

In this method, industrial processes are safeguarded using specialized ICS security appliances that function as bump-in-the-wire devices within the network. These appliances efficiently monitor and filter traffic, eliminating unauthorized data and alerting operators to any abnormal behavior. Regulated traffic is meticulously supervised and permitted to flow, providing operators with precise control over communication among critical resources. This segmentation technique does not require device reconfiguration or IP address assignment, minimizing the impact on existing infrastructure. It seamlessly scales in complex industrial environments with multiple devices, offering comprehensive protection for each process, safety system, and RTU. Operators can achieve a thorough network lockdown by implementing this specialized segmentation approach, allowing only authorized traffic internally and externally. Additionally, the same ICS security equipment can be used to restrict remote access for suppliers and third parties, further strengthening overall security.

This ICS segmentation technique provides operational teams with greater control and security while avoiding the complexity and costs associated with traditional IT segmentation solutions. It presents a practical approach to enhancing cybersecurity in industrial settings while reducing costs and preventing downtime.

6.7 Benefits of Network Segmentation

a. **Enhanced Security:** By dividing a network into smaller segments, attackers are prevented from spreading malware and gaining unauthorized access to sensitive systems, thereby reducing the attack surface. Organizations can strengthen their overall security posture by enforcing the principle of least privilege and inspecting devices for potential threats within each segment.
b. **Improved Performance:** Network segmentation reduces congestion, ensuring optimal performance for resource-intensive activities such as online gaming, media streaming, and video conferencing. By prioritizing traffic and minimizing bottlenecks, enterprises can guarantee that critical applications receive a consistent quality of service.
c. **Streamlined Response and Monitoring:** Network segmentation simplifies traffic monitoring, enabling quick detection of suspicious activities and unauthorized access attempts. By dividing subnets and implementing granular traffic monitoring, organizations can enhance security oversight and respond rapidly to potential threats. This approach reduces the risk of data breaches and ensures compliance with regulatory standards such as PCI DSS.

6.8 Differentiation of Different Segmentation

Different Segmentation are provided in below Table 6.1.

Table 6.1 Differentiation of different segmentation

Type	Microsegmentation	Internal segmentation	Intent-based segmentation	Zero-trust model
Description	Adopts a detailed approach to segmentation utilizing VLANs and access control lists. Policies are enforced on individual workloads, enhancing resilience against attacks	Segments network and infrastructure assets regardless of their location, whether on-premises or in multiple cloud environments. Establishes dynamic and granular access by continuously monitoring trust levels and adjusting security policies	Identifies where segmentation is applied, establishes trust, and applies security inspections to traffic. Integrates traditional segmentation with zero-trust principles for a cohesive security architecture	Disregards the notion of a trusted network perimeter and embraces the "never trust, always verify" philosophy. Access is continually evaluated without adding friction to users. Managed by identity and access management solutions for role-based access control
Features	a. Creates smaller, more secure zones on the network b. Minimizes traffic between workloads c. Restricts a hacker's movement between compromised applications d. Simplifies segmentation management	a. Segments network and infrastructure assets b. Enables dynamic and granular access control c. Isolates critical IT assets d. Detects and mitigates threats using analytics and automation system	a. Covers the entire network and assets b. Utilizes existing mechanisms such as identity-based tools c. Grants or denies access based on risk and trust evaluations d. Encrypts all traffic for enhanced security	a. Ensures only authorized users have access to resources b. Reduces friction in access management c. Implements IAM solutions for role-based access control d. Improves user experience e. Enhances visibility of devices and automates responses

6.9   Security Risks of Segmentation in Industrial Operations

The existing flat network configuration in industrial operations presents several security vulnerabilities because it lacks segmentation. All assets in this setup are linked to a single switch, causing all machines to share the same subnet and IP address range. This interconnection leads to vulnerabilities, where an attack on one process may readily spread to others, increasing the operational impact of each incident.

Moreover, the lack of specialized firewall regulations for different processes hinders the customization of security measures to meet particular requirements. Enforcing multiple regulations with a single firewall is difficult in practice, resulting in a uniform approach that does not cater to the specific needs of each operation. Additionally, the firewall installed at the plant level is usually overseen by IT staff and may have liberal configurations, including permitting outbound traffic on port 80 for web browsing. While this allows unrestricted internet connectivity, it also makes the ICS network vulnerable to attacks. Users in the ICS system may unintentionally download malicious malware from websites, creating a substantial security threat.

Unauthorized devices, such as laptops or wireless devices, further exacerbate the vulnerability of the ICS network. These security risks make industrial processes susceptible to cyberattacks and highlight the need for improved segmentation and security measures.

6.10   Domain Name System Leak

Domain Name System leaks may occur when using virtual private networks and other privacy-enhancing technologies. The Domain Name System is responsible for converting human-readable domain names into computer-recognizable Internet Protocol addresses, allowing users to access websites without needing to remember numerical addresses. Ideally, when a virtual private network is in use, all Domain Name System queries should be routed through the encrypted virtual private network tunnel to ensure privacy and security. However, sometimes, devices may bypass the virtual private network and use the internet service provider's Domain Name System servers instead. This unintended behavior results in a Domain Name System leak, allowing internet service providers or other third parties to monitor browsing activity despite the presence of an encrypted connection.

Domain Name System leaks expose sensitive browsing habits, revealing the websites visited and potentially compromising online anonymity. This leakage can be particularly concerning for users in regions with internet censorship, surveillance, or restrictive regulations. Additionally, cybercriminals can exploit Domain Name System leaks to conduct man-in-the-middle attacks, phishing schemes, or domain hijacking. Even if the user's internet protocol address remains hidden through the virtual private network, their browsing activity may still be traceable if their Domain Name System requests are not securely routed.

To prevent Domain Name System leaks, it is crucial to ensure that all system and application settings enforce the use of virtual private network Domain Name System servers. Many modern virtual private networks feature built-in leak prevention mechanisms that automatically block or reroute unsecured Domain Name System requests. Setting up devices to use encrypted Domain Name System protocols, like Domain Name System over Hypertext Transfer Protocol Secure or Domain Name System over Transport Layer Security, can also make them safer by stopping people from reading Domain Name System traffic. Regularly testing for Domain Name System leaks using online tools or network diagnostic utilities can help verify that all queries remain within the encrypted virtual private network tunnel. Figure 6.2 is adapted from [59].

Figure 6.2 illustrates the VPN traffic with a DNS leak. It shows DNS requests through different servers, like the VPN service, the ISP's DNS server, and the OpenDNS server.

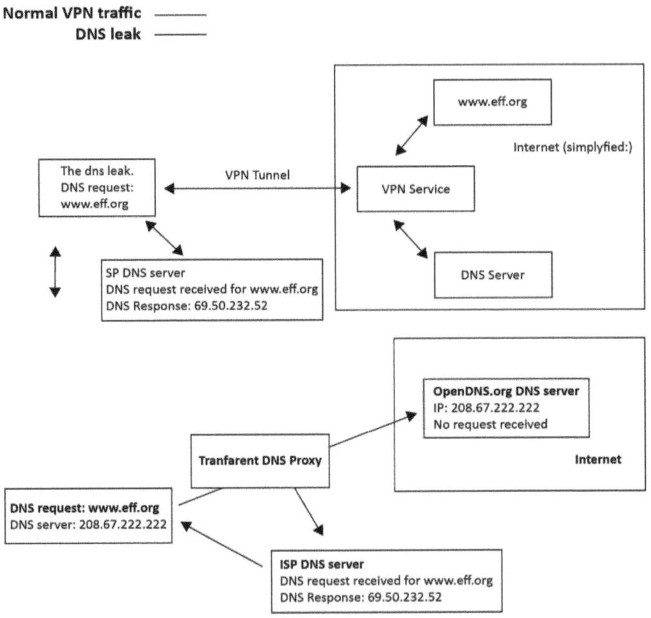

Fig. 6.2 Phenomenon of DNS leak

6.10.1   DNS Leaks in ICS Security

DNS leaks have the unintended consequence of broadening the attack surface, particularly within Industrial Control Systems ICS environments. These leaks occur when domain name searches bypass the secure tunnel of privacy tools like VPNs. This exposure reveals which websites and services devices in the ICS network are accessing.

This seemingly minor vulnerability has significant implications. By providing a window into the domains accessed within the ICS network, DNS leaks equip malicious actors with insights into the network's layout, communication patterns, and potentially sensitive systems. This information enables attackers to map out potential avenues of exploitation more accurately, identifying vulnerabilities and critical systems that could be targeted for maximum impact. Moreover, these leaks can expose internet-facing services and applications, which attackers could use as entry points. Armed with knowledge of accessed domains, adversaries can tailor their attacks to exploit known vulnerabilities associated with specific services or software used in the ICS environment.

The depth of reconnaissance is enhanced through DNS leaks, allowing attackers to gather intelligence that goes beyond traditional network scans. This helps them build a comprehensive understanding of the ICS network, aiding in the development of multi-stage attacks that are difficult to thwart. Essentially, DNS leaks heighten the risk by expanding the potential points of entry for an attacker into an ICS environment. Addressing DNS leaks is therefore essential as part of an all-encompassing cybersecurity plan for ICS. This involves not only technical solutions, such as properly configuring VPNs and DNS settings, but also fostering a culture of cybersecurity awareness to minimize the risk of human error that could be exploited due to DNS leak-related information.

6.10.2   Domain Name System Leak Process

A vulnerability that allows Domain Name System leaks to occur arises when a user utilizes certain types of virtual private networks, such as "split-tunnel" virtual private networks. Split tunneling is a configuration that allows some traffic to be routed through the encrypted virtual private network tunnel while other traffic accesses the internet directly through the local network. While this approach can enhance performance and reduce bandwidth usage, it can inadvertently lead to security risks, including Domain Name System leaks. In this scenario, the browser's Domain Name System queries bypass the virtual private network and are forwarded directly to the Domain Name System server of the internet service provider. This exposure reveals the websites the user visits to the internet service provider and any potential on-path eavesdroppers, undermining the intended anonymity and security of the virtual private network.

Another common cause of Domain Name System leaks stems from operating system features that prioritize speed over security. "Smart Multi-Homed Name Resolution," introduced in Windows 8, changed the way Domain Name System queries are processed. This feature was designed to improve Domain Name System resolution speed by sending queries across all available network interfaces and accepting the fastest response. While this mechanism enhances browsing efficiency, it increases virtual private network users' susceptibility to Domain Name System leaks. When enabled, queries may be routed outside the virtual private network tunnel, exposing the user's default Domain Name System servers within the local network or those provided by the internet service provider. This behavior can result in a privacy breach, as websites visited by the user could become visible to third parties, potentially allowing surveillance, tracking, or even data interception.

In addition to operating system vulnerabilities, improper virtual private network configurations, outdated network protocols, or software conflicts can contribute to Domain Name System leaks. Some virtual private network clients may not enforce strict Domain Name System routing policies, allowing traffic to slip through unprotected pathways. Furthermore, certain public Wi-Fi networks and corporate environments use custom Domain Name System settings that override virtual private network configurations, leading to unintended information exposure. Addressing these issues requires both technical safeguards, such as virtual private network-based Domain Name System servers and encrypted Domain Name System protocols, and user awareness to regularly test for and mitigate leaks.

6.10.3   DNS Prevention Measures

The Domain Name System is essential for converting human-readable domain names into IP addresses, enabling effective communication between internet clients and servers. Therefore, securing the DNS infrastructure is crucial to maintaining the integrity, availability, and confidentiality of network services. Below are several DNS prevention methods that can enhance security:

a. **Encrypt DNS Requests:**
   DNS queries can be encrypted using either DNS over HTTPS (DoH) or DNS over TLS (DoT). This prevents on-path eavesdroppers from intercepting sensitive information.
b. **Use VPN Clients with DNS Leak Protection:**
   Select Virtual Private Network clients that ensure DNS queries are routed through the VPN tunnel. However, not all VPN applications effectively prevent DNS leaks, so additional research is necessary before selection.
c. **Change DNS Servers:**
   DNS servers can be manually configured on local network adapters, or third-party applications such as NirSoft QuickSetDNS can be used to set them to alternative servers.

d. **Firewall Configuration:**
   DNS resolution can be disabled entirely by modifying firewall settings or assigning DNS servers to non-existent addresses such as 127.0.0.1 or 0.0.0.0. This strategy may require proxy assistance applications like Proxifier or ProxyCap, along with alternative domain resolution methods [23].

e. **Use Anonymous Browsers:**
   Employ fully anonymous web browsers, such as Tor Browser, which anonymize user activity and eliminate the need for manual DNS configuration on the operating system.

f. **Utilize Proxies or VPNs:**
   Implement proxies or VPNs across the entire system using third-party software like Proxifier or web browser extensions. However, be cautious of false positives in extensions and use external websites to verify IP and DNS leak protection.

6.11   Data Leakage in Machine Learning

Data leakage occurs when training data contains information about the target that will not be available during prediction. This issue leads to high performance on the training set (and possibly even the validation data), but the model performs poorly in production. In other words, leakage causes a model to appear accurate until it is deployed, at which point it becomes highly inaccurate. There are two main types of leakage: target leakage and train-test contamination.

Target Leakage

Target leakage occurs when predictors include data that will not be available at the time of making predictions. It is crucial to consider target leakage in terms of the timing or chronological order in which data becomes available, rather than merely whether a feature improves predictive performance.

An example will be helpful. Imagine somebody wants to predict who will get sick with pneumonia. The top few rows of raw data look like this:

Example

got_pneumonia	age	weight	male	took_antibiot ic_medicine
False	65	100	False	False
False	72	130	True	False
True	58	100	False	True

People take antibiotic medicines after getting pneumonia in order to recover. The raw data shows a strong relationship between those columns, but `took_antibiotic_medicine` is frequently changed after the value for `got_pneumonia` is determined. This is target leakage.

To prevent this type of data leakage, any variable updated (or created) after the target value is realized should be excluded.

Train-Test Contamination

A different type of leakage occurs when training data is not properly distinguished from validation data. Recall that validation is intended to measure how the model performs on data it has not encountered before. This process can be inadvertently corrupted if validation data influences preprocessing behavior. This issue is sometimes referred to as train-test contamination.

For example, imagine running preprocessing (such as fitting an imputer for missing values) before calling `train_test_split()`. The result of the model may achieve high validation scores, giving confidence in its performance, but it may perform poorly when deployed for real-world decision-making.

Example

This example will help to learn one way to detect and remove target leakage.

```python
import pandas as pd

# Read the data
data = pd.read_csv('../input/aer-credit-card-data/
    AER_credit_card_data.csv',
                   true_values=['yes'], false_values=['no'])

# Select target
y = data.card

# Select predictors
X = data.drop(['card'], axis=1)
print("Number of rows in the dataset:", X.shape[0])
X.head()
```

Listing 6.2 Loading data

Output:

```
Number of rows in the dataset: 1319
```

6.11.1 Running a Pipeline with Cross-Validation

```python
from sklearn.pipeline import make_pipeline
from sklearn.ensemble import RandomForestClassifier
from sklearn.model_selection import cross_val_score

```

```python
# Pipeline for model
my_pipeline = make_pipeline(RandomForestClassifier(n_estimators
    =100))
cv_scores = cross_val_score(my_pipeline, X, y,
                            cv=5,
                            scoring='accuracy')

print("Cross-validation accuracy: %f" % cv_scores.mean())
```

Listing 6.3 Cross-validation with RandomForest

Output:

```
Cross-validation accuracy: 0.981052
```

6.11.2 Removing Leaky Predictors

Leaky predictors are variables in a machine learning model that inadvertently
provide the model with information that would not be available at the time of
actual predictions. This usually leads to artificially high performance during training
and can severely degrade the model's effectiveness when applied to real-world
data. Leaky predictors can result from various sources, such as using future data,
information that's directly correlated with the outcome variable, or any data that
would not be available at the time of making predictions. These features essentially
"leak" information that should not be available during the decision-making process,
and they can make the model seem far more accurate than it actually is.

```python
potential_leaks = ['expenditure', 'share', 'active', '
    majorcards']
X2 = X.drop(potential_leaks, axis=1)

cv_scores = cross_val_score(my_pipeline, X2, y,
                            cv=5,
                            scoring='accuracy')

print("Cross-val accuracy: %f" % cv_scores.mean())
```

Listing 6.4 Handling data leakage

Output:

```
Cross-val accuracy: 0.830919
```

Data leakage can be a multi-million-dollar mistake in many data science
applications. Careful separation of training and validation data can prevent train-
test contamination, and pipelines can help implement this separation. Likewise, a
combination of caution, common sense, and data exploration can help identify target
leakage.

Table 6.2 Comparison of network segmentation and DNS leak prevention

Aspect	Network segmentation	DNS leak prevention
Definition	The process of dividing a network into less extensive subnetworks or segments	Precautions taken to eliminate the possibility of DNS requests being made public
Purpose	Improvements in the network's manageability and security are the purpose of this project	Preventing the visibility of DNS queries while also protecting the privacy of users
Implementation	VLANs, firewalls, and access control lists are examples of implementation and logical or physical methods	It is also possible to encrypt DNS requests, VPN client setups, and other such things
Benefits	Improved security lower attack surface, improved management	Improved privacy, as well as the prohibition of tracking based on DNS

6.12 Network Segmentation vs DNS Leak

They are two separate topics in network security: Network Segmentation and Domain Name System. Network segmentation is about breaking up a network into isolated segments for the purpose of controlling traffic flow or enhancing security, and Domain Name System leaks refer to an incident where browsing activity is inadvertently exposed because DNS queries are improperly routed. While network segmentation decreases the ability for attackers to move laterally throughout a network— thus lessening the propensity for broad breaches—DNS leaks compromise user privacy and can enable surveillance or tracking, as they manifest by indicating which websites are accessed (even when deployment involves privacy tactics such as VPNs). Given below are some of the key points of differences between them (Table 6.2):

6.13 Challenges in ICS Segmentation

Segmentation in ICS poses distinct challenges due to the critical nature of these environments and the need to balance security with operational requirements. Several difficulties arise when implementing ICS segmentation:

a. **Equipment Compatibility with Legacy Systems:**
   Legacy ICS equipment may not support modern security protocols or configurations. Businesses can address this challenge by utilizing network wrappers and protocol gateways. These tools help bridge the gap between outdated systems and current security standards.

b. **Operational Disruptions:**
Implementing segmentation within ICS networks may cause disruptions to operational services. To minimize interruptions to critical processes and activities, it is essential to carefully design the segmentation strategy and execute it in phased stages.

c. **Maintenance Access Requirements:**
Vendors often require remote access to ICS equipment for maintenance purposes. However, this access introduces potential security risks. Organizations must ensure that remote access is granted temporarily, properly monitored, and implemented using effective security mechanisms to prevent unauthorized access and mitigate security threats.

6.14 Summary

Network segmentation is a key concept that should be practiced to improve security, optimize performance, and achieve simpler network management. By segmenting a network into smaller, easier-to-manage sections, organizations can reduce the attack surface, prevent threats from moving laterally, and establish least-privilege access controls. Network segmentation also increases efficiency by prioritizing critical applications and reducing congestion. It simplifies monitoring and response, expediting the identification or rollback of malicious activity. In short, network segmentation benefits both operational continuity and efficiency, while providing a robust defense against cyber attacks.

Chapter 7
Comprehensive Overview of Field Devices in ICS: Protocol Management, Security Challenges, and Lifecycle Optimization

Abstract The security of Industrial Control System (ICS) field devices is vital for ensuring industrial safety, protecting critical infrastructure, safeguarding data privacy, enhancing cyber resilience, and maintaining regulatory compliance. This chapter examines various types of field devices, including protocols such as Modbus, Distributed Network Protocol (DNP), OPC UA, and HART (Pereira et al. (2003) Hart protocol analyser based in labview). It also delves into the anatomy of field devices, lifecycle management, sensor properties, and their comparison to SCADA systems (Anees Ara (2022) Security in supervisory control and data acquisition (scada) based industrial control systems: Challenges and solutions). Additionally, the chapter addresses the challenges of securing field devices in industrial environments, offering practical guidance on mitigating common threats and implementing robust hardware-based security solutions to protect ICS from cyberattacks. By understanding field device security systems, anyone will gain valuable insights into effective security strategies to ensure the safe and reliable operation of industrial processes.

Keywords ICS security · Field device security · Industrial control systems · Modbus protocol · Distributed network protocol · OPC UA · HART protocol · Field device lifecycle management · SCADA systems · Cyber resilience · Hardware-based security · Industrial cybersecurity

7.1 Introduction

Field devices are a broad category of specialized hardware components and equipment used in ICS to coordinate, track, and manage physical operations across various industries. These devices are integral to bridging the gap between digital control systems and the real-world environment, facilitating the seamless integration of automation technologies into industrial processes. By gathering and transmitting critical data from the physical environment, field devices allow operators to monitor and control operations more effectively. Manufacturers such as analog devices have pioneered the development of advanced field instruments that incorporate sophisticated sensors, actuators, cognitive computation, and robust communication technologies. Whether in harsh mechanical environments, high pressure, or extremely high temperatures, these devices are designed to function at their best in the most difficult industrial settings. As intelligent edge devices, field instruments perform real-time processing and analysis at the point of data collection, playing a critical role in driving automation and optimizing industrial workflows.

Field devices contribute to a variety of key operational improvements within industrial sectors. They enhance product quality by ensuring precise measurements, increase productivity by enabling automated control of systems, and reduce maintenance costs by providing real-time diagnostics and early detection of equipment failures. Additionally, they help lower energy consumption by enabling efficient control of machinery and processes and contribute to reducing unplanned downtime through predictive maintenance capabilities. These devices achieve these outcomes by converting physical signals, such as temperature, flow, pressure, and level, into actionable data and insights that can be used for monitoring, control, and optimization purposes.

As a trusted partner and leading supplier of enabling technologies, companies like analog devices provide solutions that meet the evolving needs of Industry 4.0. Their offerings span low-power sensing, precision measurement, digital processing, protection, power management, security, and communication technologies. With decades of experience and a deep understanding of industrial requirements, these companies are at the forefront of advancing industrial automation and digitalization, ensuring that businesses can fully harness the potential of smart manufacturing and connected systems. Figure 7.1 is adapted from [166].

Figure 7.1 illustrates various industrial and technical equipment in field devices.

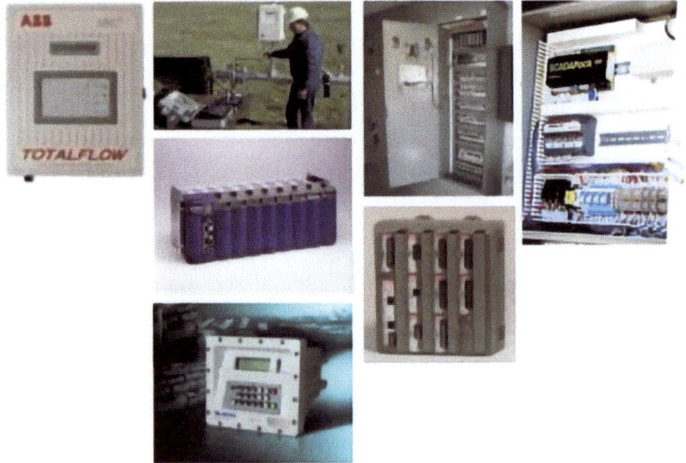

Fig. 7.1 Field devices

7.2 Common Types of Field Devices

Field devices are equipment or devices used in field applications or industrial settings to measure, interact with, monitor, or control physical processes. Field devices do many things in a multitude of industries and help make operations safe and efficient. A few examples of typical field device types are discussed as follows:

1. **Sensors:** They sense and measure physical characteristics.
2. **Actuators:** They execute instructions.
3. **Programmable Logic Controllers (PLCs):** Provide precise control.
4. **Human-Machine Interfaces (HMIs):** Enable simplicity of operator interactions.
5. **Communication Devices:** Enhance connectivity of industrial systems.
6. **Remote Terminal Units (RTUs):** Automate and monitor distant and scattered assets.

7.3 Anatomy of a Field Device

Field devices are small embedded computers typically installed in industrial environments such as factories or transportation systems. They are directly connected to actuators and sensors, allowing them to monitor and control physical processes. These devices act as the "eyes, ears, and hands" of automation systems, providing real-time data collection, decision-making, and process control.

Each field device consists of several key components. The core component is the processing unit, usually a microcontroller (MCU) or microprocessor (MPU), which runs control algorithms and logic. Common processors include ARM Cortex-M series or Intel Atom, chosen for their reliability in industrial settings. Input/output (I/O) interfaces handle communication with sensors and actuators. Analog I/O converts physical signals like 4–20 mA or 0–10 V into digital data, while digital I/O deals with binary signals such as on/off inputs. Specialized industrial interfaces such as RS-485, CAN bus, or Ethernet are also used. Memory components include flash memory to store firmware and RAM for processing data. Power supplies are designed for rugged conditions, typically using 24V DC with protection against electrical surges.

Field devices interact with their environment using sensors and actuators. Sensors measure variables such as temperature, pressure, and flow rate—examples include thermocouples, resistance temperature detectors (RTDs), and pressure transducers. Actuators carry out control commands, such as adjusting valve positions or motor speeds. Common actuators include solenoid valves, servo motors, and pneumatic devices.

To communicate with control systems like SCADA or DCS, field devices use a variety of industrial communication protocols [8]. Wired protocols include HART (which supports both analog and digital signals), Modbus (a simple master-slave protocol), and PROFIBUS or Profinet (used in high-speed automation networks). Wireless protocols such as WirelessHART and LoRaWAN enable remote and energy-efficient communication, especially useful in Industrial Internet of Things (IIoT) applications.

Firmware embedded in the field device contains the logic for real-time control, often including PID controllers. This firmware is typically written in C/C++ or standard languages such as IEC 61131-3, which includes ladder logic. Field devices also include diagnostic features like self-tests, fault logging, and predictive maintenance capabilities. To ensure cybersecurity, features such as authentication, encryption, and secure boot processes are commonly integrated.

Finally, field devices are engineered for environmental durability. They can operate in extreme temperatures ranging from −40°C to +85°C and are housed in enclosures with high ingress protection (e.g., IP67) to resist dust and water. They are also designed to withstand vibrations and shocks typically found in heavy machinery environments.

Examples of field devices include Programmable Logic Controllers (PLCs), which manage the logic of industrial machines, and smart transmitters that measure and send data on variables like pressure and fluid levels.

7.4   Field Devices in Industrial Systems

Field devices are small computers placed in areas like factories or transportation systems. They connect directly to sensors and actuators to help monitor and control physical activities, like machines running or temperatures changing.

Some of these devices are simple and only allow a Human-Machine Interface (HMI) or a SCADA system to adjust settings. Other systems are more advanced. For example, a Programmable Logic Controller (PLC) can make decisions on its own based on pre-set instructions.

Field devices can also be called by other names, like Programmable Automation Controllers (PACs), real-time automation controllers, substation controllers, remote telemetry units, relays, and communication processors. Even though they may do different tasks or have different features, they all work directly in the field to support control and automation.

For example, a PLC may be connected to a protection relay and a transformer monitor, which together control a circuit breaker. These devices are placed directly on the equipment they manage. The PLC may include an Ethernet module so it can communicate with a central control room through a network or the Internet.

Field devices are often built using standard parts on custom-made circuit boards. They can use different types of computer chips, like ARM, PowerPC, Motorola 68k, or embedded x86. Their operating systems also vary—common ones include VxWorks, embedded versions of Windows, and different types of Linux. Some devices still use custom real-time operating systems, but this is less common today.

Many field devices don't have strong security features. Often, the only protection is a password, and some have no protection at all. In the past, people thought these devices were safe from cyberattacks because they weren't connected to the main company network or the Internet. As a result, security wasn't a priority. But now, because of modern business needs and remote access, these devices are becoming more exposed to cyber threats.

7.5 Field Device Protocols

ICS field devices communicate with central control systems using various protocols. To connect with the ICS, sensors, actuators, and other devices utilize the following protocols:

1. **Modbus**
2. **DNP3 (Distributed Network Protocol)**
3. **OPC UA (Unified Architecture)**
4. **HART (Highway Addressable Remote Transducer)**
5. **Foundation Fieldbus**
6. **PROFINET**

7.6 Overview of the Modbus Protocol

The Modbus protocol follows a master-slave model where communication happens in pairs. The master initiates communication by sending a request, and the slave responds to it. In Modbus/TCP, the protocol separates the Application Data Unit (ADU) from the Protocol Data Unit by adding a Modbus Application Protocol (MBAP) header at the start of each message. This addition also introduces some basic security features. Modbus is widely used because it is simple, reliable, and effective for monitoring and controlling devices remotely.

Each ADU contains a Unit ID (or address) that helps route the message to the right application. After the header, the PDU follows, which includes a function code and the data related to the specific request.

a. **Type:** Modbus is a widely used communication protocol that works over both serial connections and Ethernet networks. For serial communication, it supports RS-232 and RS-485 interfaces, enabling point-to-point or multi-device connections. For Ethernet communication, it uses TCP/IP to operate across Local Area Networks (LANs) and Wide Area Networks (WANs), making it compatible with modern industrial systems.

b. **Description:** Modbus is popular and easy to use, but it lacks built-in security features. One major issue is the absence of message encryption, which makes it possible for attackers to intercept and modify data transmitted over Modbus networks. Additionally, Modbus does not provide strong user authentication, relying only on addresses and predictable commands. As a result, unauthorized users can access the system and cause harm, such as downtime, equipment damage, or safety risks.

c. **Vulnerabilities:** Modbus has a simple design without built-in protections, leaving it vulnerable to attacks. Without encryption, sensitive data like process values and control commands can be intercepted or altered by attackers. Modbus also lacks robust authentication mechanisms, making it easier for unauthorized individuals to manipulate critical systems. This could lead to equipment failures, production loss, and safety hazards.

d. **Security Measures:** To secure Modbus communication, organizations should implement strong security measures designed for industrial environments.

Encrypting all transmitted data is an essential step to prevent unauthorized access. Stronger authentication methods, such as mutual authentication and digital certificates, should be used to verify device identities and prevent impersonation attacks. Access control can limit communication to approved devices, reducing the risk of abuse. Intrusion detection systems (IDS) can monitor network traffic and detect suspicious activity [140]. Regular security checks, vulnerability assessments, and device software updates are also crucial for maintaining a secure and reliable system.

7.6.1 Distributed Network Protocol

The Distributed Network Protocol 3 (DNP3) is mainly used in automation systems, especially in water and electric utility sectors. It was originally developed to improve communication between control systems and field devices like Remote Terminal Units (RTUs), Intelligent Electronic Devices (IEDs), and SCADA (Supervisory Control and Data Acquisition) master stations [8].

DNP3 is mainly used for communication between master stations and RTUs or IEDs. For communication between master stations themselves, a different protocol called ICCP (IEC 60870-6) is typically used [155]. DNP3 competes with other protocols like Modbus and IEC 61850 [105].

The protocol was originally developed by Harris, Distributed Automation Products, and in 1993, responsibility for maintaining and updating DNP3 was passed to the DNP3 Users Group. This group is made up of utility companies and equipment vendors who ensure the protocol remains compatible, up-to-date, and widely supported.

DNP3 is an open standard. That means its full documentation is available to the public. The protocol's Document Library includes everything from the basic definitions and system requirements to information on secure authentication, XML device profiles, and testing guidelines.

a. **Type:** DNP3 is a communication protocol that works over both serial lines and Internet Protocol (IP) networks.
b. **Description:** DNP3 is used to connect different parts of Industrial Control Systems (ICS), such as RTUs, IEDs, and master stations. It allows these systems to communicate and exchange data reliably in real time.
c. **Vulnerabilities:** Although DNP3 is widely used and reliable, older versions of the protocol have security weaknesses. These include a lack of encryption and weak authentication, which make the system vulnerable to attacks like unauthorized access, data tampering, and spying. Using default settings or weak passwords can increase these risks. Also, because SCADA systems are often connected to the internet and many devices are linked together, a single security breach can have serious consequences. That's why strong protection methods are important.
d. **Security:** To protect DNP3-based systems, it's important to use security features designed specifically for industrial environments. Secure versions of DNP3 include features like authentication and encryption, which help protect against cyber attacks. Technologies like Transport Layer Security (TLS) and Advanced Encryption Standard (AES) ensure that data sent across the network cannot be read or changed by attackers. Strong authentication methods—such as digital certificates or two-step verification—help confirm the identity of connected devices. Regular security monitoring, vulnerability testing, and system audits are also important to catch and fix any new problems and keep the system safe and reliable.

7.6.2 OPC UA Protocol

OPC Unified Architecture (OPC UA) is an open-source, platform-independent protocol developed by the OPC Foundation. It is designed for secure and reliable data exchange between industrial devices and cloud-based systems. OPC UA follows the IEC 62541 standard and is widely used in modern industrial applications, including cloud and edge computing for tasks such as device diagnostics, asset management, monitoring, and reporting [48].

A key advantage of OPC UA is its plug-and-play capability. Any compatible device can connect to the platform without needing extra code. Once connected, the system automatically detects the device's features and data format. VPNs are not required, as OPC UA creates secure, encrypted connections, enabling communication both locally (at the edge) and remotely (in the cloud).

a. **Protocol Type:** OPC UA is an IP-based communication protocol.
b. **Description:** OPC UA is a modern communication standard created to improve industrial data exchange. It replaces older OPC versions and adds more features, such as support for different data types, detailed metadata, and consistent communication methods. Because it is platform-independent, devices with different

hardware and software can still communicate smoothly. One of its main benefits is strong built-in security, including encryption, user authentication, and access control, which ensures data is safe and reliable.

c. **Vulnerabilities:** Even though OPC UA includes strong security by design, it can still be at risk if not set up correctly. Common problems include using weak passwords, insecure network configurations, or incorrect access permissions. Also, if the system is not updated regularly or installed properly, it can be exposed to cyberattacks. Therefore, it is important to follow good security practices and monitor the system for risks.

d. **Security:** To protect OPC UA systems, organizations should use proper industrial security measures. One effective method is Role-Based Access Control (RBAC), which limits what each user can do based on their assigned role. Communication should always be protected using encryption and data integrity checks. Regular security audits, software updates, and vulnerability testing are also essential to keep the system secure, address new threats, and meet security standards.

7.6.3 HART Protocol

HART (Highway Addressable Remote Transducer) is an open communication protocol used in industrial automation systems. Its main benefit is that it allows digital communication to be added on top of traditional 4–20 mA analog signal lines without needing extra wiring. This makes it easier for industries to upgrade their systems without changing existing infrastructure. HART is widely used in both small and large industrial process control systems [115].

There are three types of HART commands:

- **Universal Commands**—Work on all HART-compatible devices. Used for basic tasks like identifying a device or reading its data.
- **Common Practice Commands**—Used in most HART devices for common operations.
- **Device-Specific Commands**—Unique to each manufacturer and used for advanced features of specific devices.

a. **Type:** HART supports both analog and digital communication.
b. **Description:** HART is commonly used in industrial automation to connect field devices such as pressure, temperature, or flow sensors to control systems. It uses Frequency Shift Keying (FSK) to send digital signals over the same wire as the 4–20 mA analog signal. This lets users program devices, get diagnostic information, and read real-time data without adding new wiring. It is a low-cost way to improve older analog systems using modern digital features.
c. **Vulnerabilities:** Although HART is useful and flexible, it has some security issues. It doesn't use encryption, so anyone with physical access to the wires might read or change the data. Attackers could fake being a valid device (spoof-

ing) or send incorrect data. Older HART systems often do not use authentication, which increases the risk of unauthorized access and data manipulation.

d. **Security:** To protect HART systems, strong security practices should be followed. Encryption methods like AES (Advanced Encryption Standard) can be used to prevent data theft or tampering. Network segmentation (dividing the network into parts) can stop the spread of attacks. Intrusion Detection Systems (IDS) can monitor traffic and alert users if something unusual is happening. It's also important to keep device firmware up to date and carry out regular security checks to fix any weaknesses. These steps help make HART-based systems more secure and trustworthy.

7.6.4 Foundation Fieldbus

Foundation Fieldbus is a base-level, industrial network used in plant or factory automation. It is an all-digital, serial, and two-way communication protocol designed to improve process control and real-time data exchange between field devices and control systems. This open architecture, developed and managed by the FieldComm Group, ensures flexibility and interoperability across various industrial systems, facilitating enhanced communication and efficiency in modern industrial environments [159].

a. **Type:** Foundation Fieldbus is a digital communication protocol specifically designed for industrial automation and process control. It is serial in nature, meaning data is transmitted sequentially in a single stream, and two-way, allowing devices to not only send but also receive data for interactive communication.

b. **Description:** Foundation Fieldbus is a digital protocol that supports real-time process control field device communication. It enables the seamless integration of a wide range of intelligent field devices such as sensors and actuators. The protocol is capable of providing detailed diagnostics, status updates, and control signals, which are vital for monitoring and optimizing processes in industries like oil and gas, chemical, and manufacturing. Its design allows for significant data exchanges, with capabilities like device configuration, process monitoring, and real-time control, which enhance operational efficiency. Additionally, Foundation Fieldbus supports both continuous and event-driven communication, making it adaptable to various industrial needs.

c. **Vulnerabilities:** Although Foundation Fieldbus provides robust features for industrial automation, improper configuration of devices and lack of adequate authentication can expose the system to security risks [159]. For instance, if devices are not securely configured, attackers could potentially access or manipulate sensitive process control information. Additionally, if authentication mechanisms are not enforced, unauthorized devices may gain access to the network, potentially compromising the integrity of the entire system. Misconfigured network connections or weak password protection can also make the system vulnerable to external threats or internal errors, risking operational disruptions or even safety hazards.

d. **Security:** To mitigate security risks associated with Foundation Fieldbus [159], it is essential to implement secure device configurations, robust authentication measures, and network monitoring protocols. Ensuring that each device connected to the Foundation Fieldbus network is properly configured with strong security settings and encryption techniques helps protect against unauthorized access. Authentication should be enforced for both device-to-device and human-to-device interactions to ensure that only authorized personnel and devices can communicate with the system. Furthermore, periodic audits, regular firmware updates, and intrusion detection systems can provide an additional layer of security, safeguarding against vulnerabilities that may be exploited by malicious actors.

7.6.5 PROFINET Protocol

PROFINET is a communication protocol used in industrial automation for fast, Ethernet-based data exchange between field devices. It works on a consumer-provider model and is especially useful in modern manufacturing and processing environments. An IO-Proxy can be used to connect lower-level PROFIBUS devices to a PROFINET network, helping reduce complex wiring and integration challenges [58].

PROFINET supports high-speed and secure data communication across different devices and platforms. It follows the four layers of the Internet model and typically does not use TCP/IP for sending real-time data. Instead, TCP/IP is only used for tasks like configuration and diagnostics.

PROFINET performs four main functions:

1. Supports real-time communication for efficient industrial operations.
2. Enables safe data transfer through PROFIsafe, ensuring worker and equipment safety.
3. Simplifies troubleshooting and fast setup, which increases system uptime.
4. Allows easy integration with existing fieldbus systems, helping preserve previous investments.

Additional details about PROFINET are as follows:

a. **Protocol Type:** Ethernet-based industrial communication protocol.
b. **Description:** PROFINET is a widely used protocol in manufacturing and automation systems. It allows fast and reliable communication between devices and supports smooth coordination across the factory floor. It also promotes device interoperability, so equipment from different manufacturers can work together without issues. The protocol is designed for high performance and reliability, which are crucial in time-sensitive industrial environments.

c. **Vulnerabilities:** Like all networked systems, PROFINET can be vulnerable to attacks if not properly secured. Common issues include network misconfiguration, weak passwords, and poor access control. These vulnerabilities can allow unauthorized users to access or disrupt the system, leading to potential data breaches or downtime in industrial processes.
d. **Security Measures:** To protect PROFINET systems, several security practices should be followed. Network segmentation can be used to isolate sensitive areas of the network. Strong access controls, such as user authentication and role-based permissions, can prevent unauthorized use. Device hardening [62] practices—including regular firmware updates, disabling unused features, and encrypting communication—help further protect the system from cyber threats.

7.7   Field Device Lifecycle Management

Industrial control systems and other critical infrastructure require lifecycle management of field devices to ensure maximum performance, System Security, and System Reliability. Lifecycle management—the complete process of overseeing an industrial field device from original planning to final disposal. This includes selecting and purchasing the correct devices, configuring them in a secure manner, keeping them up to date and strong enough for our use case, regularly updating those devices for additional security patches as necessary, and securely decommissioning devices when they are no longer needed.

It means that device lifecycle is managed effectively, reducing the time and expense of developing new industrial goods. It can be used to inform decisions about production increases and direct targeted marketing programs. All stages of life are important to ensure devices operate efficiently, safely, and in compliance with standards.

Important phases and their factors include:

- System requirements assessment and planning
- Procuring reliable vendors and products
- Device configuration and customization for operational needs
- Device deployment and installation
- System performance monitoring
- Maintenance measures such as patching
- Security steps like access control
- Upgrades or replacements strategy
- End-of-life recovery and responsible disposal
- Record keeping for audits, compliance
- Human training in device interaction

Management of all these aspects ensures that field devices will meet the immediate demands of industrial systems and also support long-term safety, scalability, and sustainability.

7.8   Field Device Sensor Parameters

Field device sensor parameters refer to the specific types of data that sensors collect in their surroundings. These variables depend on which sensor we want to use and how we want to apply it. Typical examples include temperature, pressure, humidity, and vibration. It is very important to keep track of them because all these conditions are closely related and have a significant impact on the safe, reliable, and efficient operation of industrial systems.

The continuous and real-time monitoring feature of infrared sensors permits the operator to act immediately on the event of any change in system conditions. Should a sensor read abnormally , that may mean something is wrong like an equipment malfunction or danger—alerting teams right away so they can fix the problem before anything goes awry. On the efficiency side, the information can be used to help systems like a closed-loop control of heating or cooling processes in order to save energy.

Predictive Maintenance Getting streaming data from sensors provides an early indicator of wear or failure. It thereby allows for the repair or replacement of equipment in a controlled and systematic fashion before failure, and hence reduces downtime, maintenance costs, etc. Furthermore, sensor parameters are essential to sense hazards in hazardous areas, which increase safety at work.

Sensor parameters are one of the critical components for optimal operations in intelligent industrial environments. They enable efficient operations, ensure safety, and support data-driven decision-making in a multitude of industrial applications (Table 7.1).

7.9   Comparison of Field Devices vs SCADA Systems in ICS

Field devices and SCADA systems are two ICS components that work together to ensure that industrial processes are performed effectively, safely, and in control. They allow central control and monitoring, providing operators with a broad view of industrial processes. They gather information from dozens or even hundreds of field devices and sensors, so that operators can supervise and control entire production lines or processes from a single point. SCADA systems are critical for enabling real-time data visualization, alarm handling, and remote control to allow industries such as energy, water treatment, manufacturing, and oil and gas [8] to operate efficiently and optimize.

Field devices, on the other hand, operate at the "edge" of the system, collecting critical information close to where physical events take place in a field. These devices are sensors, actuators, and intelligent tools that are designed to measure various physical factors such as temperature, pressure, flow, and level. Field devices

Table 7.1 Description of various sensor parameter types

Parameter type	Description	Common applications
Temperature	Measures temperature levels	HVAC, chemical processing, food
Pressure	Measures gas or liquid pressure	Oil and gas, automotive, aerospace
Humidity	Measures moisture content in the air	Climate control, agriculture, pharma
Flow Rates	Monitors the flow of liquids or gases	Water treatment, chemical processing
Level	Detects the level of liquids or solids	Inventory Management, Wastewater
pH	Measures acidity or alkalinity	Water quality monitoring, chemicals
Force and Load	Measures force or weight applied to objects	Material Testing, Automotive
Vibration	Detects vibrations in machinery and structures	Equipment healthmonitoring
Light and optical	Measures aspects of light, color, or wavelength	Color sorting, quality control
Gas concentration	Detects the concentration of specific gases	Environmental Monitoring, Gas Detection
Position and proximity	Determines object position or proximity	Robotics, automation, conveyor
Acceleration and motion	Measures acceleration and motion	Vehicle stability control, aerospace

gather real-world data and send it to SCADA systems for analysis and control. They are the sensory and actuator components in the industrial environment that ensure that the processes run as designed, by providing first-hand information on machinery and systems.

SCADA systems focus on monitoring and controlling industrial activities comprehensively, while field devices take up data collection and localized level control from the operational side. These two parts together make a consistent working system where operators can monitor and manage multiple operation statuses, intervene when necessary to prevent failures, and improve the overall performance. SCADA systems with field devices are used to improve the efficiency, reliability, and safety of industrial processes that ultimately help secure production outcomes while reducing operational costs (Table 7.2).

Table 7.2 Comparison of field devices and SCADA systems in ICS

Aspect	Comparison
Attack surface	SCADA systems have a larger attack surface than ICS field devices [166]. Due to their limited capability and communication, field devices are less vulnerable to certain attacks than SCADA
Physical access	Industrial field equipment is generally scattered, making it tougher to attack. SCADA systems are centralized and may be easier to access
Function	Field devices gather data or operate actuators, limiting attack vectors. SCADA systems handle several processes, creating greater attack options [8]
Criticality	Field devices directly govern physical processes, making assaults on them instantly dangerous and disruptive. SCADA systems are important but may not have direct physical effects
Communication	Field devices connect with SCADA systems and other components, providing ICS-wide attack vectors. Attackers may target them to breach the system
Attack Objectives	Field device cyberattacks frequently interrupt or influence physical processes. SCADA system assaults may target data integrity, availability, or control, creating broad disruptions
Mitigation	Physical access, firmware integrity, and communication routes protect field devices. Keeping SCADA systems safe requires robust security measures

7.10   Enhancing Field Device Security in ICS Environments

Manufacturing, energy, and utilities are a selection of industries where it is essential to have an Industrial Control System (ICS). While advancing systems with the integration of networking and computing make them more sophisticated, those same technologies compound the challenge to secure. Cloud 7 IT Services Inc. recognizes the importance of protecting ICS environments and delivers hardware-based solutions that increase security strength. These strategies encompass:

1. **Network Segmentation:** Isolate field devices from main corporate network to minimize the spread of potential attacks.
2. **Security Patch Management:** Establish a reliable method to apply both security patches and updates to field devices on a regular schedule.
3. **Access Control:** Enable rigorous access controls that will ensure only the relevant, authorized persons can interact with vulnerable field hardware.
4. **Security Training:** Train staff to recognize, respond, and report security threats effectively.
5. **Incident Response Plans:** Develop and periodically review how the response to security events can be best handled with minimal impact.
6. **Security Audits and Assessments:** Identifying the vulnerabilities within their system is always a priority.
7. **Security by Design:** Start with security mechanisms to build sensing devices and also choose robust sensors that can provide some level of security.
8. **Supply Chain Security:** Protect the entire device supply chain from manufacturing to deployment, preventing tampering or compromise.

7.11 Power of Hardware-Based Approaches

Hardware-oriented security plays a critical role in securing and strengthening Industrial Control Systems (ICS). Unlike purely software-based approaches, hardware-based methods provide a more robust level of protection by reducing the risk of attacks that target software vulnerabilities in advance.

These hardware mechanisms help build strong defenses that prevent threats from spreading across ICS networks, thereby limiting the damage an attacker can inflict. Hardware systems also enable rapid responses to incidents, which can be crucial in time-sensitive industrial environments—potentially making the difference between safety and disaster.

Moreover, hardware-based defenses are significantly more difficult for remote attackers to bypass, as they often require direct physical access to the system. This inherent requirement of physical proximity adds an extra layer of security, even when software safeguards are compromised.

Integrating hardware-based security methods enhances the system's resilience against sophisticated threats and ensures the safe and uninterrupted operation of industrial processes.

7.12 Benefits of Hardware-Based Approaches

Information technology infrastructure can be protected by incorporating security features directly into hardware, such as servers, processors, and communication devices. This approach adds another level of protection that is intrinsically hard for hackers to get past. The following lists a number of advantages of hardware-based approaches:

1. **Isolation:** Hardware-based security prevents the spread of attacks laterally within the ICS network by isolating segments. It isolates any potential threats and stops them from moving laterally through the network.
2. **Real-Time Defense:** These hardware security components help with the early identification and thwarting of a possible threat. Due to their ability to constantly surveil network system and activity, these components are able to detect anomalies immediately, making them capable of quickly taking action as a response in order from possible threats and subsequently improve the security posture of the overall infrastructure.
3. **Physical Security:** Physical access is required to breach the security protections of embedded hardware solutions. This physical barrier adds an extra layer of defense against remote attacks, ensuring that attackers must first overcome physical security measures to access the system.
4. **Resilience:** Hardware-based protections deliver tough security, even if other security measures are compromised. The hardware-integrated and reinforced security mechanisms help protect the infrastructure continuously, even if other defense layers are breached.

7.13 Implementing Hardware-Based Security

Hardware-based security implementation is essential. Here are some key points:

1. **Embedded Security Modules:** Utilize hardware security modules (HSMs) to protect cryptographic keys and conduct secure operations. HSMs provide a specialized hardware environment for cryptographic tasks, ensuring the security and authenticity of sensitive information.
2. **Network Segmentation:** Implement network segmentation solutions to divide the ICS network into multiple segments, each designated for specific components of the infrastructure. By isolating critical assets and operations into distinct network segments, organizations can reduce the risk of unauthorized access and mitigate the impact of security breaches.
3. **Secure Hardware Devices:** Selecting hardware devices for ICS components that incorporate robust security features is crucial. Choose reputable manufacturers and suppliers that prioritize security in their product design and development. Secure hardware devices provide built-in defenses against common cyber threats and vulnerabilities, enhancing the overall security posture of the ICS.

7.14 Common Field Device Security Threats

ICS are susceptible to various threats due to vulnerabilities and malicious activities. These risks are referred to as common field device security threats. Some examples include:

1. **Malware and Viruses:** Malicious software can disrupt, manipulate, and damage field devices.
2. **Unauthorized Access:** Intruders may gain access to field equipment to modify settings, implant malware, and disrupt operations.
3. **Network Attacks:** Cyber attackers exploit field device network vulnerabilities to compromise communication and network integrity.
4. **Denial of Service Attacks:** Excessive traffic can disable field equipment, halting industrial operations.
5. **Man-in-the-Middle Attacks:** Hackers secretly intercept field device communication to alter data or commands.
6. **Physical Tampering:** Unauthorized access to field devices may lead to manipulation, reconfiguration, or operational disruptions.
7. **Supply Chain Vulnerabilities:** Compromises in the supply chain may expose field equipment to exploitation.
8. **Lack of Patch Management:** Outdated software and inadequate patching leave field equipment vulnerable to known threats.
9. **Weak Authentication and Encryption:** Insufficient authentication and encryption allow attackers to manipulate or eavesdrop on field equipment.

10. **Human Factor:** Unintentional insider threats can result in security breaches, misconfigurations, and data exposure.
11. **Legacy Systems:** Older field equipment lacking modern security measures may be exploited due to known vulnerabilities.

7.15   Exploiting Passwords in Field Devices

The improper use of passwords in field devices poses a serious threat to the security of ICS and may result in various vulnerabilities and malicious activities [5]. The following are key factors to consider:

1. **Attempting to Predict a Password:** An attacker may attempt to predict a password using brute force or dictionary attacks. This involves trying numerous character combinations or commonly used passwords until the correct one is found.
2. **Reverse Engineering:** An attacker may reverse engineer the firmware or software of the field device to uncover hardcoded passwords. This can be achieved by analyzing the device's binary code or firmware updates.
3. **Device Access:** If an adversary gains physical access to a field device, they may directly extract the password from the device's memory or configuration files.

7.16   Risks and Consequences of Exploiting Passwords

Exploiting passwords in field devices presents considerable risks and may lead to severe consequences for ICS and the organizations that rely on them. Some of the most significant risks and consequences include:

1. **Unauthorized Access:** Attackers may manipulate field equipment, disrupt operations, or steal data using hardcoded passwords.
2. **Data Manipulation:** Attackers may alter field device data, affecting sensor readings.
3. **Process Disruption:** Unauthorized access may interrupt critical industrial processes.
4. **Safety Hazards:** Field devices control industrial physical processes, and attackers with access may cause accidents.
5. **Compromised Network:** Exploited field devices can provide attackers access to the ICS network, enabling lateral movement and potential compromise of SCADA systems.
6. **Data Theft:** Attackers may steal industrial designs, formulas, or process data, impacting the organization's competitiveness.

7.17   Mitigation Strategies for Field Device Security

Several mitigation measures are recommended to enhance the security of field devices in ICS:

1. **Changing Default Passwords:** To prevent unauthorized access, change default usernames and passwords to strong, unique credentials.
2. **Implementing Strong Access Controls:** Use role-based access control and MFA to restrict system access to authorized users, preventing unauthorized configuration changes and malicious activities.
3. **Regular Firmware Updates:** Keep field device firmware up to date with manufacturer-provided security patches to address known vulnerabilities and ensure security.
4. **Security Assessments:** Conduct regular security audits of field devices to identify and mitigate vulnerabilities, reducing the risk of exploitation.

7.18   Security Challenges in Field Devices

Ensuring the security of field devices in ICS presents several challenges, including:

1. **Legacy Systems and Limited Resources:** Many field devices run outdated operating systems without modern security features. Their low processing power and memory make it difficult to implement complex security solutions.
2. **Physical Accessibility:** Field devices are often located in remote or hard-to-access areas, making them susceptible to theft or tampering.
3. **Network Connectivity:** Field devices may have unstable network connections, complicating security monitoring and patch management.
4. **Lack of Security Awareness:** Field device operators and maintenance personnel may lack cybersecurity training, increasing the risk of human error and social engineering attacks.
5. **IT-OT Convergence:** Malicious actors can exploit vulnerabilities in both IT and OT systems through novel attack vectors.
6. **Evolving Threats:** New vulnerabilities and exploits emerge frequently, requiring continuous security monitoring and updates.
7. **Regulatory Compliance:** Field device installations must comply with data security and privacy regulations across various industries.
8. **Integration with Cloud Systems:** As field devices connect to cloud services, enforcing data encryption and access controls becomes essential.
9. **Managing a Heterogeneous Device Landscape:** Organizations use field devices from multiple vendors, making it challenging to standardize security policies and procedures.

10. **Cost Constraints:** Large-scale field device deployments can make implementing robust security measures expensive. Balancing security investments with budget limitations remains a challenge.

Addressing these challenges requires a multi-layered approach, including security awareness training, vulnerability management, physical security, network segmentation, secure communication protocols, continuous monitoring, and incident response planning.

7.19 Summary

Numerous security issues affect field devices, which are crucial to critical infrastructure and industrial control systems. Performance, reliability, and safety in industrial processes all depend on understanding these risks.

Supply chain attacks, sensor spoofing, and other security issues pose significant threats to field devices. These dangers highlight the importance of routine vulnerability assessments, patch management, and security updates.

Field devices must remain secure throughout their lifecycle to mitigate these risks, starting with careful design and procurement and continuing through cautious deployment, continuous monitoring, and thorough maintenance. Additionally, addressing both technological and human shortcomings requires fostering cybersecurity awareness and collaboration.

As technology evolves, malicious actors continuously adapt their strategies. Therefore, organizations must remain vigilant, monitor emerging threats, and regularly update their field device security procedures. This ongoing effort is essential to safeguarding industrial control systems and critical infrastructure from ever-evolving threats.

Chapter 8
Exploring Supervisory Systems Security Threats: Legacy SCADA Vulnerabilities, Communication Protocols, and System Security Strategies

Abstract The security of supervisory systems is vital for securing critical infrastructure, preventing cyberattacks, meeting regulatory requirements, and managing risk and resilience within system. This book will cover the introduction of SCADA (Supervisory Control and Data Acquisition) systems, the components within communicate with each other, and system design or architecture. Decoding the security vulnerabilities in these systems dealing both physical and cyber threats. The chapter provides a taxonomy of security threats such as viruses, worms, Trojan horses but also describes remedies by means of firewalls, access control, and encryption. It also discusses the difficulties of securing legacy SCADA systems. The chapter will provide readers with an overview of successful approaches to protecting these essential systems.

Keywords SCADA systems · ICS security · System resilience · Cybersecurity · Threat mitigation · Firewalls · Access control · Encryption · Legacy systems · Cyber threats

8.1 Introduction

In the digital age, where technology touches every part of life, supervisory system security is essential. Our complex industrial processes are monitored and coordinated by these systems, the silent guardians of vital infrastructure. They are susceptible to an array of increasingly complex and potent security risks due to their digital pulse. The relationship between creativity and susceptibility in this digital battleground is examined in supervisory systems security threats. Cybercriminals pose a threat to supervisory systems because we depend on them more and more; they could disrupt, impair, or deceive them. A security threat is a deliberate and harmful action intended to compromise, steal, or disrupt data inside an organization's systems or the company as a whole. A security event is an incident where firm data or a network may have been compromised.

8.2 Supervisory Systems

Supervisory systems are an important part of modern industry and infrastructure. A supervisory system is a type of control system software that is used to watch over or manage industrial processes. It is installed as an operating system program on computers. A supervisory control system may directly change different meter settings. The system helps automate data processing and formula execution, and it has ways to check the accuracy of measured values online.

These systems generally consist of two essential components:

1. Supervisory Control and Data Acquisition (SCADA)
2. Industrial Control Systems (ICS)

These components play a crucial role in several industries, including manufacturing, energy production, water treatment, transportation, and more. A comparative overview of SCADA and ICS is presented in Table 8.1.

8.3 Supervisory Control and Data Acquisition

Supervisory Control and Data Acquisition (SCADA) systems serve as the cornerstone of modern industrial efficiency and operational management. By leveraging real-time data acquisition, processing, and visualization from a network of sensors, programmable logic controllers (PLCs), and remote terminal units (RTUs) deployed across industrial facilities, these systems integrate fragmented processes into a

Table 8.1 Comparison of SCADA and ICS

Aspect	SCADA	ICS
Scope of control	Focuses on supervisory control and data acquisition	Provides supervisory and process control for industrial automation
Functionality and complexity	Used for data visualization and remote monitoring. Suited for less complex processes	Designed for intricate and suited for complicated, continuous process control
Attack surface	Mainly monitors, hence may have a lower attack surface	Its control and process management skills increase its attack surface
Security concerns	Typically related to unauthorized access, data breaches, and monitoring disruptions	Extends to unauthorized access, data integrity attacks, control manipulation, and potential physical consequences
Impact of compromises	Compromises may impact data collection and monitoring but might not directly affect industrial processes	Industrial processes may be disrupted, manipulated, or damaged by compromises

unified platform. This convergence enables seamless interaction between hardware (e.g., actuators, valves) and software (e.g., HMIs, analytics engines), providing operators with intuitive control over complex industrial workflows.

SCADA systems act as diligent observers of operational parameters, continuously monitoring variables such as pressure, temperature, flow rates, and voltage. Through advanced data historization and trend analysis, they provide a comprehensive overview of industrial processes, enabling stakeholders to:

- Detect anomalies via threshold-based alarms
- Optimize resource allocation through predictive maintenance
- Enhance safety protocols by mitigating risks in hazardous environments

The system's decision-support capabilities and remote control functionality are pivotal in critical infrastructure sectors. For instance:

In **energy distribution**, SCADA balances grid loads and isolates faults during outages.

In **water treatment**, it automates chemical dosing and monitors pipeline integrity.

In **manufacturing**, it synchronizes robotic assembly lines and tracks OEE (Overall Equipment Effectiveness).

As industries transition toward Industry 4.0, SCADA evolves by integrating with IoT edge devices, cloud computing, and AI-driven predictive analytics [91]. This digital transformation ensures compliance with stringent regulatory standards while improving uptime and reducing operational costs.

8.4 Industrial Control Systems

The Framework of Modern Industrial Automation
Industrial Control Systems (ICS) include more features than standard SCADA systems. They combine supervisory control, process management, and automation across all industrial operations. These systems connect machines, control devices (such sensors and actuators), and computers in a way that makes sure that industrial processes run smoothly, efficiently, and without errors.

There are several subsystems that work together in the ICS framework:

- **Distributed Control Systems (DCS):** Allows to control of a whole refinery process using equilibrium supervision and decentralized execution.
- **Programmable Logic Controllers (PLCs):** Execute localized, real-time control tasks (e.g., conveyor belt speed regulation).
- **Remote Terminal Units (RTUs):** Provide data collection and control in geographically distributed sites.
- **Human-Machine Interfaces (HMIs):** Real-time display and control of the process for operators

ICS ensures the seamless and precise continuation of activities in critical sectors such as:

Oil and Gas Oversees drilling operations, monitors pipelines, and ensures the safety of refineries.

Electricity Generation This encompasses load demand balance, turbine regulation, and grid synchronization.

Chemical Manufacturing Automates the management of hazardous substances, batch processing, and the regulation of reactor temperatures.

Transportation Manages airport logistics, railway signaling, and traffic control systems.

While SCADA focuses on data acquisition and high-level supervision, ICS provides a holistic ecosystem for:

- Closed-loop control of physical processes (e.g., PID tuning)
- Integration with enterprise resource planning (ERP) systems
- Compliance with industry-specific standards (e.g., ISA-95, NIST SP 800-82)

As ICS integrates the Industrial Internet of Things (IIoT) and edge computing technologies, robust cybersecurity becomes crucial. Modern ICS architectures now incorporate:

- Secure communication protocols (e.g., OPC UA, Modbus TCP/IP)
- Anomaly detection using machine learning algorithms
- Zero-trust frameworks to protect against supply chain attacks

8.5 Differentiation Between SCADA and ICS Security

SCADA (Supervisory Control and Data Acquisition) and ICS (Industrial Control Systems) are two of the most important terms in the OT field. Even though they seem like they are the same thing, they are not the same thing when it comes to security! Both SCADA and ICS systems help manage and control industrial processes, but they are very different in terms of size, features, and security issues.

SCADA stands for Supervisory Control and Data Acquisition. SCADA systems are made up of both software and hardware that allow for centralized monitoring and control of industrial processes. SCADA is a form of ICS that is only for supervisory control and data collection tasks. These systems take data from the various field devices and display it to operators. They can even do remote control operations so that operators may keep an eye on other functions. SCADA systems are vital for industries like energy, water treatment, and manufacturing, where real-time data is needed to manage operations that can be improved.

ICS, on the other hand, is a general word for any control system used in factories, like SCADA, Distributed Control Systems (DCS), and Programmable

Logic Controllers (PLC). ICS encompasses the entire infrastructure, connecting both Supervisory and Process Control tasks. ICS is more than just SCADA; it also contains various sorts of control technologies that work together to control the whole industrial system.

Security-wise the difference between SCADA and ICS is that ICS security has a wider range as compared to SCADA. While SCADA security is more focused on supervisory elements such as data transmission, remote monitoring, operator interfaces, and other SCADA operations, ICS security is concerned with protecting the whole infrastructure encompassing the process control, communication networks, and also critical field devices. The broader range of control levels from high-level supervisory systems to low-level field devices interacting with the physical world makes ICS security inherently more complex.

Because the SCADA and ICS networks are core parts of both systems, it is absolutely imperative that these industrial operations be safe and reliable via secure SCADA and ICS networks. SCADA systems can be at risk from threats targeting the control systems on which they rely, various sensors and cameras; the information transport systems that connect SCADA devices (including electronic forms of use of transducer signals); or else threat actions attempting to interfere with these SCADA systems. However, ICS security also must address threats at a higher level that encompasses everything from device-level vulnerabilities to network security to risks associated with connecting different types of control systems. With the growing linkage and interconnection of ICS networks to external networks, including corporate IT systems as well as the internet, it further complicates the tasks of protecting both SCADA and ICS infrastructures. Security-wise the difference between SCADA and ICS is that ICS security has a wider range as compared to SCADA. While SCADA security is more focused on supervisory elements such as data transmission, remote monitoring, operator interfaces, and other SCADA operations, ICS security is concerned with protecting the whole infrastructure encompassing the process control, communication networks, and also critical field devices. The broader range of control levels from high level supervisory systems to low level field devices interacting with the physical world makes ICS security inherently more complex.

Because the SCADA and ICS networks are core parts of both systems, it is absolutely imperative that these industrial operations be safe and reliable via secure SCADA and ICS networks. SCADA systems can be at risk from threats targeting the control systems on which they rely, various sensors and cameras, the information transport systems that connect SCADA devices (including electronic forms of use of transducer signals); or else threat actions attempting to interfere with these SCADA systems. However, ICS security also must address threats at a higher level that encompasses everything from device-level vulnerabilities to network security to risks associated with connecting different types of control systems. With the growing interconnection of ICS networks to external networks, including corporate IT systems and the internet, it further complicates the task of protecting both SCADA and ICS infrastructures (Table 8.2).

Table 8.2 SCADA vs ICS security differences

Aspect	SCADA	ICS
Scope of control	Focuses on supervisory control and data acquisition	Provides supervisory and process control for industrial automation
Functionality and complexity	Used for data visualization and remote monitoring. Suited for less complex processes	Designed for intricate and continuous process control. Suited for complicated processes
Attack surface	Mainly monitors, hence may have a lower attack surface	Its control and process management skills increase its attack surface
Security concerns	Typically related to unauthorized access, data breaches, and monitoring disruptions	Extends to unauthorized access, data integrity attacks, control manipulation, and potential physical consequences
Impact of compromises	Compromises may impact data collection and monitoring but might not directly affect industrial processes	Industrial processes may be disrupted, manipulated, or damaged by compromises

8.6 SCADA Architecture

SCADA is a tool that collects and analyzes data from industrial field equipment. It uses computers, networked data transfer, and a graphical user interface (GUI) to keep an eye on machinery and regulate processes. It is also a type of analytical process that deals with industrial control systems used in manufacturing, fabrication, and deployment. The four tiers of SCADA's design make it work. For instance, SCADA is often the system that keeps an entire region safe and in control. Software that is completely separate from the hardware. A supervisory system is used to gather and organize process data. SCADA is known as an RTU.

1. Process (Level 0)
2. Basic Control (Level 1)
3. Supervision (Level 2)
4. Operations Management (Level 3)

Data propagates upstream from Level 0 to Level 3 in SCADA. Data analysis at the Operations Management level leads to decision-making, which is then sent to Level 0 devices as control instructions. SCADA is a critical center for industrial automation, managing information and control instructions hierarchically.

Figure 8.1 illustrates the hierarchical structure of industrial automation and process control systems. This system has multiple tiers, including field devices, RTUs, PLCs, HMIs, and OPC (OLE for Process Control).

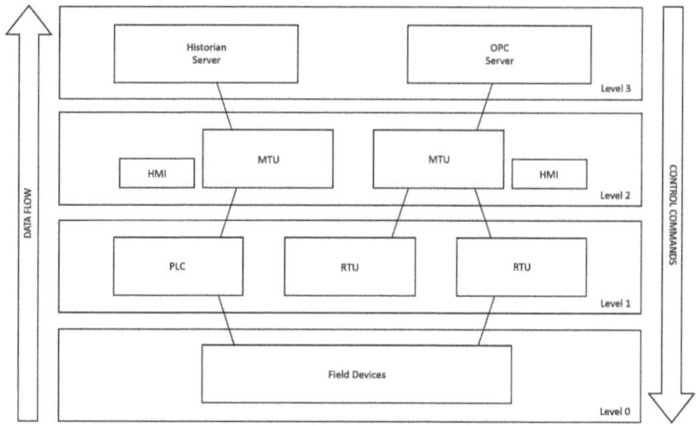

Fig. 8.1 SCADA architecture

8.7 Components of SCADA

SCADA systems are composed of several key parts to work collectively in monitoring and controlling the industries activities [8]. These components usually consist of:

1. **Field Devices**: Gather data and perform control actions.
2. **Remote Telemetry Units (RTUs)**: Deliver control instructions and transfer field device data to the MTU.
3. **Programmable Logic Controllers (PLCs)**: For data sharing and logic-based control commands.
4. **Master Terminal Unit (MTU)**: Collect data and transmit control instructions.
5. **Human-Machine Interface (HMI)**: Enable operator interaction and data analysis.
6. **Servers**: Provide data storage and intermediary OPC servers.
7. **Communication Network**: Utilize various technologies for seamless communication.

8.8 Communication Protocols in SCADA

SCADA communication index communication techniques make sure that system data and control go to their destination. These protocols make sure that SCADA system parts like field devices, RTUs, and Programmable Logic Controllers can communicate data correctly. PLC, HMIs (Human-Machine Interfaces), servers, and other things. Some common SCADA communication methods include:

1. Modicon Communication Bus (Modbus)
2. Inter-Control Centre Protocol (ICCP)
3. Distributed Network Protocol (DNP3)
4. Ethernet for Control Automation Technology (EtherCAT)
5. Open Platform Communication (OPC)
6. Common Industrial Protocol (CIP)

8.8.1 *Modicon Communication Bus*

Modicon programmable controllers can connect effectively with one another as well as with other devices in an industrial machine network. It supports Ethernet, Modbus, Modbus Plus, and Modbus for communication between controllers, allowing the user to customize his network design and integration. Ethernet is widely used for high-speed and dependable communication, making it an ideal choice for large-scale distributed systems. Modbus and Modbus Plus. Supporting serial connection choices, this is commonly found in industrial applications for more basic devices that require a low-cost manner of connecting to the network.

Modicon's built-in ports, network adapters, option modules, and gateways can increase communication by allowing controllers to easily integrate into a variety of network designs, including local area networks (LANs) and wide area networks (WANs). These modules enable old systems and new digital infrastructure to interface with Modicon controllers, resulting in high interoperability with a wide range of industrial manufacturers' technology.

Modicon also offers ModConnect partner agreements, which allow Original Equipment Manufacturers (OEMs) to add Modbus Plus capability into their designs, hence increasing the system's interoperability and access. Because of its flexibility, Modicon-based systems can be easily altered and customized to meet the needs of various industrial automation solutions, including monitoring, control, and data collection (across multiple devices and subsystems). A figure is shown below in Fig. 8.2.

Fig. 8.2 Overview of Modbus was from [31]

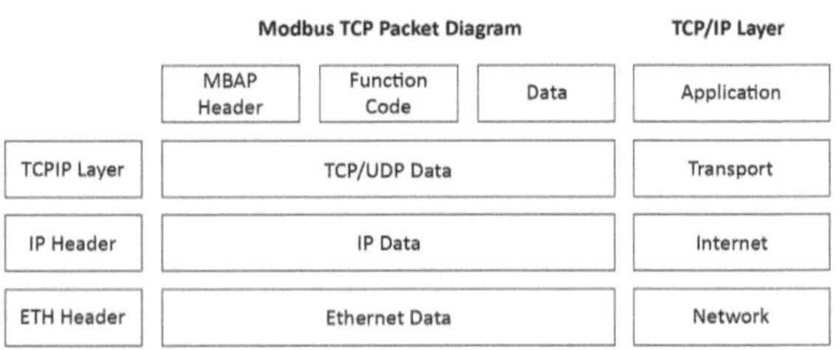

Figure 8.2 shows devices linked in a hierarchy of networks using different communication methods. Each network's packet structure contains the Modbus protocol, which devices use to communicate data. The industrial machine network's Modicon PLCs are able to communicate effectively with one another and other devices. In terms of network architecture and integration, it offers Ethernet, RS485, Modbus Plus, or the option for the user to select other controllers. Ethernet's high speed and dependable connectivity enabled it to be chosen quickly in large-scale, dispersed applications. Both Modbus and Modbus Plus. Serial connection techniques: Most frequently utilized in industrial settings (for cost-effective network connections for low-end devices; frequently, though not solely, found in specific Digi product series).

The controllers themselves can be used to expand communication to almost any network design, including local area networks (LANs) and wide area networks (WANs), by integrating the built-in ports, network adapters, option modules, and gateways of Modicon devices. Because these modules are very compatible with the technology of many industrial manufacturers, systems have access to modern digital infrastructure and may work with Modicon controllers.

The system's accessibility and interoperability are further increased by Modicon's ModConnect partner agreements, which enable Original Equipment Manufacturers (OEMs) to include Modbus Plus functionality in their designs. A system built on the Modicon platform is adaptable and may be readily customized to meet the needs of certain industrial automation applications, including data gathering (from various devices and subsystems), control, and monitoring.

8.8.2 Inter-Control Centre Protocol

IEC 60870-6 (ICCP) is the primary standard for WAN-based utility control center messaging and grid management [155]. This allows utility control centers to communicate important information, improving the efficiency of the power network. Although significant, ICCP contains security weaknesses such as spoofing attacks, session hijacking, weak authentication, and trust-building concerns in WAN. Control hubs and data sources are geographically scattered. A Fig. 8.3 is taken from [106].

Figure 8.3 represents the Inter-Control Center Communications Protocol Server Object. It shows details about requests, responses, reports, actions, events, data objects, and control center data objects related to the Inter-Control Center Communications Protocol. This also includes various operations and data objects relevant to the Inter-Control Center Communications Protocol Server Object.

To improve communication, the ICCC will require many protection measures to combat the threat of tactical denial-of-service attacks, which cannot be addressed by current methods alone [155]. Defense-in-depth tactics must be used, servers must be diligently patched, and vulnerabilities must be examined. Implement network segmentation and access rules to keep any attacker behind a sandbox environment

Fig. 8.3 ICCP objects

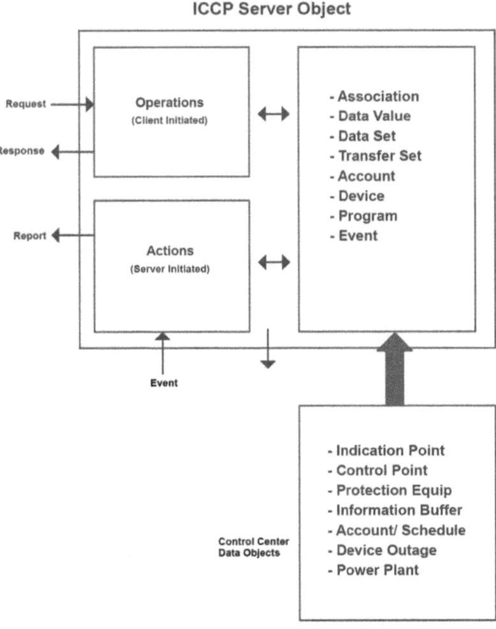

that offers minimal capabilities for observing the data stream and prevents them from interacting directly with the system.

8.9   Types of Inter-Control Center Communications Protocol

The Inter-Control Center Communications Protocol (ICCP), also known as Inter-Control Center Communications Protocol, is a set of communication protocols that are standardized in the electric utility sector to exchange real-time data between control centers. There are many instances of usage in the industry of many different Inter-Control Center Communications Protocol protocols.

1. **Server Items:**

   a. **Definition**: These items are defined by IEC 60870-6-503.
   b. **Characteristics**:

      a. Client-initiated operations or server-initiated actions, such as reacting to breaker status changes, are considered Server Objects.
      b. Implementation of the Inter-Control Center Communications Protocol protocol requires the use of internal objects.

2. **Data Objects:**

  a. **Definition**: These items are defined by IEC 60870-6-802.
  b. **Characteristics**:

    a. Data objects encompass control messages, status information, analog values, schedules, text, and basic files.
    b. New data objects may be defined as required.
    c. Data objects are external and are not mandatory for Inter-Control Center Communications Protocol implementation.

8.10 Security System of Inter-Control Center Communications Protocol

Inter-Control Center Communications Protocol boasts robust security measures. In addition to peer entity authentication, data integrity, and data secrecy, Inter-Control Center Communications Protocol governs connection establishment and individual object operation permissions. These enhanced security measures are provided by Bilateral Tables, which are Bilateral Agreements established for distant control centers linked to the central control center. These tables secure relationships and object actions against unauthorized access. With its robust security features, Inter-Control Center Communications Protocol is a reliable and secure option for important control center communications, particularly in high-security contexts.

8.11 Distributed Network Protocol

General Electric was the original developer of the Distributed Network Protocol 3 telecommunications standard, which was published in 1993. The Distributed Network Protocol Users Group then took control of it. Distributed Network Protocol 3 was first designed for Supervisory Control and Data Acquisition, but it is currently utilized in many different fields worldwide, including security, oil and gas, water infrastructure, power, and more. A Fig. 8.4 is taken from [97].

Figure 8.4 illustrates Distributed Network Protocol 3, which includes Supervisory Control and Data Acquisition Master Station/Control Center, communication lines, distant substations, intelligent devices, actuators, radios, and meters. Radio, fiber optics, twisted-pair, and data transmission rates are shown. This also shows Programmable Logic Controllers and Human-Machine Interface/Supervisory Control and Data Acquisition interfaces for process monitoring and control.

Distributed Network Protocol 3 is responsible for sending transient information payloads and makes sure that each one reliably and predictably sends the same

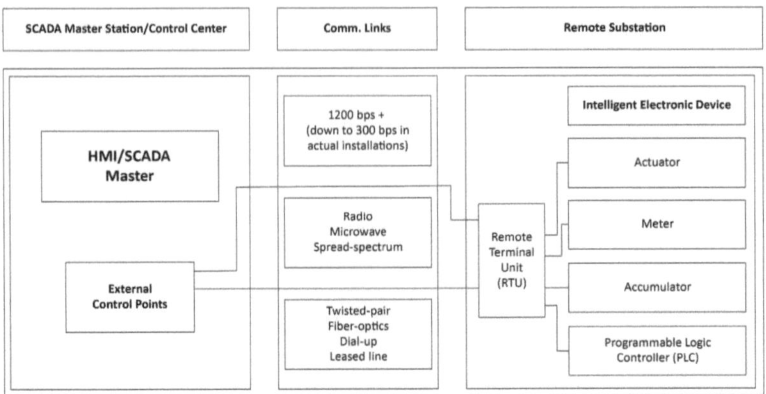

Fig. 8.4 Distributed Network Protocol 3

quantity of little data packets. Application data might vary in size and is always communicated in an organized manner. Information is separated into fragments, also known as Application Protocol Data Units, which can contain up to 2048 bytes apiece. Error control is enhanced by segmentation. To fit into the Data Link frame, Transport Protocol Data Units are 250 bytes in size. It is marked with a Frame (Link Protocol Data Unit, or LPDU), which is created by adding CRC bytes and a link layer header. The LPDU is transmitted to this physical layer. It contains one additional start/stop bit in addition to eight data bits. Distributed Network Protocol 3 file transfer is dependable and effective when this methodical methodology is used.

8.12 Structure of Distributed Network Protocol

In order to improve range, Distributed Network Protocol 3 now uses Ethernet and the Internet Protocol. Originally, it used RS232, RS422, and RS485, which could be used in many contexts. Transport layer frames of the Internet Protocol Suite encapsulate DNP3 data-link layer packets. Three layers are necessary for error detection and distributed network protocol addressing: Ethernet local area networks, User Datagram Protocol, Internet Protocol, and Transmission Control Protocol. UDP is used for LANs or regular broadcasts, whereas TCP is used for WANs or LAN + WAN. Distributed Network Protocol 3 can be hierarchical, polled, report by exception, multi-server, multiple-master, or peer-to-peer. A Fig. 8.5 is from [106].

Figure 8.5 displays a technical layout for ASDU, APDU, APCI, TSDU, TPDU, LPDU, and LH data transmission components, including data sizes, bits, and transmission architectures. This highlights the technological communication or networking system.

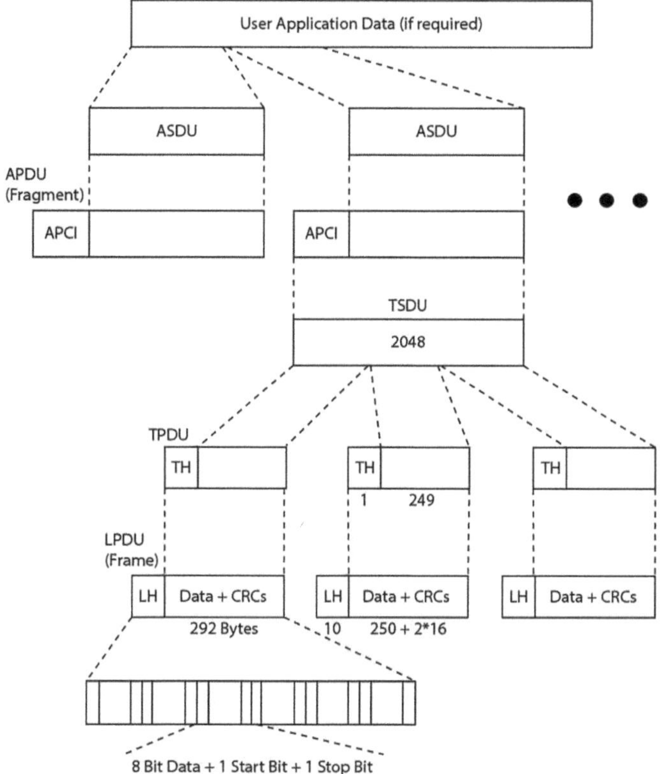

Fig. 8.5 Structure of DN3 protocol

8.13   Ethernet for Control Automation Technology

EtherCAT is used by the industrial Ethernet system to connect automation sector field devices. It is set up to provide instantaneous and efficient communication. EtherCAT-based hardware-based synchronous motion with the quickest, most accurate robotic and motion control systems. Due to serious flaws in EtherCAT authentication and encryption, lower-level packet delays and packet injections are possible. It lowers these risks by providing intrusion detection in the industrial network. Figure 8.6 is from [141].

Figure 8.6 represents a simple computer network with an Ethernet switch, 100BaseT hub, router, server, and gateway-connected Ethernet network.

It's adaptable network topology options make it a cutting-edge solution for large and complicated networks. An EtherCAT network, of course, has a single master device that controls the network and the initial slave device. Since all slave devices have at least two ports, they can usually be connected.

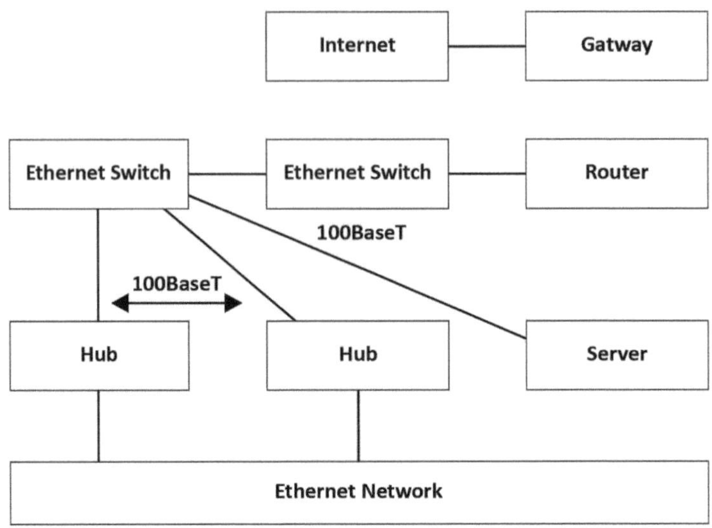

Fig. 8.6 EtherCAT

8.14 Open Platform Communication

OPC standards are important to industrial Windows-based control systems. It enables the intercommunication of data between industrial software programs and devices providing a universal protocol creating compatibility. OPC enables integration of multiple industrial systems and ensures seamless operation between hardware and software, regardless of the manufacturer. Figure 8.7 is from [13]. However, despite being versatile, OPC can also become a security vulnerability if not configured and protected appropriately. Hardened OPC Servers, as OPC-based systems, must be secured. This method would require tight security perimeters to prevent unauthorized access and communication by intruders. Companies can thus protect their control systems while exploiting the full potential of their OPC installations.

Figure 8.7 represents a system architecture with OPC client applications, servers, and hardware devices. OPC client applications link to OPC servers, which connect to hardware devices. Interfaces on hardware devices include USB, Ethernet, and RS-232.

8.15 Common Industrial Protocol

In industrial environments, this enables peer-to-peer communication between supervisory control systems and field-level equipment. Numerous industries, particularly the manufacturing and process sectors, use CIP in their automation and control sys-

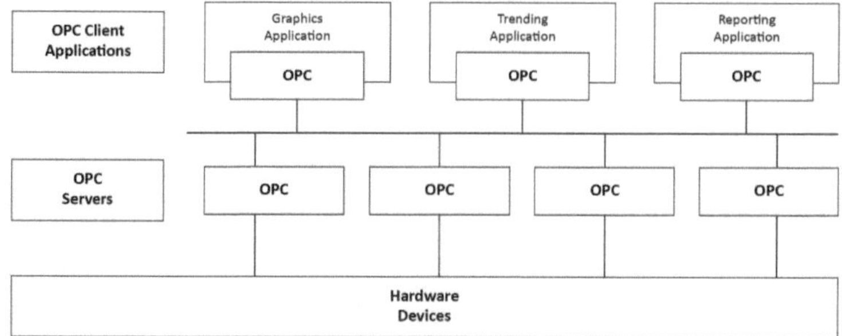

Fig. 8.7 OPC architecture

Fig. 8.8 Common industrial
protocol

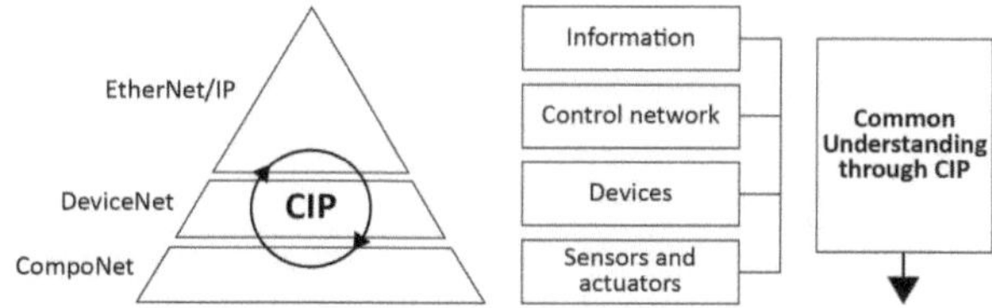

tems. This facilitates direct, end-to-end data interchange between external systems and almost any kind of field device, including sensors, actuators, and controllers. This improves process visibility through monitoring, control, and optimization. CIP was developed with the ability to integrate with various industrial networks, including EtherNet/IP, DeviceNet, and ControlNet, and it functions with multiple network topologies. Since it provides a common method of communication between various equipment types and genres, this is a crucial component in achieving interoperability and system integration in complex industrial environments. CIP's peer-to-peer connectivity facilitates direct device-to-device connection, which speeds up process automation and decision-making. Figure 8.8 is from [29].

Figure 8.8 illustrates the CIP within the context of industrial automation and control systems, including EtherNet/IP control networks, DeviceNet, and CompoNet device communication.

There are security dangers associated with CIP's advantages. Potential dangers include session hijacking, CIP data eavesdropping, and message alteration. Strong security is required for these systems. This included hashes, MACS (message authentication codes), digital signatures, and more. These security methods assist industries in protecting their entire environment and communications pertaining to CIP.

8.16   Security in SCADA Systems

SCADA and ICS are heavily used in whole range of critical businesses that provides society's essential services, e.g., energy, water, transportation, manufacturing, and others by monitoring and controlling the energy flow or service availability. These systems are essential to automate industrial processes and sit at the nexus of IT and OT. When referring to the large-scale industrial systems, SCADA systems receive and store data concurrently, including sensors as well as control devices that help in the timely decision-making by operators to effectively manage all operations.

Industrial Control Systems, including SCADA but also DCS and PLC to control continuous (SCADA, DCS) and controlled discrete (PLC) physical processes safely. The incorporation of SCADA and ICS right in the heart of critical infrastructure is inevitable for energy, water supply, manufacturing, and transportation, among other sectors that depend upon de facto availability, reliability, and smooth operation. Ensuring that they work together seamlessly is critical in order to keep these essential services up and running.

However, organizations face both opportunities and challenges as a result of the convergence of OT and IT with the rapid advancement of technology. On one hand, technology offers greater operational efficiency, real-time data analysis, and improved decision-making, enabling businesses to optimize processes, reduce costs, and increase productivity. On the other hand, as SCADA and ICS equipment become more remotely accessible and interconnected with other devices and networks, they are increasingly exposed to cyber threats, making them vulnerable to a wider range of security issues.

While improving efficiency and response time, this also opens up a whole new range of attack vectors for bad actors to exploit, allowing them to particularly target industrial systems with cyberattacks, including unauthorized access attempts or manipulation. The risk landscape for SCADA and ICS environments is becoming increasingly complex, which further complicates security measures. With an ever-growing list of new cyber threats that are more sophisticated every day, a multi-faceted approach is required to secure SCADA and ICS systems from both internal and external threats and protect our critical infrastructure as well as the integrity and continuity of their operations.

8.17   Physical System Vulnerabilities

The security of essential infrastructure and industry is at risk due to physical system vulnerabilities. Unauthorized access to field equipment, hardware, and software, communication networks, and OT/IT environments can offer serious risks, which can lead to a variety of hazards. Below is a summary of these vulnerabilities and recommended security measures:

1. **Unauthorized Access to Field Devices:**

   a. **Vulnerability:** Unauthorized users who have physical access to field devices are vulnerable.
   b. **Risk:** Weak security in field devices can lead to unauthorized tampering or disruption.
   c. **Solution:** Implement mandatory authentication and authorization through proper access control policies, limiting physical access to authorized personnel.

2. **Lack of Security Hardening and Open Ports:**

   a. **Vulnerability:** Security holes and unprotected network ports in hardware and communication networks.
   b. **Risk:** Malicious insiders may exploit insecure ports and devices.
   c. **Solution:** Implement security hardening measures and close unnecessary open ports to prevent unauthorized access and malware uploads.

3. **Lack of Security Perimeters:**

   a. **Vulnerability:** Lack of physical security perimeters in hardware and communication networks.
   b. **Risk:** Physical security breaches can allow unauthorized trespassing.
   c. **Solution:** Establish safe boundaries between software and hardware (e.g., firewalls to prohibit unwanted access).

8.18 System Security

System security is the technical and process-oriented safeguards that ensure that a system is secure from any possible harm, unauthorized access, or loss of data. Its function is to ensure all digital resources are kept confidential, available at all times, and their integrity is intact with various methods and approaches. Intrusion detection, encryption, firewall deployment, and antivirus software (among others) act as defenses. In a highly wired world that is more vulnerable than ever to digital predators, system security stands as an active sentinel protecting the computer-reliant society from unwelcome systems disruptions.

Cyberattacks

There are frequent cyberattacks against supervisory systems, the most dangerous of which is malware and ransomware. When used to breach critical systems, these malicious tools will disrupt regular operations and expose sensitive data, as cybercriminals frequently do. And even worse is the issue of zero-day vulnerabilities, which means that attackers are attacking systems before security patches have been released, leaving many with unauthorized access and ruined systems. Increasingly sophisticated attacks and changes in threat profiles highlight the importance of robust cybersecurity measures to prevent supervisory systems from being compromised by crippling attacks. Cyberattacks can manifest in various ways:

1. **Unauthorized Entry**

   Unauthorized entry poses a significant threat to control systems security [5]. Key concerns include:

   a. **Weak Authentication:** Poor authentication mechanisms allow attackers to access control systems using default or improperly managed credentials, leading to unauthorized access and manipulation of critical systems.
   b. **Insufficient Access Controls:** Inadequate access controls may lead to privilege escalation, allowing attackers to gain higher levels of system control than authorized. This can compromise system integrity and security.

   Addressing these vulnerabilities requires robust authentication mechanisms and well-defined access control policies to prevent unauthorized entry and safeguard control systems.

2. **Data Handling**

   Effective data handling is crucial for maintaining the security and integrity of systems and processes. Key concerns include:

   a. **Data Falsification:** Cybercriminals may attempt to manipulate data, leading to incorrect decisions and potentially disastrous consequences. Measures must be implemented to detect and prevent data tampering.
   b. **Data Theft:** Company procedures and consumer data are prime targets for cybercriminals. Unauthorized access to sensitive data can result in theft, misuse, financial losses, and reputational damage.
   c. **Protective Measures:** Organizations should implement strong data integrity checks, access restrictions, encryption protocols, regular audits, employee training programs, and a robust incident response strategy. These measures help mitigate the risks associated with data manipulation and theft, ensuring the reliability and trustworthiness of supervisory systems.

   By addressing these concerns proactively, organizations can strengthen their systems against data breaches.

3. **Physical Attacks**

   Physical attacks on supervisory systems pose severe risks to operations and finances. Key considerations include:

   a. **Disruption of Operations:** Physical attacks can cause operational interruptions, leading to financial losses for organizations.
   b. **Methods of Attack:** Attackers may exploit vulnerabilities by hacking devices or cutting critical wires to disrupt operations and damage equipment.
   c. **Consequences:** Theft of equipment or data resulting from physical attacks can further impact business continuity and financial stability.

d. **Protective Measures:** To protect vital infrastructure and industrial processes from physical attacks, organizations should prioritize security through surveillance systems, access controls, and perimeter defenses. These measures can help deter attackers and mitigate the impact of physical threats.

A proactive approach to physical security can reduce operational risks and safeguard organizational assets.

4. **Insider Threats**

Insider threats in supervisory systems can stem from both intentional and accidental actions. Key concerns include:

a. **Intentional or Accidental Actions:** Insider attacks may be deliberate, with malicious intent, or accidental, due to negligence or ignorance of security protocols.

b. **Damage to Critical Infrastructure:** Security breaches caused by insider threats can inflict significant harm on vital infrastructure, compromising operations and leading to disruptions.

c. **Risk from Trusted Individuals:** Even trusted employees or contractors may pose security risks, either intentionally or unintentionally, highlighting the need for robust security protocols and ongoing personnel training.

By implementing comprehensive security measures and continuous training, organizations can mitigate insider threats and protect their supervisory systems.

5. **Security Implications**

Security breaches in critical infrastructure systems can have far-reaching consequences. Key implications include:

a. **Operational Disruption:** Attacks on critical services can lead to downtime, financial losses, and potential risks to public safety.

b. **Environmental Damage:** Industries such as energy and water treatment are vulnerable to cyberattacks that may cause pollution or other ecological hazards.

c. **Financial Losses:** The costs associated with cyberattack investigations, system repairs, and legal penalties can be substantial.

d. **Loss of Trust:** Security breaches can damage public trust and harm the reputation of critical infrastructure organizations.

e. **Legal Consequences:** Violations of security regulations may result in severe fines, lawsuits, and legal actions.

Mitigating these security implications requires robust cybersecurity measures, proactive risk management, and continuous vigilance to protect critical infrastructure from evolving threats.

8.19 Types of Cybersecurity Risks to SCADA Systems

SCADA systems control and monitor critical infrastructure across sectors such as energy, water, transportation, and manufacturing. As these systems become more interconnected, they are increasingly vulnerable to a variety of cybersecurity threats [125]. Cyber attackers may target SCADA systems to disrupt operations, steal sensitive data, or cause physical damage to infrastructure. Given their critical role, ensuring the security of SCADA systems is paramount to prevent breaches that could compromise public safety, operational continuity, and national security. Common cybersecurity concerns for SCADA systems include threats such as unauthorized access, malware attacks, denial of service, and vulnerabilities within legacy systems that have not been adequately updated or secured. A list of these concerns is provided in Table 8.3.

8.20 Supply Chain Vulnerabilities

The supervisory system's supply chain encompasses various components, including hardware, software, manufacturing, distribution, and installation. However, this chain is susceptible to several vulnerabilities, including:

1. **Malware-Infected Hardware:** PLCs and sensors may become infected with malware during manufacturing or delivery. Once integrated into supervisory systems, this malware can compromise their functionality and security.
2. **Tainted Software:** Hackers may tamper with firmware upgrades and drivers before their deployment in supervisory systems. Such modifications introduce vulnerabilities that enable unauthorized access to sensitive data.
3. **Backdoored Devices:** Attackers may embed hidden weaknesses in components used in supervisory systems. These backdoors provide covert access to critical infrastructure, potentially compromising system integrity and operations.
4. **Counterfeit Components:** Substandard or counterfeit components may infiltrate the supply chain, leading to insecure integrations within supervisory system

Table 8.3 Cybersecurity risks to SCADA systems

Risk	Description
Malware	Attacking and controlling data via malware
DoS attacks	System crashes from network overload
Data theft	Unauthorized access to confidential data
MITM attacks	Intercepting data for access [79]
Phishing	False emails to steal data
Buffer overflow attacks	Accessing memory buffer without authorization

equipment. Such components can cause malfunctions and compromise overall system security.

To reduce these supply chain risks, it's important to have strong risk management plans, good quality checks, and work closely with reliable suppliers to keep the supervisory system parts safe and secure.

8.21 Mitigation of Supply Chain Vulnerabilities

We recommend the following measures to address supply chain vulnerabilities in supervisory systems:

1. **Assessing Suppliers and Manufacturers:** Evaluate suppliers and manufacturers for compliance with security and quality standards to ensure that integrated components meet the required security and reliability levels.
2. **Secure Distribution Channels:** Utilize secure distribution channels to prevent tampering with hardware and software during transportation. Implementing robust encryption and authentication mechanisms safeguards components from unauthorized modifications.
3. **Testing and Validation:** Conduct thorough testing and validation of components before integration into supervisory systems. This process helps identify abnormalities and security risks early, enabling timely remediation.
4. **Digital Signatures and Hash Checks:** Implement digital signatures and hash checks to verify the authenticity and integrity of software packages and updates. This ensures that only authorized and unaltered software is installed on supervisory systems.
5. **Regular Auditing:** Perform regular audits of the supply chain and associated procedures to identify and mitigate vulnerabilities. A proactive approach prevents security breaches and ensures the ongoing integrity of supervisory system components.

By implementing these mitigation measures, organizations can enhance the security and resilience of their supervisory systems against supply chain vulnerabilities.

8.22 Legacy Supervisory Systems

Legacy supervisory systems pose significant security challenges due to outdated technology and the absence of modern security features. Key issues include:

1. **Outdated Technology:** Legacy systems lack contemporary security protections, making them vulnerable to cyber threats and attacks. They often use obsolete software and hardware components that vendors no longer support, increasing exploitation risks.

2. **Unsupported Software:** These systems may rely on outdated operating systems or applications that no longer receive security updates, heightening their exposure to cyber attacks.
3. **Incompatibility:** Due to outdated architecture, legacy systems may struggle with encryption and access control. Integrating modern security measures can be challenging and costly.
4. **Limited Visibility:** Older supervisory systems often lack sufficient monitoring and logging capabilities, making it difficult to detect and respond to security incidents, leading to delayed threat detection.
5. **Obsolete Hardware:** Upgrading legacy hardware is expensive and resource-intensive. Consequently, these systems may not support modern security features or technologies.

To reduce security risks in old supervisory systems, it's important to plan carefully and invest in updates. This will help keep important infrastructure safe and strong.

8.23   Security Threats

System security is the process of protecting digital assets, including computer systems, networks, and data, from potential harm, unauthorized access, and cyber threats. Its primary goal is to ensure the confidentiality, availability, and integrity of digital resources through various methods and approaches. These include intrusion detection, encryption, firewall deployment, antivirus software, and access control. In an era of increasing digital vulnerability, system security acts as a vigilant safeguard, ensuring the reliability and smooth operation of digital environments.

Types of Security Threats
Organizations and individuals face numerous security threats in the digital age. To mitigate risks, continuous monitoring and effective cybersecurity measures are essential. Below are different types of security threats:

1. Virus
2. Worms
3. Trojan Horse
4. Logic Bombs
5. Trapdoors (Backdoors)
6. Spoofing
7. Email Viruses
8. Macro Viruses

8.23.1 Virus

In the realm of computer security, a virus is a type of malicious software (malware) designed to replicate itself and spread across systems or devices. Infected systems may suffer from file corruption, data theft, or operational disruption. Below are key characteristics of viruses:

1. **Behavior:** A virus attaches itself to executable files or documents, disguising itself as legitimate software. It spreads similarly to a biological virus.
2. **Propagation:** Viruses often require human action to spread, typically through infected email attachments, compromised file sharing, or downloads from untrusted sources.
3. **Payload:** Once activated, viruses can execute various malicious actions, including file alteration, data theft, or system damage.
4. **Examples:** Notable viruses that caused significant damage in the early days of the internet include Melissa and ILOVEYOU.

8.23.2 Worms

A worm is a type of malware specifically designed to autonomously replicate itself and spread to other computers or devices across a network. Unlike viruses, worms do not require human intervention or the execution of host software to propagate [151]. Below are some fundamental characteristics of worms:

1. **Behavior:** Worms are a class of malware that can autonomously spread across computers and networks. Unlike viruses, they do not require a file to attach to in order to spread rapidly.
2. **Propagation:** Worms exploit vulnerabilities in computer networks to spread without user assistance, often depleting resources and slowing down network performance.
3. **Payload:** Worms can perform various malicious activities, including installing additional malware, creating backdoors for unauthorized access, and launching Distributed Denial-of-Service (DDoS) attacks [33].
4. **Examples:** Notable worms include the WannaCry ransomware worm and the Conficker worm [149].

8.23.3 Trojan Horse

A Trojan horse, or simply a "Trojan," is a type of malware that disguises itself as legitimate software or files to deceive users and gain unauthorized access to their systems. Unlike viruses and worms, Trojans do not have the capability to self-replicate. Instead, they rely on social engineering tactics to trick users into executing them. Below are some essential attributes of Trojan horses:

1. **Behavior:** A Trojan horse is a type of malware that masquerades as legitimate or desirable software, drawing its name from the legendary wooden horse. Users unknowingly install or execute it, often leading to harmful consequences.
2. **Payload:** Trojans can facilitate unauthorized access, steal sensitive data, or cause system damage.
3. **Examples:** Banking Trojans like Zeus and TrickBot are notable examples.

8.23.4   Logic Bombs

In cybersecurity, the term "logic bomb" refers to a type of malicious code designed to trigger and execute under specific conditions, such as on a particular date or after a predefined event. Below are key characteristics:

1. **Behavior:** A logic bomb is a piece of malicious code embedded within a system or program, remaining dormant until activated by a predefined trigger.
2. **Payload:** When activated, logic bombs can disrupt systems, delete data, or disable functionality.
3. **Examples:** A "time bomb" is a variant designed to activate on a specific date, potentially causing data loss or system failures.

8.23.5   Trapdoors (Backdoors)

Trapdoors, also known as backdoors, are hidden access points intentionally placed in software, hardware, or networks to bypass standard authentication or security measures. They allow unauthorized access to systems or data, often for malicious purposes. Below are key characteristics:

1. **Behavior:** A backdoor is a deliberately concealed access mechanism left in a system or software, bypassing normal authentication or security protocols.
2. **Purpose:** While developers sometimes use backdoors for debugging and maintenance, attackers can exploit them to gain unauthorized access.
3. **Examples:** In 2015, a backdoor vulnerability in Juniper Networks allowed attackers to access VPN connections.

8.23.6 Spoofing

Spoofing is a deceptive technique in computer networking that involves forging or altering data to impersonate another entity or conceal one's identity. It manipulates data packets, IP addresses, or other network identifiers to mislead recipients or evade security systems. Below are key aspects of spoofing:

1. **Behavior:** Spoofing involves fabricating the source of data or communication to deceive the recipient.
2. **Types:** Common spoofing methods include email spoofing, website spoofing, and IP spoofing, where attackers create fake websites to mislead users.
3. **Examples:** Email spoofing is widely used in phishing attacks [80] to make fraudulent emails appear legitimate.

8.23.7 Email Virus

An email virus is a type of malware that spreads via email communication, typically through infected attachments or malicious links. Once activated, it can infect the user's system and potentially spread to contacts in their address book. Below are key characteristics:

1. **Behavior:** Email viruses propagate through infected attachments or links. When opened, they execute malicious code on the user's system.
2. **Payload:** These viruses can steal data, encrypt files (ransomware), or spread further by sending infected emails to additional contacts.
3. **Examples:** The Mydoom worm and the ILOVEYOU virus are well-known email-based malware.

8.23.8 Macro Virus

A macro virus is a type of malware written in a macro language, such as Visual Basic for Applications (VBA), and embedded in documents or files that support macros. These are the primary characteristics:

1. **Behavior:** Macro viruses target files that allow macros, such as Microsoft Office documents (Word, Excel, PowerPoint), and execute malicious actions when the file is opened.
2. **Propagation:** They spread through infected documents, often via email attachments or downloads.

3. **Payload:** Macro viruses can install additional malware, grant unauthorized access, or corrupt data.
4. **Examples:** The Melissa and W97M/Melissa viruses exploited Microsoft Office documents.

8.24   Firewall in Supervisory System Security

Firewalls are essential in protecting the supervisory control systems. They keep a safe distance between the outside world, trying to get in and out of the non-essential signals across network systems, using intelligent filtering, transparent data packets. Firwalls function in the sense that they check each data packet as per their defined rules and access control list to permit only authorized traffic.

Capabilities such as stateful inspection and comprehensive monitoring allow firewalls to prevent attacks more effectively. This allows firewalls to get advance notice about suspicious behavior and traffic patterns, and actively stop this behavior using intrusion detection and prevention systems (IDPS).

Network activity logging and auditing should be robust as well. These features enable efficient incident response, and detection of policy breaches that allow continuous auditing to maintain visibility and enforce responsibility.

Firewalls are typically deployed in high-availability and redundant configurations to provide continuous filtering. This arrangement makes sure that the supervisory control systems also stay secure and in operation even when there are technical failures or cyber threats.

8.25   Access Control

Access control is a key part of keeping management systems and sensitive data safe. It means allowing only the right people or systems to use important systems or data.

To build strong access control, the following main ideas should be followed:

1. **Strong Authentication:** Use strong methods to verify identity before giving access. This can include passwords, biometrics, or multi-factor authentication (like a password and a code sent to a phone).
2. **Clear Permission Levels:** Set different access levels for users based on their role or job. Give only the access they need, which helps lower the chance of security risks.
3. **Least Privilege Rule:** Give users or systems only the minimum access they need to do their work. This limits the chances of someone misusing the system.

By following these rules, access control becomes stronger, making the system more secure and less likely to be hacked or misused.

8.26 Cryptography and Encryption

Cryptography for Securing Data and Communication in Industrial Control System (ICS). It relies on mathematics to secure information from the likes of integrity, authenticity, and confidentiality. It is used to stop any form of unauthorized access that leads with sensitive operational data. This section goes into detail about cryptography in ICS, detailing the encryption methods, and some of these complexities, along with its common challenges and how it is specifically applied to certain industries.

Encryption Algorithms in Industrial Control Systems Today's ICS incorporates robust encryption standards for securing data at rest or over a network. Some popular encryption types include:

- **AES (Advanced Encryption Standard)**: A symmetric-key algorithm widely used to secure communication between programmable logic controllers (PLCs) and human-machine interfaces (HMIs). AES-256 is often used for its strong security and good performance.
- **RSA and ECC (Elliptic Curve Cryptography)**: These are asymmetric algorithms used for secure key exchange and digital signatures, especially in supervisory control and data acquisition (SCADA) systems.
- **Lightweight Cryptography**: Algorithms such as ChaCha20 or PRESENT are designed for devices with limited resources, like field sensors or embedded systems [138].

Key Management Challenges The success of encryption depends on how well the cryptographic keys are managed. Key management in ICS brings several challenges:

- **Legacy System Compatibility**: Older systems may not support modern key exchange methods, requiring customized or hybrid solutions.
- **Secure Storage**: Hardware security modules (HSMs) are used to keep keys safe, even in tough industrial environments.
- **Lifecycle Management**: Keys need to be rotated or updated regularly, and this must happen without interrupting critical operations.

Real-Time Constraints and Encryption Overhead Industrial networks must operate in real time, so encryption must be efficient:

- **Latency Sensitivity**: Encryption in protocols like OPC UA is designed to keep delays low, which is essential for tasks like robotic control.
- **Bandwidth Considerations**: Protocols such as Modbus Secure use compact encryption to avoid reducing data transmission speed.
- **Hardware Acceleration**: Many routers and gateways now include special hardware to handle encryption, reducing the load on the main system.

Use Cases in Industrial Environments

- **Secure Remote Access**: VPNs using IPsec/IKEv2 encryption allow safe access to distributed control systems (DCS) for remote maintenance.
- **Firmware Integrity**: Asymmetric encryption ensures that only trusted firmware can be installed on devices like PLCs.
- **Data Historian Protection**: Encryption at the column level in databases protects stored telemetry and process data.

Compliance and Standards Several regulations and standards require encryption in critical infrastructure:

- **NERC CIP**: Requires encryption of sensitive data in the bulk electric system.
- **IEC 62443**: Defines encryption and security guidelines for industrial automation and control systems.
- **GDPR**: Affects ICS that process EU citizens' data, requiring personal data encryption in manufacturing systems.

Emerging Trends and Future Directions

- **Quantum Resistance**: Preparing for future threats by adopting post-quantum algorithms like CRYSTALS-Kyber.
- **Zero-Trust Architectures**: Using end-to-end encryption and mutual TLS authentication between system components.
- **Secure Protocol Updates**: New industrial protocols like MQTT over TLS are being used in Industrial Internet of Things (IIoT) environments.

Risks of Improper Implementation Improper use of cryptography can create serious security issues in ICS:

- Using outdated encryption algorithms (e.g., DES) on legacy equipment
- Poor random number generation for keys on embedded devices
- Weaknesses in custom or proprietary encryption protocols

By using encryption strategies that match the needs of industrial environments, organizations can keep critical systems secure without hurting performance. As ICS becomes more connected with cloud services and edge computing, it's important to keep reviewing and improving cryptographic practices to deal with new security risks.

8.27 Security Challenges in Legacy Supervisory Systems

This system is designed with outdated systems and Infrastructure making security unique to the supervisory system. These should be addressed with careful planning and selection of security measures. Some of the most critical security concerns in legacy supervisory systems include:

1. **Isolation:** Legacy systems should be isolated from other networks to reduce the risk of unauthorized access and exploitation. Limiting external connectivity helps lower the attack surface.
2. **Virtual Patching:** Updating unsupported software in legacy systems can be difficult. Virtual patching techniques—such as intrusion detection and prevention systems—can help reduce risks by adding extra security controls to block known threats.
3. **Regular Assessment:** Perform regular security checks to find weaknesses and potential threats in legacy systems. Knowing about these vulnerabilities makes it easier to apply effective security measures.
4. **Modernization:** When possible, upgrade or replace legacy systems with more secure and supported technologies. These improvements strengthen security, boost performance, and ensure long-term system reliability.
5. **Security Awareness:** Provide training for system administrators to teach them best practices and risks related to older systems. Raising awareness helps them protect legacy systems better and respond effectively to security issues.

By addressing these security challenges early, organizations can improve the safety and strength of their legacy supervisory systems.

8.28 Summary

Their criticality importance to supervisory systems that control power and emergency (and other) infrastructure is obvious. He said the other safety-related risks included cyberattacks, physical vulnerabilities, outdated systems and insider threats including supply chain breaches. Last defense strategy additionally requires cybersecurity measures, physical security protocols as recently as solutions for legacy systems. Regular audits, security awareness, and cross sector collaboration is key to keeping up with emerging threats. Maintaining a vigil is key to protecting the heartbeat of both critical infrastructure and society today.

Chapter 9
Assessing Supervisory Systems Security Threats: Mitigating Sectoral Risks, Addressing Insider Threats, and Designing Human-Centric Security Solutions

Abstract This chapter provides an in-depth exploration of supervisory systems, focusing on risk assessments at both sectoral and entity levels and the identification of various risk types and sources. It emphasizes the importance of addressing cross-cutting issues such as consistency, technological adaptation, and interagency collaboration. The chapter covers critical aspects, including data privacy, cybersecurity resilience, transparency, and accountability (Chougule (2023) Ics cybersecurity resilience and the importance of remote laboratory, Retrieved from https://gca.isa.org/blog/ics-cybersecurity-resilience-and-the-importance-of-remote-laboratory). It also examines common supervisory frameworks and addresses potential threats such as insider attacks, vulnerabilities from third-party sources, and data breaches. Special attention is given to advanced persistent threats (APTs) and the necessity of security awareness training for supervisors. Readers will gain valuable insights into threat intelligence analysis and security measures, including real-time monitoring, threat data analysis, and the development of customized threat profiles.

Keywords Supervisory systems · Risk assessment · Cybersecurity resilience · Data privacy · Threat intelligence · Advanced persistent threats · Insider attacks · Third-party vulnerabilities · Security awareness training · APT

9.1 Introduction

Many security issues in today's interconnected digital world are closer to supervisory systems. In a variety of applications, such as banking, energy, and communication, they oversee vital infrastructure. The hazards stated will be examined in further depth in part 2 of this article, along with their potential effects on supervisory systems around the world.

However, those systems have frequently been proven to be deficient. That latter element highlights how urgently improved security is needed. The significance of the threat is illustrated by hundreds of actual cases, such as bank hacks and private

M. A. Rahman et al., *Securing Industrial Control Systems*,
https://doi.org/10.1007/978-3-032-03018-4_9

data breaches. Hacking of critical weapons systems can have catastrophic effects on confidence and finances.

The main dangers to oversight structures are described in this section of the article. It also describes the causes, mechanisms, and potential harm of these hazards. Stakeholders can detect vulnerabilities more quickly if they have a more thorough awareness of these threats. In a similar vein, they may react faster and virtually in real time, protecting the supervisory systems from intrusions.

9.2 Supervisor's Risk

Supervisors that are responsible for risk-based supervision should have a clear understanding of money laundering, terrorist financing, and proliferation financing risks in the businesses and organizations they supervise. General risk assessments help managers provide them with the necessary information on such risks. The higher risks are listed through their severity and impact, supervisors plan similarly hence transmit the valuable load over there.

A comprehensive risk assessment involves analysis of the risks and vulnerabilities, as well as potential consequences, associated with money laundering, terrorist financing, and proliferation financing. Complying with the Financial Action Task Force (FATF) recommendations ensures that the process is formal and stringent.

On October 2020, the FATF included PFR into regular ML-TF risk assessments. Those risks now have to be baked into risk-based decisions, and integral to supervision.

New guidance from the FATF will also look to enhance methods to detect and address proliferation financing risks. Being safe in the New Financial Era is going to require vigilance, adaptability and informed discernment on the part of monitors.

9.3 Sectoral and Entity-Level Risk Assessment

At the entity level, a sophisticated risk monitoring solution that is sectoral-focused for combating money-laundering and countering terrorist-financing provides a better insight on risks than any rule-based AML/CFT mechanism. Supervisors may assess the riskiness of specific industries, companies or business segments to form a more complete picture of risk. The specifics of the approach may be unique to the sector and different in various jurisdictions, depending on its characteristics (the number and type of entities) and on the supervisory level.

For instance, bank supervisors may decide to conduct separate reviews of every institution or they may correlate institutions to be reviewed together, considering their size and the risk of money laundering and terrorist financing they present. This is about aligning supervisory efforts more precisely and effectively. Similarly, supervisors of designated non-financial businesses and professions (DNFBPs) may wish to commence with sectoral reviews. Supervisors can improve the way they group and prioritize risks associated with entities by classifying them based on activities, structure, customer base, and geographic risks.

Sectoral risk analysis is a broader concept, and the manner of application to determine entity-specific risk comprehension will need some detailed structured methodology according to different types of entities as they face different threats. In some cases, people performing the role of company formation agents in the trust and company service provider sector would continue to pose a higher risk of money laundering or terrorist financing than other providers in this sector.

9.4   Entity-Level Risk Assessment

Risk assessment is not only an important component of an organization's internal control system, but also identifies its internal and external threats (Table 9.1).

9.5   Security Risks in the Supervisory Information Systems

In this digital world, supervisory information systems are often needed to monitor industries such as finance or healthcare and manufacturing. These systems gather and examine data in real-time to enable organizations meet certain rule-based conditions, thereby ensuring that they will be able to maintain proper operations. At the same time, however, they are also exposed to various cybersecurity risks that can affect their operations as well as steal privileged information. These include

Table 9.1 Key aspects of entity-level risk assessment

Aspect	Description
Sectoral assessment	Identify sector-specific AML/CFT control risks and weaknesses
	Group entities by subsector for risk assessment
Entity-level assessment	Assess individual entity ML/TF risk considering business type, size, client profile, and high-risk jurisdictions
Risk mitigation	Tailor supervisory engagement based on entity-level risk assessment
	Rate the quality of mitigation measures considering sectoral and entity-level inherent risks
Residual risk rating	Develop a risk matrix considering inherent ML/TF risks and the quality of AML/CFT mitigation
Aggregation of assessments	Identify common ML/TF risks by aggregating entity-level risk assessments at the sectoral level
Impact	Informs supervisory actions and prioritizes efforts
	Basis for sectoral and national regime improvements

risks such as malware, ransomware, phishing attacks, social engineering, insider threats, and vulnerabilities in software and hardware.

Malware and ransomware can block access to important data by locking it and asking for money to unlock it. Phishing and social engineering tricks can fool employees into giving away passwords or secret information. Insider threats are even harder to deal with because they come from people inside the organization, whether by mistake or on purpose. Also, hackers can take advantage of weaknesses in software and hardware to break into systems and steal or change data.

When critical or private information is exposed, it can cause big problems. Organizations may lose their good reputation, face fines, and lose the trust of their customers. This is even more serious in industries with strict rules and regulations. A data breach can affect not only the company but also its clients.

To lower these risks, organizations must take active steps. They should use multi-factor authentication, check their security regularly, and test their systems to find and fix weak spots before hackers can use them. Following important security rules, like the NIST Cybersecurity Framework or ISO/IEC 27001, is important to stay protected from new threats.

Similarly, learning from previous incidents through case studies also dramatically supports in escalating cybersecurity blueprints. Organizations can take preventative action by analyzing what happened before, both positively and negatively, to face anything that may potentially come at them. Remaining vigilant, informed about new threats and continuously enhancing security parameters are fundamental to safeguarding supervisory information systems and securing critical industries.

9.6 Types and Sources of Supervisor's Risk

Supervisors deal with different types of risks when making decisions. These risks can come from how we think, outside pressures, personal interests, or issues with technology. Understanding these risks helps supervisors make fairer and better decisions.

1. **Cognitive Biases**

   Cognitive biases are errors in thinking that can lead supervisors to make poor decisions. These errors happen because of how our brains work.

   Sources:

   a. **Confirmation Bias:** This happens when a supervisor looks for information that supports what they already believe, ignoring anything that disagrees.
   b. **Overconfidence Bias:** Sometimes, supervisors feel they know more than they do, which leads them to take unnecessary risks or ignore helpful advice.
   c. **Anchoring Bias:** This happens when supervisors place too much importance on the first piece of information they receive, even if it's wrong or outdated.

2. **External Pressures**

   External pressures are things outside the supervisor's control that can affect their decisions. These pressures can make decisions less fair.
   Sources:

   a. **Political Interference:** Sometimes, politicians or government groups try to influence supervisors to make decisions that benefit them.
   b. **Industry Lobbying:** Business groups may try to change rules to make things easier for their companies, even if it's not in the public's best interest.
   c. **Economic Interests:** Big companies or powerful people might pressure supervisors to make decisions that help them make more money, even if it harms others.

3. **Conflicts of Interest**

   Conflicts of interest happen when a supervisor's personal interests get in the way of doing their job fairly. This can affect how they make decisions.
   Sources:

   a. **Financial Interests:** If supervisors have money in the companies they oversee, they might make decisions that benefit them personally.
   b. **Personal Relationships:** If supervisors are friends with people in the companies they oversee, they might not be strict enough with them.

4. **Technology and Cyber Risks**

   Technology and cyber risks happen when problems with computers or the internet affect the work of supervisors.
   Sources:

   a. **Cyberattacks:** Hackers may try to attack supervisory systems, steal information, or cause disruptions.
   b. **Data Breaches:** If someone gains access to sensitive information without permission, it can cause serious problems for supervisors and the entire organization.

9.7 Cross-Cutting Issues in Supervisory Systems

Cross-cutting issues are big challenges that affect many parts of supervisory systems. These issues are important for effective monitoring because they apply to different industries and organizations. Solving these issues is crucial for making sure the supervisory system works well and is trustworthy.

1. **Consistency and Standardization**
2. **Adapting to New Technologies**
3. **Collaboration Between Agencies**
4. **Skilled Workforce and Expertise**

5. **Protecting Data and Keeping It Private**
6. **Staying Strong Against Cyber Threats**
7. **Being Agile and Flexible**
8. **Ensuring Transparency and Accountability**

9.8 Consistency and Standardization

The ICS must be utilized in a consistent and standardized manner to ensure its proper functioning. They ensure that engines and components from various parts of the system are reliable, compatible, and safe. It also refers to keeping everything stable and the same throughout all systems, protocols, and processes. Standardization is a need to follow certain rules and guidelines in order not to unfit together or fit for the handreich part, whatever what desired.

1. Making supervisory procedures, methods, and criteria the same across different sectors and organizations.
2. Inconsistencies can lead to some sectors getting unfair advantages or disadvantages, which can harm fairness and control.

9.9 Adaptability to Technological Advancements

Adaptive technology modifies existing technologies or devices to allow people to interact more easily with the technology. This is particularly useful for handicapped people, assisting them to work easily.

1. Supervisory systems will need to evolve to remain current with the new and ever-changing technology and digital advancements.
2. New technology spell new risks and issues for regulatory systems; it also means that regulatory systems must innovate to address these challenges.
3. It's important to monitor and manage risks quickly as technology changes, especially in industries like finance.

9.10 Inter-agency Collaboration

Inter-agency collaboration is coordinating or collaborating between two or more federal agencies, and also within parts of the same institution. True collaboration is a group of organizations coming together to produce greater public value than any one organization alone can achieve. It provides consolidated oversight of ICS within and across sectors to resolve common vulnerabilities that can affect critical infrastructure, including the power grid or water treatment plants.

Cross-Agency Coordination
- **Example:** For Instance, collaboration between the Department of Energy (DoE) and CISA in USA to exchange ICS-specific threat intelligence (like ransomware campaigns against SCADA systems).
- **Framework:** Adoption of ISA/IEC 62443 standards securing industrial automation and control systems.

Knowledge and Resource Sharing
- **Tools:** Industrial Cyber Threat Sharing Platforms (e.g., E-ISAC for energy sectors) enable real-time alerts about zero-day exploits in PLCs or HMIs [6].
- **Impact:** Joint incident response drills (e.g., simulated attacks on oil pipelines) improve recovery times for OT (Operational Technology) systems.

Risk Mitigation
- **Case Study:** After the Ukraine power grid hack (2015), agencies collaborated to develop air-gapped backups for ICS networks.
- **Challenge:** Harmonizing regulations like NERC CIP (North America) with the EU's NIS Directive for cross-border infrastructure.

9.11   Data Privacy and Confidentiality

Important safeguards to protect ICS must be data privacy and confidentiality [46]. This data generated must be protected to engender trust and orchestrate safe operations, for companies enabling functions in energy, transportation. In order to prevent things like sabotage or accidental misconfiguration, ICS needs personnel which are both knowledgeable about cybersecurity and industrial processes.

Specialized Training
- Certifications: GICSP (Global Industrial Cybersecurity Professional) or Certified SCADA Security Architect (CSSA) certify ICS-specific abilities.
- Gap: only 23% of industrial firms have dedicated OT security teams (SANS 2023), creating a clear need for workforce development.

Risk Assessment and Response
- Example: MITRE ATT&CK for ICS defines adversary behaviors (i.e., manipulating sensor data to evoke equipment failure)
- Tools: Nozomi Networks or Clarity to monitor OT network traffic anomalies.

Systemic Performance
- Case Study: A German automotive plant reduced downtime by 60% after training engineers in ICS patch management and intrusion detection

9.12   Cybersecurity Resilience

Cybersecurity resilience means being able to withstand, adapt, and recover from cyberattacks while keeping operations running safely and securely [30]. A proactive approach to cybersecurity helps minimize damage and maintain business continuity. Protecting ICS data (e.g., process parameters, equipment configurations) prevents industrial espionage and sabotage.

Regulatory Compliance
- Standards: NIST SP 800-82 guides encryption of ICS data at rest (e.g., historian databases) and in transit (e.g., MQTT protocols).
- Penalty: Unauthorized access to ICS blueprints could violate trade secrets laws (e.g., Defend Trade Secrets Act).

Trust-Building
- Example: Anonymized data sharing among utility companies to benchmark operational efficiency without exposing proprietary workflows.

Preventing Misuse
- Technique: Role-Based Access Control (RBAC) limits engineers' access to critical systems like turbine controllers or DCS (Distributed Control Systems).

9.13   Agility and Flexibility

Industrial Control Systems (ICS) global flexibility and agility are vital to maintain ICS strong, efficient and safe. These criteria allow ICS to adapt fast enough as new threats emerge. Strength (resilience) means that the system can continue to operate even under circumstances as drastic as Stuxnet or Triton, both of which targeted nuclear power plants and safety systems.

Proactive Defense
- **Redundancy:** Adding backup systems (like manual controls for valves) so the plant can keep running even if there's a cyberattack.
- **Segmentation:** Keeping the IT and OT networks separate (like using firewalls between office computers and machines) to stop hackers from moving around easily.

Incident Response
- **Example:** In 2021, after a ransomware attack, Colonial Pipeline used manual controls to get the system running again while fixing the problem.
- **Tools:** Systems like Dragos and Tenable.ot help find and fix security problems in ICS.

9.14 Transparency and Accountability

Transparency and accountability are all about being open and responsible. It builds confidence and controls the ICS. This flexibility extends also to the advent of new threats, including ransomware screen- (HMI) based attacks or vendor software hacking.

Adaptive Strategies
- **Example:** Protecting old ICS devices (like Siemens PLCs) by blocking attacks using tools that stop known tricks (virtual patching).

Regulatory Adaptation
- **Framework:** Updates to rules like the NIST Cybersecurity Framework (CSF) now include cloud-connected ICS and smart devices (Industrial IoT) [91].

Innovation
- **Case Study:** A chemical factory used "digital twins" (virtual models) to safely test how cyberattacks might affect their machines without risking real damage.

9.15 Common Supervisory Frameworks

Supervisory frameworks are used to control and secure ICS. They provide step-by-step guidance for monitoring, optimizing and securing industrial processes.

1. **Risk-Based Approach (RBA)**

   - Focuses supervision and resources on the most dangerous areas.
   - Makes sure resources are spent wisely.

2. **Three Lines of Defense Model**

   - Explains how risks are handled at three levels: daily work, risk management teams, and internal audits.
   - Helps companies follow rules better and manage risks properly.

3. **Twin Peaks Model**

   - Splits supervision into two parts: one protects customers, the other keeps the financial system stable.

4. **Consolidated Supervision**

   - Looks at the risk of a whole financial group, including its smaller companies.
   - Helps spot problems between different parts of a group.

5. **Supervisory Colleges**

- Groups where regulators from different places share information about big financial institutions.
- Helps to understand and better manage global risks.

6. **Financial Action Task Force (FATF) Standards**

- FATF which creates worldwide rules to fight money laundering and terrorism funding.
- Countries prioritize the most significant threats through a risk-based approach.

9.16   Insider Threats to Supervisory Systems

Insider threats happen when a current or former employee misuses their position of authority, such as when they have authorized access to a system within the company. They damage systems and steal data, among other things. This can only be avoided by implementing strict access controls, training employees, keeping an eye on usage, and conducting frequent security audits.

Example An unhappy engineer changes PLC programs to dangerously raise gas pressure in a pipeline.

Prevention Tools like Fortinet FortiSIEM can pick up on suspicious commands and shut them down before they escalate. External actors can even exploit trusted insiders through phishing or social engineering to gain control of their credentials. Hence, layers of prevention such as multi-factor authentication and security awareness training are essential ways to protect against these attacks.

Example A maintenance worker's email account is compromised, enabling an attacker to upload a malicious program to the SCADA server during scheduled maintenance activities.

Prevention Robust Strong password policies, social engineering simulations and network security enforcement can reduce the impact of compromised accounts. They add that regular vulnerability scanning of internal systems are also part-and-parcel in identifying and remediating threats before they can be taken advantage off.

9.17   Third-Party Vulnerabilities in ICS

Industrial Control Systems (ICSs) commonly rely on third-party software and tools. Therefore, if the system of one supplier is breached, it leads to a dangerous strategy being applied to industrial control systems. One of these examples is the SolarWinds cyberattack. That was the case in which attackers implanted malicious code into a software update. Hackers then compromised those networks by using the unsuspecting customers' unwitting installation of an update.

Another example is the erroneous software upgrades released for GE Simplicity, which is a popular ICS program. These upgrades looked quite genuine, but the code they distributed was dangerous. ICS operators installed the virus, they may have used it to infect and take over the systems that were running it.

Why This Is a Problem

They are widely used for managing critical services like power, water supply and industrial machines. A trusted software update that can be used to infiltrate could cause widespread mayhem from hackers. That could mean shutting down devices, breaking equipment or taking intellectual property. The age of systems is such that few to none may readily be updated, leaving many ICSes even more susceptible to this type of attack.

Mitigation

To reduce this risk, organizations can use a Software Bill of Materials (SBOM). An SBOM is a detailed list of all the components used in a piece of software, including code from third parties and open-source libraries. It helps in the following ways:

- Tracks the origin of each software component.
- Quickly identifies if any part has known security issues.
- Helps avoid or replace risky components before damage occurs.
- Supports faster responses during a cyberattack by showing which systems might be affected.

Using an SBOM improves transparency and makes it easier to manage and secure software, especially when many vendors are involved.

9.18 Dependency Risks

Monitoring systems often use tools from other companies (third-party tools). This creates risks, especially when these tools are very important for daily operations. If the company providing the tool has money problems or cannot support its tool properly, it can slow down or stop important work. If their technology fails or they stop offering the service, it can cause major problems, making it hard to do important monitoring jobs and hurting regulatory work. Also, if the third-party tool has security problems, hackers might attack, risking data accuracy and the privacy of sensitive information.

To manage these risks, companies should use tools from different vendors instead of depending on just one. They should carefully check a vendor's background before choosing them to make sure the tools are reliable and safe. Contracts should clearly say what services the vendor must provide and what will happen if services are interrupted. Companies should also keep checking the security of these tools regularly to fix any new problems quickly. Managing third-party risks is very important to keep monitoring systems safe, stable, and reliable over time.

9.19 Malicious Inclusions

Malicious inclusions are a serious danger to monitoring systems. Hackers or bad actors might hide harmful software (like malware or secret access doors) inside third-party tools while they are being built or shared. Once these hidden threats get into the system, they can steal private data, damage the system, or commit fraud. Such attacks can cause huge financial losses and damage trust in the system. It can be very hard for an organization to fix the damage and win back trust after such an attack.

To prevent malicious inclusions, companies must use strong cybersecurity practices. They should do regular security checks, watch for strange activities, and quickly block any threats they find. Good security helps protect systems from hidden dangers.

9.19.1 Malicious Inclusion Detection Script

```python
import re
from collections import import defaultdict

# Sample file contents
file_contents = [
    "This is a normal file content.",
    "This file contains a malicious signature: SIGNATURE_1234."
    ,
    "Another file content with no issues.",
    "Warning! Malicious code detected: SIGNATURE_5678.",
    "File contains a benign text."
]

# Known malicious signatures
malicious_signatures = [
    re.compile(r"SIGNATURE_1234"),
    re.compile(r"SIGNATURE_5678"),
    re.compile(r"SIGNATURE_9101")
]

# Detect malicious inclusions
def detect_malicious_inclusions(file_contents,
    malicious_signatures):
    detected_malicious = defaultdict(list)

    for i, content in enumerate(file_contents):
```

```python
25          for signature in malicious_signatures:
26              if signature.search(content):
27                  detected_malicious[signature.pattern].append((i
    , content))
28
29      return detected_malicious
30
31  # Main process
32  detected_malicious = detect_malicious_inclusions(file_contents,
        malicious_signatures)
33
34  # Output the detected malicious inclusions
35  if detected_malicious:
36      print("Detected Malicious Inclusions:")
37      for signature, entries in detected_malicious.items():
38          print(f"\nSignature: {signature}")
39          for entry in entries:
40              print(f" - File {entry[0]}: {entry[1]}")
41  else:
42      print("No malicious inclusions detected.")
```

Listing 9.1 Malicious inclusion detection

9.19.2 Example Input

```python
1  file_contents = [
2      "This is a normal file content.",
3      "This file contains a malicious signature: SIGNATURE_1234."
    ,
4      "Another file content with no issues.",
5      "Warning! Malicious code detected: SIGNATURE_5678.",
6      "File contains a benign text."
7  ]
```

Listing 9.2 Sample file contents

9.19.3 Known Malicious Signatures

```python
1  malicious_signatures = [
2      re.compile(r"SIGNATURE_1234"),
3      re.compile(r"SIGNATURE_5678"),
4      re.compile(r"SIGNATURE_9101")
5  ]
```

Listing 9.3 Malicious signatures

9.19.4 Example Output

```
Detected Malicious Inclusions:

Signature: SIGNATURE_1234
 - File 1: This file contains a malicious signature: SIGNATURE_1234.

Signature: SIGNATURE_5678
 - File 3: Warning! Malicious code detected: SIGNATURE_5678.
```

9.19.5 Explanation

- **Sample File Contents:** The script starts with a list of file contents, each representing a different file. Code is referred from: https://github.com/ Sunzidasiddique1/ICS.
- **Malicious Signatures:** Defines known malicious patterns using regular expressions.
- **Detection Function:** The `detect_malicious_inclusions` function scans each file for known signatures and records matches.
- **Main Process:** The script processes files, identifies malicious inclusions, and prints results.

Future Enhancements For more advanced detection, machine learning algorithms and sophisticated pattern-matching techniques can be integrated.

9.20 The Inadequacy of Security

When monitoring systems are constructed using third-party tools, there may be security flaws that can be taken advantage of. It's possible for sensitive information to be stolen, changed, or even hidden. Third-party tools frequently have security flaws because of poor design, a lack of funding, or a lack of in-depth expert knowledge that prevents them from making the right choices. Businesses affect systems that are left mainly and remain very vulnerable if they do not invest enough time or resources in these fixes. If this continues, it leads to data leaks, hacking, and losing power. Businesses must make sure that any external supplier they employ for their instruments is financially stable and has safe application procedures in place. The procedures that should be taken in terms of security must be explicit in the contract. Regular security inspections are necessary, and preventative actions are advised. Maintaining system security necessitates constant labor, from protecting networks to deciding who has access to data.

9.21 Advanced Persistent Threats in Supervision

Advanced persistent threats (APTs) are stealth cyberattacks usually conducted by state actors or criminal outfits that have all the skills at their disposal. Because these attacks lay silent for such an extended period, they would become quite dangerous to monitor. APTs have the capacity to swipe sensitive data and interfere in large scale operations with considerable financial ramifications, such as those in finance or critical infrastructure. Social Engineering and zero-day attacks, malware techniques—These types of hackers enter the system in stealth mode. Once they are inside, they can take or modify data with not a spark of it been noticed. However, to combat APTs businesses must not only create multi-layered defenses. They need to monitor their networks continually, instruct their employees on how to recognize scams, and have written plans on how to respond should an attack occur. It is important when dealing with these types of potentially deadly threats to be vigilant and act quickly.

9.22 Security Training and Awareness for Supervisors

Advanced security is not only a tech thing but It also relies heavily upon people, especially on managers. Supervisors are the leaders of the team and must provide support to employees to perform their daily tasks. They ensure others know and adhere to the security policies defined by the organization.

Why Supervisors Are Important
Supervisors are the link between the management and the staff. If they understand security well, they can teach others and make sure that rules are followed. But if they are not aware of security risks, they might miss problems or fail to take action in time. This is why it is very important to train supervisors properly.

What Supervisors Need to Learn
- How to spot suspicious activity, such as strange emails or unauthorized access [24].
- What to do if they think there is a security problem.
- How to teach and remind employees about safe practices.
- How to respond quickly and correctly during a security incident.

Why Regular Training Matters
Cyber threats change all the time, so one training session is not enough. The information must be kept up to date with the latest types of attacks and how they should be handled so that, supervisors can know new ways of their prevention. This training might be in the form of workshops, online courses, videos or just a couple of friendly reminders in a team meeting.

Building a Security-First Culture

Besides formal training, it's also important to create a strong workplace culture where everyone cares about security. Supervisors should lead by example. If they follow security rules and take threats seriously, other employees will do the same. They should also encourage open communication, so team members feel safe reporting anything suspicious without fear.

Benefits of Good Training and Awareness

- Fewer mistakes by employees that could lead to security problems.
- Faster detection of issues before they become serious.
- A safer and more prepared work environment.
- Stronger protection against cyberattacks, data leaks, and other risks.

Organizations can dramatically enhance their broader safety by providing supervisors with the correct knowledge and building a security-first mindset in them. But ultimately, people are the first and last line of defense in any cybersecurity strategy.

9.23 An In-Depth Guide to Supervisor Security Training and Awareness

Protecting an organization through supervisor security training, team supervisors need to establish, enforce, and reinforce security policies with their teams. Here is a simple guide for building a strong supervisor security training program.

1. **Assessing Training Needs:** Find out what supervisors already know and where they need help.
2. **Define Training Objectives:** Set clear goals based on company rules and common threats.
3. **Developing Customized Training Modules:** Create courses for supervisors covering topics like:

   a. Security policies and procedures
   b. How to spot threats (phishing, malware, social engineering)
   c. Protecting private data
   d. How to report and respond to security incidents
   e. Using security tools

4. **Interactive and Engaging Training:** Use real-life examples, role-play, and practice sessions.
5. **Continuous Education:** Offer regular workshops, online classes, and webinars.
6. **Hands-on Practice and Simulations:** Give supervisors real practice with security tools.
7. **Encourage Reporting and Communication:** Make it easy and normal for supervisors to report problems.

8. **Reward and Recognition:** Reward supervisors who take security seriously.
9. **Integration with Performance Reviews:** Include security behavior in yearly reviews.
10. **Cross-Functional Collaboration:** Work with other departments to share security knowledge.
11. **Crisis Preparedness Training:** Train supervisors on what to do during emergencies.
12. **Legal and Ethical Considerations:** Teach supervisors about laws and ethics in security.
13. **Regular Refresher Training:** Keep skills sharp with refresher classes.
14. **Evaluation and Metrics:** Measure success by tracking fewer incidents and better compliance.

9.24 Threat Intelligence and Analysis in Supervision

Supervisors must use threat intelligence to protect company assets and data [5]. Here's a guide for adding threat intelligence to supervision:

1. **Comprehending Threat Intelligence:** Understand what threat intelligence is.
2. **Real-Time Monitoring and Alerts:** Watch for threats in real time.
3. **Analyzing Threat Data:** Study threat reports to find risks.
4. **Integration into Decision-Making:** Use threat information to guide actions.
5. **Customized Threat Profiles:** Build threat profiles based on the company's needs.
6. **Prioritizing Threats:** Focus first on the biggest risks.
7. **Collaboration and Information Sharing:** Share threat information with others.
8. **Effective Incident Response:** Act fast when a threat is found.
9. **Legal and Ethical Considerations:** Follow the law and good ethics when handling threats.
10. **Evaluating and Improving Processes:** Regularly review and improve security actions.
11. **Training on Tools and Technologies:** Teach supervisors how to use threat detection tools.
12. **Adaptive Learning and Updates:** Keep updating knowledge as new threats appear.
13. **Performance Measurement:** Track how well security efforts are working.
14. **Cross-Departmental Training:** Train across teams for better teamwork.
15. **Promoting a Security-Centric Culture:** Make security part of everyday work life.

Threat intelligence [5] helps supervisors find dangers early and react quickly. It comes from government agencies, security companies, databases, and public sources. By using this information well, supervisors can spot weaknesses, block attacks, and keep important systems safe.

9.25 Open-Source Software Security in Supervisory Systems

Software is used extensively today as an open source because it is the best way to save costs and work in collaboration. Security is required to be a major concern while implementing with the Open Source Software in supervisory systems. The security is much better since the source code is open and thus bugs are easier to find, allowing to fix it immediately. Participating in open-source communities helps companies to receive new functionality and security information.

But this openness is double-edged while other parties can analyze the code and look for security vulnerabilities, too. It is therefore necessary to maintain all software on our computers up-to-date and through careful monitoring. Ideally, organizations should have a process in place to monitor and check for updates and vulnerabilities on a regular basis. A specialized team for OSS security can help detect problems early in the process.

Staying on top of updates and patches is essential to addressing known issues in a timely manner. Organizations are also taking unorthodox approaches and using software bills-of-materials, carrying out third-party audits of open source dependencies, managing dependencies carefully, training staff to look for security issues in the code they use, and meeting certain guidelines of good OSS security practice.

Detailed control and less costBy being alert and utilizing the security aspects of open-source tools supervisory systems can keep their vital processes along with data safe and that too at a cheaper means possible. Open-Source Software, in the right planning and attention scope, is an economic way to as secure as possible.

Figure 9.1 illustrates the Risks-Controls mapping system including steps like using peer review and quality gates, security-specific unit tests, vulnerability

Fig. 9.1 Risks-Controls mapping

scanning solutions, CI/CD pipelines, and managing open-source repositories to mitigate risks and ensure compliance.

9.26 Key Aspects of a Risk-Based Supervisory Framework

A risk-based supervisory framework helps financial regulators check and control companies based on how much risk they bring to the financial system. Here are the main parts of a risk-based supervisory framework:

1. **Complete Risk Check**

    a. Uses a strong method to understand the risks related to money laundering and terrorist financing.

2. **Matching Supervisory Actions with Risk**

    a. Adjusts how supervisors work so that financial services run smoothly while managing high money laundering or terrorism financing risks.

3. **Influencing Firm Behavior**

    a. Gives advice, watches actions, and enforces rules to make sure firms follow anti-money laundering (AML) and countering the financing of terrorism (CFT) laws.

4. **Flexible Risk Management**

    a. Keeps checking risks and quickly adjusts to new and growing threats.

5. **Enough Skills and Tools**

    a. Makes sure there are enough skilled people, tools, and other resources to manage and supervise risks well.

6. **Working Together**

    a. Supports teamwork between law enforcement, regulators, and supervisors.
    b. Improves the system by sharing important information, setting risk priorities together, and working on joint supervision efforts.

9.27 Summary

In conclusion, maintaining the integrity, stability, and resilience of financial markets depends critically on protecting supervisory systems from security risks. The wide range of dangers that supervisory frameworks face–from malware infiltration and cyberattacks to insider threats and data breaches–has been clarified. Proactive risk mitigation strategies and strong cybersecurity protocols are essential as financial authorities navigate the complexities of a connected digital world. Through cooperation, information exchange, and technological innovation, stakeholders can improve the security posture of supervisory systems and lessen the impact of any threats. Vigilance, adaptation, and resilience will be crucial in preserving the integrity and reliability of supervisory frameworks worldwide as the financial regulatory landscape continues to evolve.

Chapter 10
Controller Security Threats: Mitigating Advanced Persistent Threats (APTs), Enhancing Authentication, and Securing Control Architectures

Abstract This chapter provides an in-depth examination of protecting controllers within Industrial Control Systems (ICS) from various attacks. It details the roles and types of controllers, including network controllers, access control systems, and automotive control units, and highlights the potential consequences of a controller breach, such as data manipulation, service disruptions, financial losses, and national security risks. The chapter covers essential mitigation techniques, emphasizing the importance of securing domain controllers and Active Directory. Multi-factor authentication (MFA) is discussed as a critical mitigation strategy, alongside securing DNS for domain controllers (Bright Security (2024) Dns tunneling: Techniques, tools, and prevention). Additionally, the chapter explores security metrics to help readers assess and enhance their security posture. Emerging threats, such as zero-day exploits (Kaspersky, n,d., Zero-day exploit, Retrieved from https://www. kaspersky.com/resource-center/definitions/zero-day-exploit), ransomware attacks, and the role of AI/machine learning in both attack and defense, are examined. The chapter also addresses data privacy, compliance concerns, and the security of wireless controllers, while discussing potential future trends in controller security threats. By understanding these vulnerabilities and mitigation strategies, readers will be equipped to safeguard controllers and ensure the overall security of their ICS environments.

Keywords Controller security · Industrial control systems · Network controllers · Access control systems · Automotive control units · Multi-factor authentication · Domain controllers · Active directory · Security metrics · Zero-day exploits · Ransomware · AI in cybersecurity · Data privacy · Wireless controllers

10.1 Introduction

Controller security threats mean different types of dangers that can harm network or system controllers. These threats can cause big problems for Industrial Control Systems (ICS). Some examples are: hackers getting into controllers to change how systems work, malware infections that damage controllers, denial-of-service (DoS) attacks that stop controllers from working, employees misusing their access, attackers physically tampering with devices, wrong settings that make systems weak, leaking of secret information, weak passwords that are easy to break, and bugs in software that attackers can use.

These problems can break the safety, privacy, and working ability of controllers. As a result, systems may stop, data may get stolen, and people may be put at risk. To stop these dangers, we must use strong security steps. This includes using strong passwords, checking systems regularly, updating software on time, setting up devices correctly, and training workers to spot and report problems. A full security plan is very important to keep controllers safe and working properly.

10.2   Mitigating Controller Security Threats in ICS

The goal of "Mitigating Controller Security Threats in Industrial Control Systems" is to find out what dangers controllers face and how to protect them. Controllers are important parts of ICS, but they can easily be attacked in many ways, such as hackers breaking in, malware infections, bad use by employees, and wrong settings.

To protect controllers, we must use strong methods. These include using strong login methods like multi-factor authentication (MFA) and role-based access controls (RBAC) to limit who can access the system. It is also important to update and patch controller software often to fix known problems. Training employees helps them understand security risks and how to deal with them. New technologies like Artificial Intelligence (AI) and blockchain can also help [123]. AI can find strange behaviors early, and blockchain can safely keep records that cannot be changed. Using these methods together can make ICS much stronger against cyber threats [123].

10.3   Controllers in Cybersecurity

In cybersecurity, a controller can be a device, a program, or a person who watches over the security of a network or system. Their main job is to make sure data, systems, and resources are safe from hackers, attacks, or any kind of harm.

Controllers make sure that security rules are followed. They control who can access networks, check system activities, and quickly respond when something bad happens. In automatic systems, controllers can be machines or software like firewalls or intrusion detection systems (IDS). Human controllers, like security officers, are responsible for making sure security plans are carried out, checking for risks, and keeping systems protected. Without controllers, it would be much harder to keep networks and systems safe.

10.4 Controllers and Their Importance

Controllers are like the brain of automated and semi-automated systems. They are special hardware or software parts that manage machines, systems, and processes. Controllers take inputs, use set rules or logic to decide what to do, and then send outputs to control the system.

Controllers help systems react correctly to different situations. They turn commands and inputs into actions.

In industries and factories, PLCs and DCS are common. PLCs (Programmable Logic Controllers) control machines and processes on the factory floor. DCS (Distributed Control Systems) manages bigger and more complex operations that are spread out over many locations. In computer networks, routers, switches, and gateways act as controllers. They control how data moves between devices. They also keep the network safe by managing connections, controlling data flow, and applying security rules. Modern cars also use controllers called ECUs (Electronic Control Units). ECUs control airbags, emissions, braking systems, and engines. As cars get more complex, they need more controllers to stay safe and work well.

Controllers are very important because they help systems work quickly, correctly, and safely. They reduce human mistakes, improve safety with fail-safe systems, and boost the system's performance and efficiency.

Because controllers are so important, keeping them secure is critical. Protecting controllers from attacks or unauthorized access is necessary to keep systems safe and running properly. As technology grows and systems connect more, securing controllers becomes even more important for protecting important infrastructure.

10.5 Key Types of Controllers

In industrial automation and control systems (IACS), different types of controllers help run and manage operations. They are key to safety, reliability, and smooth work. Some main types of controllers are:

1. **Industrial Control Systems (ICS):** These systems monitor and control industrial processes in areas like manufacturing, energy, and transportation. Controllers help automate and manage the work.
2. **Network Controllers:** Routers, switches, and firewalls control the flow of data in networks and help keep connections safe.
3. **Access Control Systems:** These systems manage who can enter facilities or networks and control security settings.
4. **Automotive Control Units:** Electronic Control Units (ECUs) manage important parts of vehicles like engines, brakes, and airbags.

10.5.1   Importance of Controller Security

Controllers are very important for running automated and semi-automated systems. Keeping them secure is critical. Here are the main reasons why controller security matters:

1. **Privileged Access and Control:** Controllers often have high-level access to important systems and data. This makes them a big target for hackers [111].
2. **Potential for Big Cyberattacks:** If a controller is hacked, it can cause major problems like stealing data, stopping systems from working, or even damaging machines.
3. **Data Manipulation and Tampering:** Attackers might change settings or fake data, which can make systems act in unsafe or wrong ways.
4. **Service Disruptions and Downtime:** Hacked controllers can cause systems to stop working, hurting business work and slowing down productivity.
5. **Financial and Reputation Losses:** Companies can lose money and trust if attackers break into their controllers.
6. **National Security Risks:** Controllers used in important services like power, water, and transport can affect public safety if attacked.
7. **Interconnected Systems:** Many controllers are connected to other systems. A problem in one controller can spread and create bigger security issues.

It is very important to protect controllers to lower risks, keep systems running safely, protect important data, and ensure public safety is not harmed.

10.6   Privileged Access and Control

Controllers in automated systems have special access and control over important parts of the system. They monitor, check, and manage how machines work. Because of this critical role, hackers often try to attack controllers. If attackers gain control, they can stop services, change important data, or damage systems. These privileged accounts often have fewer restrictions, which makes them attractive targets for cybercriminals [111]. If not properly managed, even a small mistake can lead to big problems across the entire system. That's why it's important to limit who can use these accounts and monitor their activities regularly. Using multi-factor authentication, strong passwords, and access logs can help improve safety.

In many cases, controllers are connected to other networked devices. This makes it easier for attackers to move through the system once they get access. Therefore, organizations should use the principle of least privilege—only giving users the minimum access needed to do their job. Privileged accounts should also be reviewed often to check for any unnecessary access or unusual activity. Training staff on how to use these accounts safely is just as important as technical controls. When people understand the risks and responsibilities, they are less likely to make mistakes that

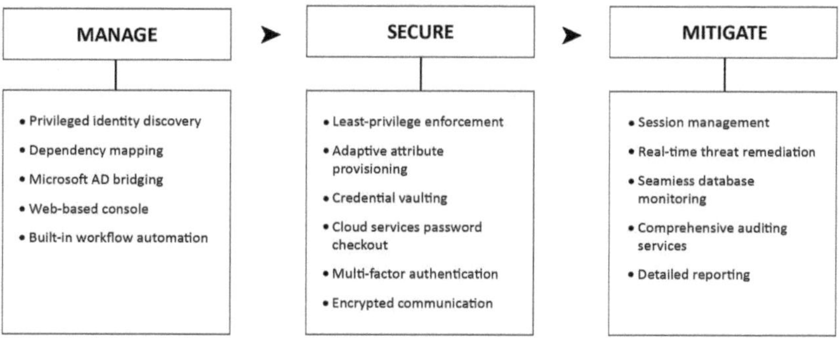

Fig. 10.1 Privileged access and control

lead to security problems. With the right mix of tools, policies, and education, organizations can better protect their controllers and reduce the chances of a serious cyberattack.

Protecting controllers from attacks and illegal access is very important to keep systems safe and working properly. A related figure is shown in Fig. 10.1, taken from [111].

Figure 10.1 illustrates the Privileged Access and Control system, including managing, securing, and mitigating identities. It provides some components and tools like least-privilege enforcement, session management, adaptive attributes, real-time threats, and tools like Microsoft AD bridging, credential vaulting, database management, and multi-factor authentication.

In automated systems, controllers play a very important role because they have special access and control over key parts of the system. They check data, monitor operations, and control machines. Because they are so important, controllers are often targeted by hackers.

If an attacker succeeds, they can take full control of the system or process. This can cause big problems like changing important data, stopping services, or even damaging important infrastructure. These attacks can have serious and wide-ranging effects. So, it is very important to protect controllers with strong defenses to stop attacks and block unauthorized access. Protecting controllers helps keep systems safe and working properly.

10.7 Potential for Devastating Cyberattacks

Today, technology is growing fast, and systems are very connected. This makes the risk of big cyberattacks even higher. Cyberattacks can badly hurt people, businesses, and even entire countries.

Several reasons make cyberattacks very dangerous:

1. If a controller is hacked, it can cause major damage.
2. Hackers can change important data, stop services, or damage physical systems.
3. In factories, a hacked controller can break machines, hurt workers, and cause big financial losses.
4. These risks show why companies must work hard to protect controllers from hackers and breaches.
5. Strong security helps keep controllers safe and ensures the system keeps running smoothly.

10.8   Data Manipulation and Tampering

Data manipulation and tampering mean changing data without permission to cheat, mislead, or gain unfair advantages. This can happen in many places like cybersecurity, financial systems, and data analysis.

Figure 10.2 represents the variety of physical and network security-related tampering attacks.

Below are some examples of data manipulation and tampering:

1. Weak controller security can lead to big changes and tampering of important data.
2. Controllers handle sensitive data and key system instructions.
3. If a controller is hacked, attackers can change this data in harmful ways.
4. In factories, this could mean changing production settings, which leads to faulty products.

Fig. 10.2 Data manipulation and tampering

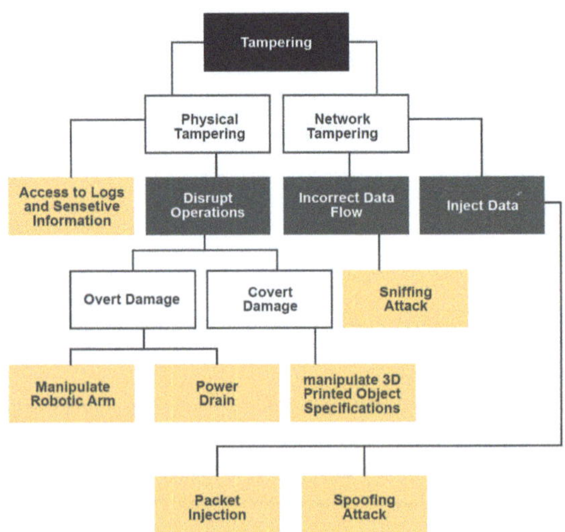

5. These problems hurt product quality, company reputation, and customer safety.
6. To protect data and stop unauthorized changes, controller security must be strong.
7. Good controller security builds trust in processes, products, and customer satisfaction.

10.9 Service Disruptions and Downtime

Interruptions and downtime in Industrial Control Systems (ICS) can cause serious problems, including losing money and creating safety risks. These disruptions can happen because of machine failures, cyberattacks, human mistakes, or natural disasters.

1. Controller attacks can cause service interruptions and downtime, especially in important areas like healthcare, banking, and utilities.
2. Controllers manage critical tasks, and an attack can stop services.
3. Service disruptions cause huge financial losses and put public safety at risk.
4. The risk of big disruptions shows why controller security must be a high priority.
5. Safe and steady services need strong controller protection.

10.10 Financial and Reputational Losses

Cyberattacks can cause financial and reputational losses that hurt companies and individuals badly [132]. Here's how it happens:

1. A controller security breach can lead to money loss and damage a company's reputation.
2. After an attack, companies must spend a lot on investigations and stronger security.
3. These costs can put pressure on a company's budget.
4. Losing money also damages the company's public image.
5. Security problems reduce customer trust and hurt brand value.
6. Over time, losing trust can drive customers away to competitors.
7. Building trust back takes time and a lot of money, so strong controller security is important to avoid these problems.

10.11 National Security Implications of Controllers

Controllers used in critical systems can have a big impact on national security. Here are some important points:

1. Weak controllers can threaten national security, especially in important areas like energy or defense.
2. Attackers could change or control defense systems or power grids.
3. A successful hack could stop important services or take over critical assets.
4. These breaches can cause big economic and social problems.
5. Strong protection, regular checks, and active defense methods are needed for controllers.
6. Protecting controllers is key to keeping the country's assets, systems, and people safe.

10.12   Controller Breach and Interconnected Systems

In today's world, many systems are connected to each other. If one controller is breached, it can create bigger problems across many systems.

1. Today's systems are highly connected.
2. Controllers play a big role in keeping these systems running.
3. If one controller is hacked, the problem can quickly spread to other connected systems.
4. This chain reaction makes it very hard to control the damage.
5. Controller security is critical because attacks can move across systems.
6. A breach can affect many industries, services, and even whole regions.
7. Securing controllers protects not just one system but the entire network, which keeps the digital world stable and safe.

10.13   Key Aspects of Controllers in Cybersecurity

Controllers are very important in cybersecurity. Their main tasks include:

1. **Access Control and Authorization:** Controllers allow only approved users to access important resources.
2. **Security Policy Enforcement:** They make sure that security rules, encryption, and other protections are properly followed.
3. **Monitoring and Detection:** Controllers watch network traffic, system activities, and user actions to find anything suspicious, using tools like IDS (Intrusion Detection Systems) and IPS (Intrusion Prevention Systems).
4. **Incident Response Coordination:** After a security attack, controllers manage the investigation, reduce the damage, fix the problems, and put steps in place to stop it from happening again.
5. **Security Compliance and Auditing:** Controllers check if security rules and industry standards are being followed and find ways to improve.

6. **Configuration and Patch Management:** They manage hardware and software settings, apply updates, and fix security weaknesses.

10.14 Controller Breach and Interconnected Systems

Today, many systems are linked together. Controllers are key parts of these connections. If a controller is attacked, the problem can quickly spread to other systems. This creates a chain reaction, making it hard to stop the attack.

- A controller breach can harm many systems at once—not just one company but entire industries or regions.
- Because systems are connected, a small problem can grow very big.
- That's why securing controllers is so important.
- Strong security protects not just one system but the whole network of systems.

To keep the digital world stable, we must take proactive steps to protect controllers and apply strong defense methods.

10.15 Roles and Responsibilities of Security Controllers

Security controllers have different jobs depending on the company and industry. But in general, their main responsibilities are:

1. **Access Control and Authorization:** Controllers decide who can see certain data or use certain systems. They make access rules to keep sensitive data safe.
2. **Security Policy Enforcement:** They make sure everyone follows the company's security rules, like how to set up devices safely and use encryption properly.
3. **Monitoring and Detection:** Controllers watch how users behave, how systems run, and how networks work. They use systems like IDS to find possible threats early [64].
4. **Incident Response Coordination:** If there's a cyberattack, controllers organize the response. They investigate what happened, fix the damage, and make changes to avoid it happening again.
5. **Security Compliance and Auditing:** Controllers check if the company is following cybersecurity laws and standards. They also look for areas that need improvement.
6. **Configuration and Patch Management:** They manage system settings and install updates to fix security gaps and keep systems strong.
7. **Cybersecurity Education and Training:** Controllers teach employees about safe practices, security rules, and how to spot risks. This helps build a strong security culture.

8. **Security Incident Handling:** Controllers set up clear steps for reporting, investigating, and solving security problems, following company rules.

10.16   Domain Controller

A domain controller is a very important server in a Windows network [84]. Manage user logins and decide who can access different parts of the network. Domain controllers are a key part of Microsoft Windows Active Directory. They help organize and manage users, computers, and network resources. Instead of each computer checking passwords on its own, the domain controller does it all in one place. This makes it easier for users to log in and for companies to apply security rules. Domain controllers also help monitor user activities and make it easier to check if security rules are being followed. They are especially important in industrial control systems (ICS), where they:

- Check the identity of operators and administrators.
- Control who can access important systems.
- Manage user accounts and their permissions.
- Help with security checks and following industry rules.

The main job of a domain controller is to keep the network safe and working properly. Whether it's in a regular office or an ICS environment, domain controllers make security better, simplify management, and help with meeting regulations.

10.16.1   Types of Domain Controllers

Domain controllers are important for network security and management. They handle centralized login and permission services for users, computers, and devices. They are responsible for authenticating user credentials and controlling access to resources across the network. Without domain controllers, managing large networks and ensuring security would be much more difficult. There are mainly two types of domain controllers, depending on their role and setup in a network: Read-Write Domain Controller (RWDC) and Read-Only Domain Controller (RODC).

- **Read-Write Domain Controller (RWDC):** This is the most common type of domain controller. It has full read and write access to the directory data, meaning it can make changes to the network's configuration and user permissions. When a user logs in or tries to access resources, the RWDC verifies their credentials and updates the network's directory as needed. RWDCs can also replicate changes made on other domain controllers to keep the network data synchronized.
- **Read-Only Domain Controller (RODC):** A Read-Only Domain Controller is used in situations where security is a concern, such as in remote offices or less

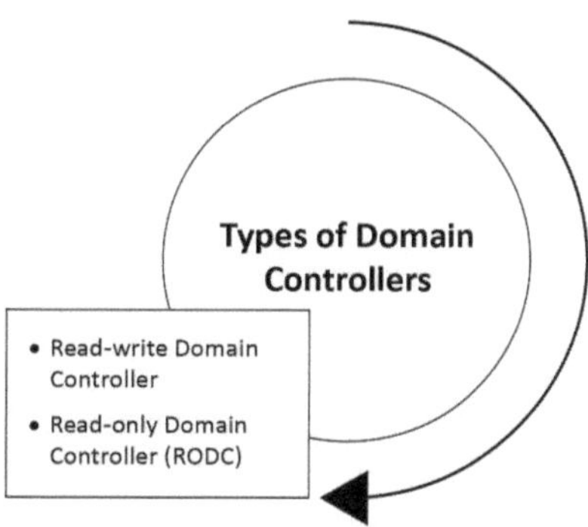

Fig. 10.3 Domain controller type

secure locations. Unlike RWDCs, RODCs can only read the directory data and cannot make changes. This means they cannot directly modify user permissions or settings, which adds an extra layer of security. However, RODCs can still authenticate users and provide access to network resources, making them useful in environments where users need to log in but changes to the network are rare or unnecessary. RODCs also replicate data from RWDCs, but only in a one-way direction, ensuring that sensitive information is protected.

Figure 10.3 illustrates different Domain Controllers, including a Read-write Domain Controller and a Read-only Domain Controller.

10.16.2 Domain Controller Work Process

A domain controller (DC) is like the network's gatekeeper [86]. It checks who can log in and who can use different resources.

When users log in, the DC checks their username and password using Active Directory. If the information is correct, the user gets access; if not, they are blocked.

Each DC has a copy of Active Directory to make the system faster and more reliable. DCs also help apply security rules across the network and protect user information.

Summary of Domain Controller
1. Users must log in to access the network.
2. DCs manage who can do what inside the network.
3. They apply security rules like password strength and access rights.

10.17   Securing Domain Controllers in Cloud Environments

In the cloud, security is shared:

- The cloud provider secures the cloud itself.
- The organization secures its data and apps inside the cloud.

  Important steps include:

1. **Identity and Access Management (IAM):** Set up strong IAM rules. Check access rights often and give users only the permissions they need.
2. **Network Encryption and Security:** Encrypt data when it is stored and when it is sent. Use firewalls and network security groups (NSGs) to control network traffic.
3. **Regular Auditing and Monitoring:** Set up logs and monitoring systems. Use cloud tools to keep an eye on the domain controller's activities [24].
4. **Regular Security Checks:** Perform tests to find and fix security problems in the cloud system [24].

10.18   Roles and Significance of Domain Control

Domain control helps keep a network safe, organized, and running smoothly. A domain controller checks users' identities and controls what they can access.

- It enforces rules such as password policies and data access limits.
- It tracks and manages user accounts and network resources.

  As a company grows, domain control also grows to support more users and technologies.

  Domain control also helps with disaster recovery by keeping backups and recovery plans ready. This helps reduce downtime if something bad happens.

  Finally, domain controllers help monitor network health and user actions. They can detect security problems early and respond quickly.

  In short, domain control is key to managing and protecting a company's network.

10.19   Primary Functions of Active Directory

Active Directory mainly helps manage identities and access in a Windows network. Its main functions are:

1. **Location of Resources:** Active Directory organizes things like printers, files, and users so they are easier to find.

2. **Centralized Security Management:** It helps apply the same security rules to all users, devices, and resources.
3. **Single Sign-On Access:** Users can log in once and then use many network resources without logging in again.

10.20 Multi-Factor Authentication (MFA)

MFA means "multi-factor authentication [47]." It asks users to prove their identity in more than one way before they can log in. For example, a user might need a password, a smartphone app code, and a fingerprint. This makes it much harder for attackers to break into accounts and improve security. Figure 10.4 is taken from [47].

Figure 10.4 illustrates MFA security control system. Security includes password, code from app or SMS, fingerprint, PIN, push notification, face scan, USB token, and iris scan.

10.20.1 MFA Mitigation of Controller Security Threats

Multi-factor authentication (MFA) is a well-known and strong way to stop unauthorized people from accessing important controllers in a network.

MFA asks users to prove who they are in more than one way:

- Something they know (like a password)
- Something they have (like a smartphone or token)
- Something they are (like a fingerprint or face scan)

Even if someone's password is stolen, MFA makes it harder for attackers because they also need the other proofs to log in. MFA protects against many common attacks like phishing, brute force attacks, and stolen login information. Since

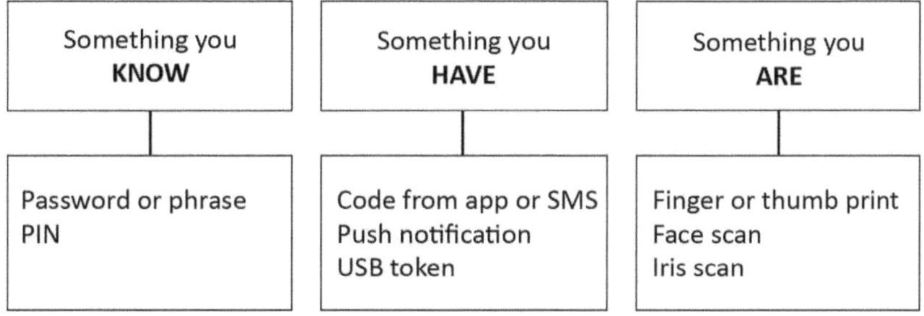

Fig. 10.4 MFA security control system

Table 10.1 Mitigation of security threats by multi-factor authentication (MFA)

Security threat	How MFA mitigates the threat
Credential theft	Even with stolen credentials, an attacker needs the second factor
Phishing attacks	Phished credentials are insufficient without the second factor
Brute force attacks	Even with a correct password, the attacker needs the second factor
Insider threats	Insiders would require an additional factor for unauthorized access
Unauthorized remote access	MFA adds a layer of security, requiring additional authentication

controllers are very important in industrial control systems, using MFA makes sure that only trusted people can change settings. This helps stop hackers or malware from damaging the system. By adding MFA to access control rules, companies can make their systems stronger, prevent break-ins, and better protect their industrial systems from both inside and outside attacks (Table 10.1).

10.20.2   Components of Multi-factor Authentication

Multi-factor authentication (MFA) makes systems safer by asking for more than one type of ID. It builds a stronger wall against people who should not get in. Here are the main parts:

1. **Knowledge Factor:** This means something to know, like a password or a PIN (Personal Identification Number). It is the first step to proving who you are. To stay safe, it is good to use strong passwords and change them often.
2. **Possession Factor:** This means something to have, like a smartphone, a special device (token), or a smart card when logging in.
3. **Inherence Factor:** This means something that uses body features like fingerprints, face, eyes, or voice. These are very hard to fake, so they make logging in even safer.

10.21   Implementation Steps for MFA on Domain Controllers

Adding MFA to domain controllers is very important to keep the network safe. Here are the steps to do it:

1. **Assessment and Planning:** Check the network and choose an MFA solution that works well with domain controllers [86].
2. **Selecting MFA Methods:** Pick how users will prove their identity—options include text codes, mobile apps, hardware tokens, or biometrics.
3. **Integration with Active Directory (AD):** Set up the MFA solution so it works smoothly with Active Directory.
4. **User Enrollment:** Help users sign up for MFA and connect their devices or biometrics.
5. **Testing and Rollout:** Test the MFA system first in a small area. If it works well, expand it to the whole organization.
6. **User Education and Training:** Teach users why MFA is important, how to use it, and help them with any problems.
7. **Ongoing Monitoring and Management:** Keep an eye on how MFA is working. Update the system when needed and manage user accounts properly.

10.22   Benefits of Implementing MFA for Domain Controllers

Using MFA for domain controllers makes the network much safer [84]. Here are some main benefits:

1. **Stronger Security:** MFA makes it harder for hackers. Even if they steal a password, they still need another proof, like a phone or a fingerprint.
2. **Less Risk from Phishing and Credential Theft:** If someone falls for a phishing email and gives away their password, MFA still protects the account by needing a second proof [24].
3. **Meets Security Rules:** Many laws and standards now ask for MFA. Using MFA helps to company stay compliant.
4. **Easy for Users:** Today's MFA tools are simple to use. Anyone can get a push notification, a code by SMS, or use fingerprint to log in.
5. **Flexible and Adaptable:** MFA systems can be set up in different ways to meet company's needs. Anyone can choose methods that are easiest for users and still meet security goals.

10.23   Securing DNS for Domain Controllers

Keeping DNS (Domain Name System) safe is key to keeping the network running well. Here are some steps to secure DNS:

a. **Update DNS Servers:** Always install updates and patches to fix security holes.
b. **Control Access:** Only allow trusted people and systems to use DNS servers. Use firewalls to control traffic.

c. **Use DNS Security Extensions (DNSSEC):** DNSSEC signs data to make sure it is real and not changed.
d. **Protect Against DDoS Attacks:** Use defenses like traffic filtering and limit requests to stop attacks.
e. **Use DNS Filtering and RPZ:** Block bad websites and set custom rules for DNS traffic.
f. **Log and Monitor DNS Activity:** Keep records of DNS actions and watch for strange behavior.
g. **Encrypt DNS Traffic:** Use DNS over TLS (DoT) or DNS over HTTPS (DoH) to hide DNS requests from attackers [23].
h. **Strong Authentication:** Use MFA and give users only the access they really need.
i. **Do Regular Security Checks:** Check DNS settings often and test security with experts.
j. **Backup and Disaster Plan:** Keep backup copies of DNS settings and make a plan to recover fast if something goes wrong.
k. **Train Staff:** Teach admins and users about DNS security and how to spot problems.

10.24   Security Metrics

Security metrics help to measure how safe the systems are [2]. They collect and analyze important data to find out to make improvements.

Metrics usually compare two or more sets of data over time. A measurement, on the other hand, shows something at one point in time.

In simple words:

- **Measurement:** A simple fact collected at a single time.
- **Metric:** A result to compare multiple measurements.

Good metrics come from good data, and they should always give the same results if measured by different people. This helps teams track the health of systems and make smart decisions.

10.24.1   *Importance of Cybersecurity Metrics*

Cybersecurity metrics help companies understand and improve their defenses [2]. A cybersecurity matrix uses these metrics to give a full view of how strong their protections are. Here are the key parts:

1. **Vulnerability Assessment Metrics:** Measure weak points in systems and networks, like old software or open ports.

2. **Attack Detection and Prevention Metrics:** Measure how well the company can spot and stop cyberattacks.
3. **Compliance Metrics:** Measure how well the company follows cybersecurity rules and policies.
4. **Performance Metrics:** Measure how well the security program is doing. Look at things like the number of incidents and costs.

10.25 Metrics Versus Measurements

Metrics are used to measure, track, and check how well a process, system, or activity is working. Metrics usually come from measurements. They help an organization understand how effective, efficient, or successful it is. Metrics are often linked to key performance indicators (KPIs) and help make better decisions using data.

Measurements mean collecting data or observing something about a feature, condition, or event. Measurements can be numbers (like height or time) or descriptions (like color or type). The raw data collected through measurements are used to create metrics. A table with more details is shown in Table 10.2.

Table 10.2 Metrics versus measurements

Metrics	Measurements
Metrics are quantitative assessments that provide insights into the performance, effectiveness, or health of a process, system, or activity	Measurements are specific data points or values obtained through observation or instrumentation, serving as the raw data used to calculate metrics
Metrics are often derived from multiple measurements and are designed to convey meaningful information for decision-making	Measurements are individual data points that may lack context and interpretation until aggregated or analyzed to form metrics
Metrics are typically used to track trends, evaluate performance against objectives, and make informed decisions	Measurements offer a detailed view of a specific attribute, allowing for precise analysis but may require aggregation for a broader understanding

10.26   Metric Lifecycle

The metric lifecycle shows the steps for making, using, checking, and improving cybersecurity metrics [2] in an organization. Here is a simple look at the stages:

a. **Create:** Collect important input data from trusted places, like store products or apps made for clients.
b. **Calculate:** Use some analysis steps on the collected data to get results. These results are saved in the metric results database and usually shown as one or more rows in a table.
c. **Communicate:** Share the metric results in different ways, like sending email alerts when a rule is broken, sending email updates, or showing the results in easy-to-read charts.

10.27   Zero-Day Exploits in Controller Systems

Zero-day exploits are serious problems in controller systems that do not have fixes yet [89]. Hackers can attack these problems before the company even knows about them or fixes them. This makes controller systems very open to cyberattacks. A figure is shown in Fig. 10.5, based on [47].

Figure 10.5 illustrates the zero-day vulnerabilities, exploits, attacks, and techniques used by malicious actors. It describes a successful exploit using a zero-day vulnerability work.

To build a strong security plan, it is important to know and check the risks of zero-day problems in controllers. Finding these problems before hackers use them is very important. Some good ways to lower the risk are using strong intrusion detection systems and network monitoring tools, updating and fixing systems as soon as updates are available, and keeping a strong focus on cybersecurity in the organization. Protecting controller systems from zero-day attacks also needs teamwork with cybersecurity experts and always learning about new dangers.

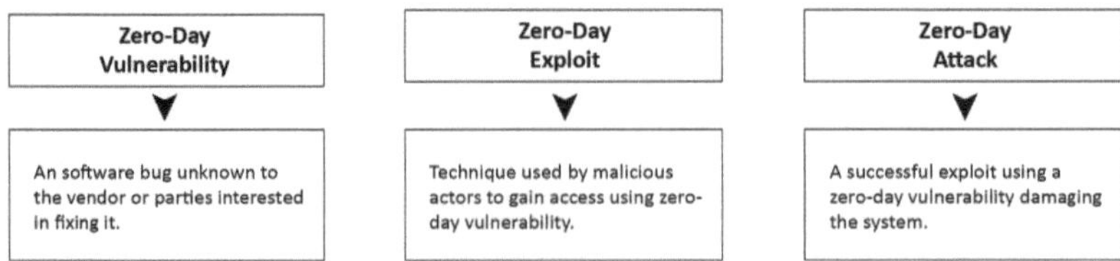

Fig. 10.5 Zero-day exploits in controller systems

10.28 Different Controller Attacks

Different controller attacks are the ways bad actors try to break or control controllers. This can hurt their security and stop them from working properly. Controllers are devices or systems that help manage work in networks, machines, and industrial settings. Here are some common types of attacks on controllers:

Ransomware Attacks on Controllers

Ransomware attacks on controllers and important systems are a big problem. Ransomware locks the data and asks for money to unlock the systems. This bad software can stop work and create safety risks. Security tools like firewalls, access controls, and regular system updates help stop phishing and ransomware attacks. A figure is shown in Fig. 10.6, based on [91].

Figure 10.6 shows how malware can infect a system and ask for ransom. The problem usually starts when someone opens a phishing email [24]. This allows malware to lock files and ask for money to unlock them and restore network access.

To stop such attacks, networks must be monitored by intrusion detection systems and strong cybersecurity tools. If a system gets attacked, it should be quickly separated from the network, authorities should be informed, and a response plan should start. Making regular, safe, and separate backups helps recover without paying a ransom. Being careful, improving system security, and working together with experts can protect controller systems.

AI and Machine Learning for Controller Security

AI and machine learning (ML) are very important for making controllers more secure. These smart technologies can find and stop threats quickly, keeping controllers and connected systems safe.

Using AI and ML lets systems watch and study large amounts of network data all the time. This helps find patterns, strange behavior, and risks that normal security

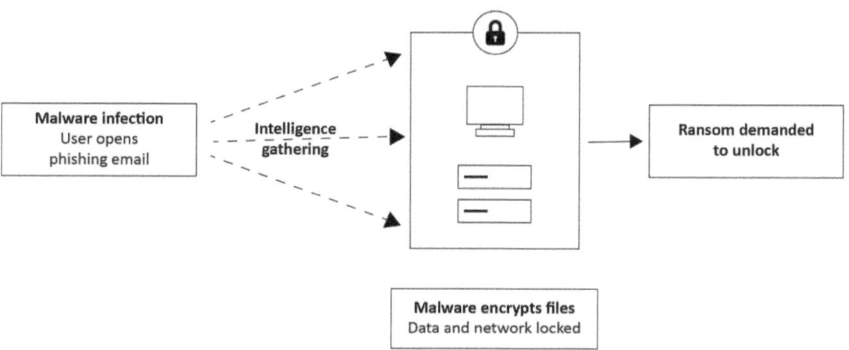

Fig. 10.6 Ransomware attacks system on controller

Fig. 10.7 The proposed block diagram of blockchain

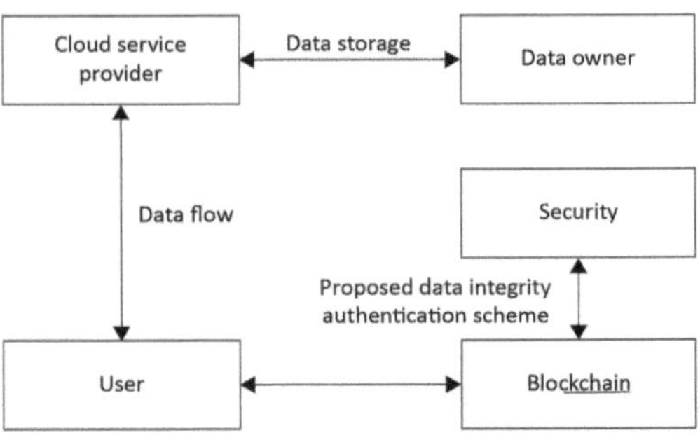

tools might miss. AI and ML can also learn from old data and get better at finding new types of attacks.

Some ways AI and ML help include detecting intrusions, spotting strange behavior, finding malware, and studying how users act. They help organizations act fast when new dangers appear, keeping important systems safe and trustworthy.

Research in this area needs teamwork between AI experts, cybersecurity teams, and industry specialists. As AI and ML improve, they will make controller systems even stronger and more secure.

Blockchain for Controller Security and Integrity

Blockchain technology can also help keep controllers safe and trustworthy. Because blockchain is decentralized and cannot be easily changed, it protects systems from hacks and errors. Figure 10.7 is from [123].

Using cybersecurity rules and best practices with blockchain can reduce the chance of attacks and fraud. Blockchain uses cryptography, decentralization, and agreement (consensus) to make sure data stays safe and trusted. Blockchain also improves privacy [123]. It keeps data secure and anonymous, and lets people control their own information. This means less trust is needed in a single organization, and personal information stays better protected. Because data is so important, using blockchain can change how we protect and manage sensitive information in the future.

Figure 10.7 shows how blockchain helps keep data safe and authentic. It involves a cloud service provider, data storage, a data owner, and security measures working together.

Smart contracts make security even better by automating tasks and following safety rules. Blockchain's use of cryptography and its clear, open design make it a strong solution for protecting critical systems where safety and trust are very important.

Phishing Attacks Targeting Controllers

Industrial control systems (ICS) are the backbone of key industries like power, water treatment, and manufacturing. These systems use controllers to manage operations and control equipment and machines.

Fig. 10.8 Phishing attack type

Cyberattackers use phishing to trick people and get unauthorized access to controllers. Phishing involves sending fake emails or making fake websites to steal passwords or login details. These attacks pretend to be trusted sources to break into controller systems. Figure 10.8 is from [24].

Figure 10.8 shows different types of email attacks, including spear phishing, malware, whaling, smishing, and vishing [24].

Phishing attacks can put important systems and data at risk, allowing unauthorized access. It's crucial to understand how these attacks work. Being aware, educating staff, and having strong security rules in place are key to spotting and stopping phishing attacks on controllers, helping protect system security.

10.29   Data Privacy and Compliance in Controller Security

Data security is about keeping information safe from unauthorized access, misuse, and leaks [46]. It also ensures that data is not altered or destroyed without permission. To protect data, organizations use a mix of rules, technology, and physical security measures while following privacy laws and managing privacy risks effectively.

ICS security focuses on keeping industrial control systems running safely. These systems rely on both hardware and software to stay secure. Controller security is closely tied to data privacy and following laws like the Health Insurance Portability and Accountability Act (HIPAA) and the General Data Protection Regulation (GDPR). Safeguarding sensitive data from unauthorized access or exposure is key to maintaining privacy. Figure 10.9 is from [46].

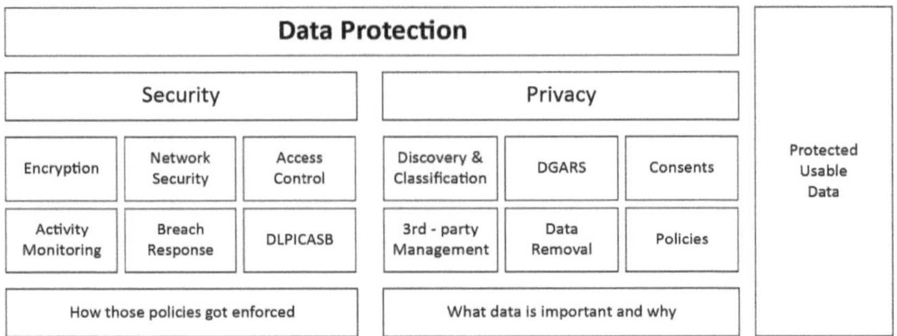

Fig. 10.9 Data protection

Figure 10.9 shows a timeline that highlights important parts of security, privacy, and data protection. It includes topics like encryption, network access, security controls, classifications, DSARS, consents, data activity, breaches, DLP/CASB, third-party data monitoring, management, removal policies, and response. It is important to meet legal requirements and follow the standards made to protect sensitive data. Following these rules helps create a safe environment for controllers, builds trust, and protects personal information.

10.30   Security of Wireless Controllers and Networks

Keeping wireless controllers and networks safe is very important for the good performance of IoT and ICS devices. Wireless networks are flexible and cheaper because they need less wiring, but they also bring new security risks. These risks come from weak communication protocols, poor encryption, and the chance that hackers may access sensitive data and systems.

To protect wireless networks, strong steps are needed, like using powerful encryption, updating software regularly, and setting strong access controls. These steps help lower the chances of attacks.

Phishing is a common cyberattack where fake emails or websites trick people into giving away passwords or other private information. Hackers act like real companies to steal information and break into controller systems. In wireless networks, phishing attacks are harder to notice because users trust wireless communication easily [24]. Phishing can cause hackers to steal data, get unauthorized access, or damage important systems. So, it is important for organizations to understand these attacks and stay alert. Training users, using anti-phishing technologies, and following strong security rules are necessary to protect wireless controllers and networks. By staying prepared and using many layers of security, organizations can lower the risk of attacks and keep their wireless systems running safely [21]. A Fig. 10.10 is taken from [21].

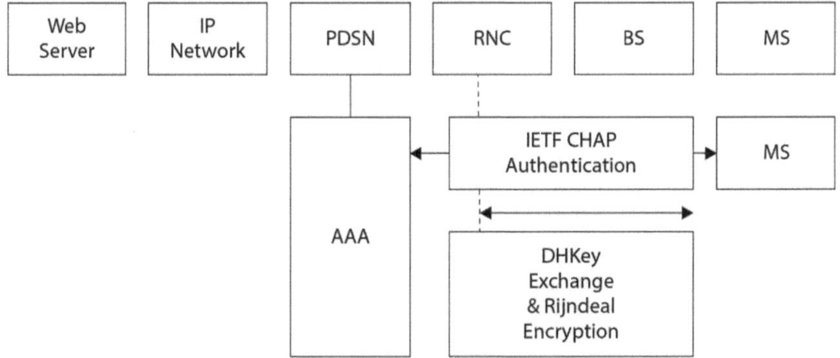

Fig. 10.10 Wireless network security

Figure 10.10 illustrates the setup of wireless network security, including components such as a web server, IP network, PDSN, RNC, BS, MS, server network, IETF CHAP authentication, and encryption.

In the modern era of interconnected IoT and industrial systems, understanding and addressing wireless network and controller security risks is essential. Proactive security measures are critical to protect data and maintain safe and reliable operations within wireless ecosystems.

10.31   Privacy-Preserving Techniques for Controllers

Today, ensuring data security is a critical requirement. Privacy-preserving techniques allow users to protect their personally identifiable information (PII) while enabling service providers and applications to continue operating effectively.

Controllers play an essential role in managing operations and processes. Privacy-focused methods such as homomorphic encryption, differential privacy, and secure multi-party computation (SMPC) are explored to enable data processing without compromising sensitive information. These techniques are particularly important in data-sensitive industries such as healthcare and finance.

A balanced approach is needed to maintain both operational efficiency and strict privacy requirements. Effective strategies for preserving privacy include using secure and encrypted communication channels and storage systems, obtaining informed consent, anonymizing sensitive data, restricting access to critical information, and complying with applicable legal and ethical standards.

10.32 Future Trends in Controller Security Threats

The future of industrial control system (ICS) security will depend on our ability to adapt, integrate, and enhance defenses against evolving threats. Advancements such as AI-powered threat detection and the resilience provided by blockchain technology will have a significant impact on protecting critical infrastructure [123]. Security threats targeting controllers are becoming more sophisticated and diverse. Malware is expected to evolve by using advanced encryption and evasion techniques to infiltrate controller systems. Industrial IoT (IIoT) devices will increasingly become targets due to their expanding use, posing risks to critical operations. AI-driven targeted attacks are anticipated to grow, making threat detection and prevention more challenging. Other emerging threats include zero-day vulnerabilities, supply chain attacks, 5G-related weaknesses, and blockchain exploitation. Social engineering and spear phishing tactics will continue to threaten controller systems by targeting individuals who grant system access [123]. Future threats may also involve hardware-level attacks, manipulation of controller data, and tampering with critical algorithms. To safeguard controller systems, organizations must adopt proactive security strategies, maintain continuous monitoring, and ensure the resilience and integrity of operational environments.

10.33 Summary

In conclusion, controller security risks are a growing concern in industrial, IoT, and critical infrastructure sectors. Advanced attack vectors include malware, social engineering, supply chain threats, and AI-driven attacks [156]. Key security challenges involve data manipulation, unauthorized access, and denial-of-service attacks. Emerging zero-day vulnerabilities and hardware-level attacks present additional risks.

A proactive approach to controller security requires strong authentication, encryption, regular updates, and workforce education. Staying ahead of the rapidly evolving threat landscape and securing critical systems and operations necessitates collaboration, real-time monitoring, and adaptive security measures.

Chapter 11
Building ICS Cyber Resilience: AI and Machine Learning Strategies, System Hygiene Practices, and Secure Smart Grid Frameworks

Abstract Industrial cyber resilience control systems are crucial for recovering from intrusions and maintaining operational continuity. This chapter focuses on cyber-security resilience within Industrial Control Systems (ICS), detailing methods to protect these systems from cyberattacks. It explores cyber resilience and defensive techniques, including system hygiene, common cyberattacks, and attack-resilient frameworks for Distributed Energy Resources (DER). The chapter also addresses cybersecurity concerns specific to power systems and SCADA, highlighting AI and machine learning solutions for resilience. Additionally, it examines advances in cyber-physical attack testing and simulation, as well as the role of blockchain technology in enhancing ICS resilience, security, and privacy. Topics covered include IT and ICS security integration, the impact of contemporary ransomware attacks, security control placement, Active Directory, SIEM systems, and SOC recommendations. The chapter concludes with a discussion of the "sliding scale" of cybersecurity, providing readers with a framework to evaluate their security posture.

Keywords Industrial Cyber Resilience · ICS Security · Cybersecurity Resilience · System Hygiene · Cyberattacks · DER Attack-Resilient Frameworks · Power System Security · SCADA Security · AI and ML Resilience · Blockchain Technology · Cyber-Physical Attack Testing · Ransomware · Security Control Placement · Active Directory · SIEM Systems · SOC Recommendations · Sliding Scale of Cybersecurity

11.1 Introduction

Cyber resilience is the ability of an organization to handle, recover from, and prepare for cyberattacks and data breaches while continuing to operate smoothly. For Industrial Control Systems (ICSes), cyber resilience means having plans and actions in place that help them survive, adjust to, and recover from cyberattacks

© The Author(s), under exclusive license to Springer Nature Switzerland AG 2026   279
M. A. Rahman et al., *Securing Industrial Control Systems*,
https://doi.org/10.1007/978-3-032-03018-4_11

without stopping important tasks [16]. Cyber resilience includes the following key areas:

1. **Threat Protection:** This refers to defending against the increasing number and types of cyber threats. DNS authentication is important for making sure that communication between Internet-connected devices is genuine. Tools like DNSSEC (Domain Name System Security Extensions) help check that DNS responses are valid, lowering the chances of attacks such as DNS spoofing or DNS cache poisoning [23]. Endpoint Detection and Response (EDR) tools are also useful as they watch devices for harmful activity and react quickly when something suspicious happens.

2. **Recoverability:** Strong backup and recovery systems are necessary to get business operations back on track after a cyberattack. This is especially important during ransomware attacks where data might be locked and used for ransom. The process includes regularly saving important data and system information in safe places, both within the company and on Internet-based storage. A clear incident response plan should also be in place to guide how to restore data and systems. Testing the backup and recovery steps regularly ensures they work properly when needed.

3. **Flexibility:** Cybersecurity plans should be able to change and adjust to new types of threats and attack methods. Staff training is very important because trained employees are better at noticing and dealing with possible risks. Tools that use machine learning and artificial intelligence can quickly look through large amounts of data to spot signs of harmful activity. Updating security settings regularly helps ensure the system stays strong against new threats.

4. **Durability:** Good management of IT systems, strong data protection methods, and reducing human mistakes all help make systems more secure. To keep sensitive information safe from leaks or unauthorized access, companies should use proper access controls, encryption, and tools that prevent data loss. Keeping systems secure also means regularly updating software and fixing known problems. Training employees and making them aware of security issues helps prevent mistakes and insider threats. Improving security procedures based on past experience and best practices makes organizations stronger against cyber threats.

Cyber resilience is a key part of modern digital safety. It shows how well systems, organizations, and people can deal with attacks and disruptions, adjust to changes, and recover quickly. In today's connected world, where cyber threats are serious, cyber resilience goes beyond simple protection [16]. It's a smart and complete approach that not only protects networks and systems but also builds a culture of readiness, fast response, and ongoing improvement.

11.2 Resilience of Cybersecurity in Industrial Control Systems

Industrial Control Systems (ICS) cyber resilience means the ability of an organization to stop, absorb, and recover quickly from cyberattacks on its control systems. ICSes are used to manage and monitor key operations in places like factories, transport systems, water plants, and power stations. Cyber resilience in ICSes uses strategies, rules, tools, and methods to protect systems from attacks and to keep them running, even during problems [16]. The main goal is to reduce the impact of cyber incidents and keep important processes safe, reliable, and always available. Figure 11.1 is taken from [30].

Figure 11.1 shows different network parts and devices. These include a Controller, Simulators, Backup Server, WSUS Node, AV Update Node, Firewall, OPC Server, CVE Scanner, and other parts. To protect important systems from new cyberattacks, we need to improve the cybersecurity strength of Industrial Control Systems (ICS). Regular risk checks, using strong security steps, training staff, and planning for problems can help companies find, stop, and fix cyberattacks. To stay ahead of attackers, companies must work together, share threat news, and keep watching for dangers. Cybersecurity resilience helps companies keep their ICS systems safe from new cyber threats [30].

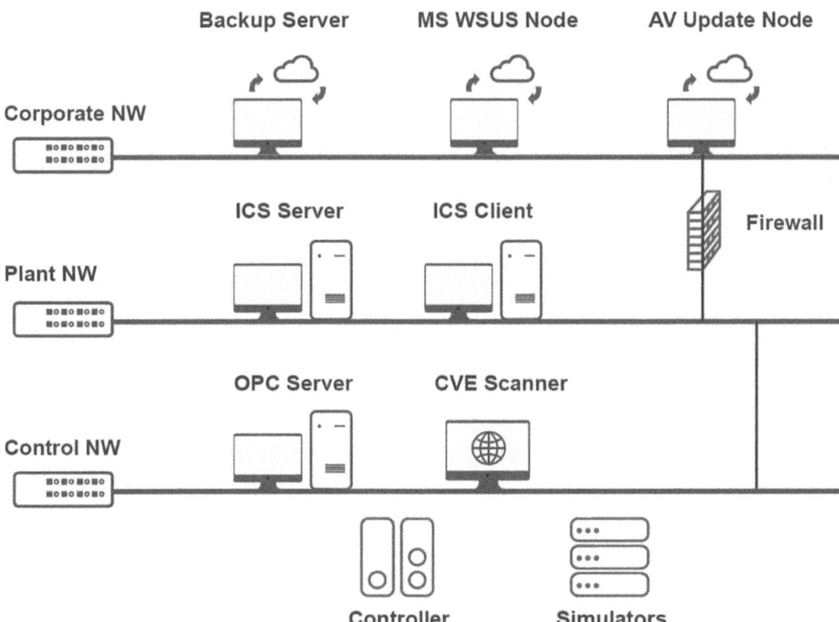

Fig. 11.1 Remote test lab for ICS cybersecurity

11.3   Attack Resiliency Analysis

Attack Resiliency Analysis checks if Industrial Control Systems (ICS) can avoid, handle, and recover from cyberattacks.

This includes guessing possible attacks, finding weak points in ICS devices and steps, checking security tools, and testing safety methods like backups and action plans. It also looks at how well teams respond to problems, how trained the workers are, and if the system uses security updates and watches for dangers. The main goal is to make ICS stronger so that key industry systems stay safe even when cyber threats increase.

Key Features of Cyber Resilience

Cyber resilience helps protect companies from human mistakes and software problems. The four main parts are:

a. **Threat Protection:** Companies can stop new cyber threats by using DNS checks and EDR tools. This helps protect important data and lowers the risk of attacks. Good security rules can block hackers and limit the harm from attacks.
b. **Recoverability:** Using backups across different network systems helps companies recover fast after a cyberattack. With extra copies of important data and systems, they can go back to normal quickly. This lowers money loss and keeps their good name.
c. **Adaptability:** Companies can reduce risks by quickly finding attacks and updating their protection. Admins watch for strange activity and adjust safety steps when new threats appear. This helps protect against new kinds of cyberattacks.
d. **Durability:** Making better devices, matching IT with business goals, and planning updates helps companies stay strong during cyberattacks. It keeps their systems safe over time. Investing in good systems and keeping them updated helps businesses stay open and ahead in the digital world.

These steps help companies build strong cyber safety. They can resist and recover from attacks while keeping their system and data safe and ready to use.

11.4   Strategies for Cybersecurity and Cyber Resilience

Organizations implement cybersecurity and cyber resilience strategies to protect their systems, networks, and data from various cyber threats [16]. These strategies also ensure rapid recovery in the event of an attack. Cybersecurity focuses on protecting systems and data from unauthorized access, data leaks, and malicious software. Common measures include firewalls, antivirus tools, secure passwords, and access control. Cyber resilience ensures that organizations can continue operations and recover quickly after an attack. This involves regular data backups, disaster recovery plans, and maintaining essential system functions during disruptions. Effective cybersecurity includes identifying and fixing system weaknesses, maintaining up-to-date software, and training employees to recognize threats such as phishing attempts. Resilience planning also requires regular testing of backup and recovery procedures. Real-time monitoring helps detect unusual activity and allows for a quick response to minimize damage. Clear communication during and after a cyber incident is critical. Organizations must keep staff, customers, and regulators informed with timely and accurate updates. Adhering to industry standards, such as GDPR or HIPAA, supports the development of secure systems and processes. Together, these practices create a strong defense that protects organizational assets, maintains trust, and ensures business continuity. Figure 11.2 is adapted from [16].

Figure 11.2 shows a step-by-step guide for improving cybersecurity. It explains how to keep systems safe by following a clear and regular process.

The steps are:

1. **Step 1: System Care:** Keep systems clean, updated, and well maintained on a regular basis.
2. **Step 2: Make a Plan:** Create a team with people from different parts of the organization to plan for possible cyber problems.
3. **Step 3: Check for Risks:** Study common attacks to make a plan that fits the company's needs.

Fig. 11.2 Seven strategic steps for cyber resilience

4. **Step 4: Measure Risk:** Use simple estimates to understand risk levels. Avoid spending too much time on small details.
5. **Step 5: Reduce Risk:** Use time and money to protect the most important systems that could be harmed.
6. **Step 6: Get Support:** Buy cyber insurance to get financial help and expert advice if an attack happens.
7. **Step 7: Start the Work:** Begin building strong protection with a clear plan and small, achievable goals. Keep improving the system step by step.

By following these steps, organizations can better protect their digital systems, lower the chance of cyberattacks, and handle problems more easily in a fast-changing threat environment.

11.5 Overview of System Hygiene

Cyber hygiene refers to a set of routine practices that individuals and organizations follow to protect their digital systems from cyber threats. It involves taking preventive actions to secure sensitive data and ensure the privacy, reliability, and integrity of digital environments. Implementing strong cyber hygiene helps organizations prevent security incidents, data breaches, and other cyber-related risks that could impact operations or damage reputation. Core practices include using strong and unique passwords, enabling multi-factor authentication (MFA), regularly updating software and firmware, and securing devices against unauthorized access. Routine tasks such as scanning for malware, reviewing access logs, and applying security patches are essential to maintaining a secure system environment. Regular data backups and disaster recovery plans are also key elements of effective cyber hygiene. Additionally, organizations should establish clear cybersecurity policies, provide ongoing employee training, and conduct regular security audits to identify and resolve vulnerabilities. Promoting a culture of cybersecurity awareness and accountability across the organization helps reduce the risk of attacks. Cyber hygiene goes beyond prevention. It emphasizes continuous monitoring, effective incident response planning, and readiness for recovery to maintain a strong and resilient security posture. Figure 11.3, adapted from [88], illustrates the essential components of cyber hygiene and highlights key practices necessary for protecting digital systems [88].

Figure 11.3 shows the key parts of cyber hygiene. These include personal, social, thinking-related, environmental, and technical factors. Some of the specific topics include confidence in using technology, social habits, trust, background, communication, awareness, and privacy.

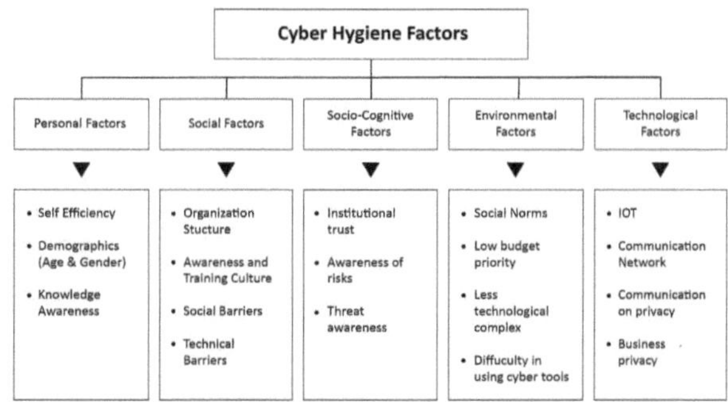

Fig. 11.3 Conceptual map of cyber hygiene factors

The main areas of cyber hygiene are [88]:

1. **Secure Password Management:**

   a. Use strong and different passwords for each account.
   b. Change passwords regularly and avoid using easy-to-guess information.

2. **Regular Software Updates:**

   a. Keep systems, apps, and security tools updated to fix known problems.

3. **Phishing Awareness:**

   a. Be careful with unknown websites, emails, or messages that might be fake.
   b. Check if messages are real before clicking on links or sharing important details [24].

4. **Firewall Protection:**

   a. Use firewalls to control network traffic and block unwanted access.

5. **Safe Wi-Fi Use:**

   a. Use strong Wi-Fi security settings.
   b. Change default router passwords and use unique passphrases.

6. **Device Security:**

   a. Install antivirus software and keep it updated.
   b. Use device encryption to protect information from theft or loss.

7. **Data Backup:**

   a. Back up important data regularly and store it in a safe place.
   b. Test recovery options to make sure backups work when needed.

8. **Two-Factor Authentication (2FA):**

   a. Turn on 2FA where available to add extra protection to logins.

9. **Privacy on Social Media:**

   a. Change privacy settings to control who can see your personal information.

10. **Cybersecurity Education:**

   a. Stay informed about common cyber threats.
   b. Learn and share good habits for staying safe online.

Cyber hygiene involves ongoing learning, staying alert to new threats, and taking part in cybersecurity efforts [88]. In today's connected world, both individuals and organizations can better protect their systems and data by following these good practices.

11.6 Most Common Types of Targeted Cyberattacks

Modern cyberattacks are complex, long-term campaigns that target an organization's IT systems to cause harm, steal private data, or disrupt normal business activities. These attacks are carefully planned and use different methods to take advantage of weaknesses in systems, networks, or human behavior. While the techniques can vary, some types of cyberattacks are more common and frequently used by attackers. These attacks typically focus on valuable assets like intellectual property, financial data, and personally identifiable information (PII).

Phishing attacks are the most common and successful type of cyberattack. They exploit human weaknesses by tricking people into revealing sensitive data like login details or financial information. Other common attacks include ransomware, advanced persistent threats (APTs), and denial-of-service (DoS) attacks. For example, ransomware attacks usually involve encrypting important files and asking for payment to unlock them. APTs are long-term efforts to steal information over time. Knowing about these attack methods is essential for creating a strong defense against cyber threats.

1. **Targeted Data Breaches:**

   a. Attackers target specific organizations to steal sensitive information.
   b. Stealing user passwords, business plans, or intellectual property is a major threat.
   c. Large data breaches can lead to legal and financial problems.

2. **Malware Attacks:**

   a. Attackers use phishing emails to send malware into systems [24].
   b. This infects the network, making it open to more attacks.

3. **Vulnerability Exploits:**

   a. Attackers take advantage of software weaknesses before they are fixed.

4. **DoS and DDoS Attacks:**

   a. Denial-of-Service (DoS) and Distributed Denial-of-Service (DDoS) attacks flood servers with fake requests, blocking legitimate access to resources.

5. **Spear Phishing:**

   a. Spear phishing is a type of phishing that targets specific employees, especially those with access to sensitive information or important tasks.

6. **Supply Chain Attacks:**

   a. Attackers target software vendors to enter the systems of the organizations they supply, like the 2020 SolarWinds attack.

7. **Man-in-the-Middle (MitM) Attacks:**

   a. Attackers use methods like ARP spoofing, HTTPS spoofing, session hijacking, and DNS spoofing to intercept and change data or impersonate trusted users.

11.7 Attack-Resilient Framework for DER Cybersecurity

Cybersecurity is an essential part of cyber resilience. It includes several key areas such as device protection, network protection, and training employees to follow safe online practices. Together, these parts create a complete cybersecurity system. On the other hand, cyber resilience focuses on keeping the organization running by using data protection, regular backups, and recovery plans [16]. For example, to keep the power grid stable and reliable, a cybersecurity plan must be built for distributed energy resources (DERs) that can resist different types of cyberattacks. A strong cybersecurity plan protects business data and equipment. It ensures that the business can continue to work, even when facing serious online threats. It also helps prevent data loss and gives full protection from cyber risks. This system is made to find and respond to cyberattacks on both the hardware and utility levels. It focuses on four main areas (Fig. 11.4):

a. **DER Resilience Planning and Measures:** This part helps find weak points by offering simple steps and rules to increase trust when using distributed energy resources.
b. **Preventing DER Attacks:** This part looks at current systems to find weaknesses and build better safety designs that fit DER features.
c. **Detecting DER Attacks:** This part helps quickly spot unusual actions or signs of misuse using real-time tracking and pattern checks.
d. **Responding to DER Attacks:** This part focuses on quick and proper responses to attacks based on how harmful the activity is and how sure the system is about the threat.

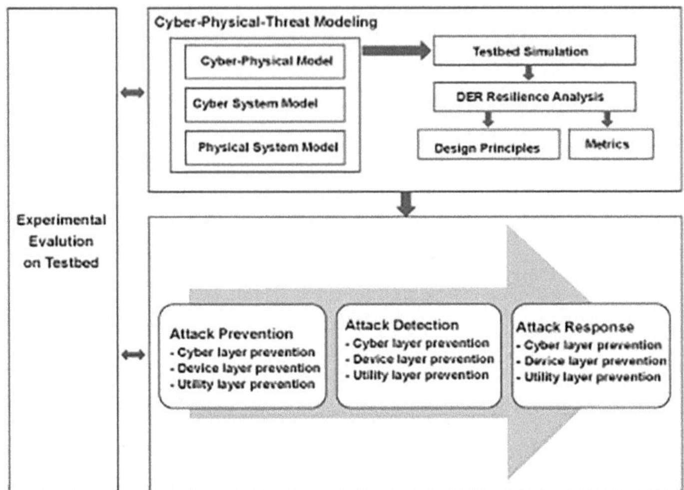

Fig. 11.4 Attack-resilient framework for DER cybersecurity

Figure 11.3, taken from [119], shows a system that combines protective steps and considers both the physical and digital parts of the power system to build a strong and safe way to connect distributed energy resources (DERs).

11.7.1 Cyberattacks on Cyber-Physical Power Systems with Distributed Energy Resources

Potential cyberattacks on cyber-physical power systems using Distributed Energy Resources (DERs) highlight important cybersecurity vulnerabilities in modern power grids. Numerous cyber threats target the digital and physical components of these cyber-physical systems. Threats to the electricity system include attacks on DER controllers, smart inverters, and interference with wide-area monitoring, security, and control systems. Understanding and modeling these risks is essential as the power system integrates an increasing number of DERs to mitigate risks and ensure system security and reliability. The research emphasizes the necessity of addressing cyber-physical threat models, assessing risks, and developing appropriate security solutions to protect the electrical grid from potential cyberattacks.

Additionally, proactive and adaptive cybersecurity measures are crucial for safeguarding the electrical grid from possible cyberattacks on cyber-physical power systems with DERs. The power sector can ensure the continuous reliability and security of the evolving energy landscape by adopting cutting-edge technologies and collaborative approaches.

11.7.1.1 DERShield: A Resilience Framework for Cybersecurity

To secure DERs in cyber-physical systems, a versatile approach is offered by the *Attack-Resilient Framework for DER Cybersecurity*. This design ensures DER asset integrity and functionality even in the face of evolving cyber threats. Strong security protocols, anomaly detection, incident response strategies, advanced threat modeling, and continuous monitoring are all incorporated into this architecture to strengthen DERs against a range of threats. DERShield provides an all-encompassing and flexible cybersecurity architecture for DERs in cyber-physical systems. Its diverse strategy, when combined with resilience and collaborative principles, makes it an invaluable tool for securing decentralized energy systems in the future.

11.8 Resilience in SCADA Systems

Cyber resilience refers to an organization's capacity to facilitate business growth (enterprise resiliency) by proactively planning for, effectively responding to, and successfully recovering from cyberattacks. A cyber-resilient company can efficiently respond to and recover from both anticipated and unanticipated crises, threats, adversities, and obstacles.

On the other hand, SCADA systems are essential for industrial and critical infrastructure management. They offer centralized monitoring and control but are vulnerable to cyberattacks. SCADA vulnerabilities may arise from outdated software, insufficient authentication, lack of encryption, and Internet exposure. To enhance SCADA resilience, organizations must understand their specific weaknesses.

a. **Legacy Infrastructure Issues:** Many SCADA systems are outdated and not security-focused. Operating systems and devices without security upgrades may be exploited.
b. **Interconnectivity Risks:** SCADA systems' Internet and business network connections expand the attack surface. Any corporate network breach could affect SCADA.
c. **Human Factors:** Insider threats and human errors pose significant risks. Employee mistakes, inadequate training, and credential mishandling can lead to security breaches.
d. **APTs Targeting SCADA Systems:** SCADA systems are critical, making them attractive targets for Advanced Persistent Threats (APTs). APTs may infiltrate networks and cause significant harm.

11.9 Strategies to Enhance Resilience in SCADA Environments

A diversified strategy is needed to improve SCADA resilience. Technologies, cybersecurity best practices, and organizational tactics are essential components.

a. **Isolate Key SCADA Assets:** Segment the network to isolate key SCADA assets. This reduces the attack surface and minimizes the impact of breaches.
b. **Multi-factor Authentication and Access Management:** Use MFA and conduct frequent access reviews to manage access. Restrict access to essential personnel only, enhancing overall security.
c. **Regular Patching and Updates:** Implement a structured patch management plan to keep SCADA network software and devices updated with security patches.
d. **SCADA Incident Response Planning:** Develop and test a comprehensive incident response plan. Quickly detect, contain, and recover from security incidents to minimize potential damage.
e. **Monitor for Anomalies:** Continuously monitor SCADA network data for unusual activity that may indicate a cyberattack. Utilize intrusion detection systems and anomaly detection technologies for early threat detection.
f. **Cybersecurity Training for Personnel:** Provide regular training for SCADA operators and personnel on cyber threats, security best practices, and incident reporting [8].

11.10 Role of AI and Machine Learning in Industrial Control System Resilience

Advanced techniques for ICS security are outlined below:

a. **Anomaly Detection Algorithms:** AI and ML play a crucial role in ICS anomaly detection. Machine learning algorithms can identify unexpected patterns in historical data that may indicate a cyberattack or system failure.
b. **Behavioral Analytics:** AI can establish a baseline of normal system activity and detect deviations. This approach is essential for identifying threats that traditional rule-based methods may overlook.
c. **Predictive Maintenance:** AI-powered predictive maintenance analyzes data to forecast equipment failures and necessary repairs. Preventing failures enhances system resilience and reduces downtime.

11.11 Cyber-Physical Response with Machine Learning

Cyber threats have evolved with the incorporation of machine learning into cyber-physical systems. The predictive, adaptive, and intelligent characteristics of machine learning are highly beneficial to cyber-physical systems, which integrate computational elements with physical processes.

a. **Automated Incident Response:** AI and ML can identify, analyze, and contain cyber incidents more rapidly. This automation enables swift responses to attacks, minimizing system impact.
b. **Enhanced Threat Intelligence:** Machine learning can process and evaluate vast amounts of threat data to uncover emerging threats. This enhances incident response strategies, allowing them to adapt to new attack vectors [156].
c. **Security Adaptation:** Machine learning in incident response enables systems to adapt to evolving threats. This adaptability is crucial for combating sophisticated cyberattacks.

11.12 Ensuring Resilience in Energy ICS

Energy ICS is fundamental to the production, distribution, and management of energy infrastructure. Ensuring resilience in these systems is critical for maintaining reliability, efficiency, and security. Understanding the energy ICS threat landscape—including cyberattacks, natural disasters, and other disruptions—is the first step toward resilience. These diverse threats necessitate multiple solutions.

a. **Redundancy and Diversity:** Incorporating redundancy and diversity into an energy distribution network is a key aspect of resilience. This involves multiple energy distribution channels and backup power sources. In the event of a failure or cyberattack, the system can seamlessly switch to alternative sources, minimizing disruptions.
b. **Rapid Restoration Strategies:** Implementing well-defined plans and strategies for the swift restoration of the distribution network is essential. A skilled workforce, automated restoration procedures, and advanced monitoring systems ensure a prompt recovery from any disruptions.
c. **Investment in Technology:** Strengthening energy distribution networks through advanced technologies enhances resilience. These technologies reduce vulnerabilities to cyberattacks, enable predictive maintenance, and facilitate rapid intervention.

11.13   Protecting Smart Grids from Cyber Threats

Cyberattacks can target smart grids, which use advanced technologies to distribute electricity efficiently. Protecting these vital infrastructures requires a holistic approach that addresses vulnerabilities and ensures smart grid security. Developing smart grid protections requires understanding the evolving cyber threat landscape. Communication network cyberattacks, data manipulation, unauthorized access, and energy service disruptions are significant threats.

a. **Network Segmentation and Access Control:** Smart grids must have robust network segmentation and access restrictions in place. This ensures that critical systems remain secure even if a portion of the network is compromised.
b. **Frequent Security Audits and Upgrades:** Conducting regular security audits and timely hardware and software upgrades for smart grids is crucial. Identifying vulnerabilities and promptly addressing them helps prevent future cyberattacks.
c. **Employee Awareness and Training:** Educating staff members about cybersecurity risks and best practices is a key component of safeguarding smart grids. Since human error is often a weak point, training reduces the likelihood of accidental security breaches.

11.14   Resiliency Testing and Simulation

Resilience testing of Industrial Control Systems (ICS) involves evaluating their ability to withstand risks, disturbances, and challenges. Assessing the resilience of industrial processes and critical infrastructure is a systematic and proactive procedure.

a. **Identifying Vulnerabilities:** Penetration testing, also known as pen testing, is an essential method for assessing an ICS's security. It helps identify system vulnerabilities, including weaknesses in hardware, software, or operational procedures.
b. **Mimicking Real-World Attacks:** Penetration testing simulates real-world cyberattacks by potential adversaries. This approach provides a practical evaluation of the system's security posture and identifies possible entry points for attackers.
c. **Improving Defenses:** Insights gained from penetration testing enable organizations to strengthen their defenses against cyberattacks by updating procedures, implementing necessary changes, and enhancing security measures.

A comprehensive cybersecurity and risk management plan must include ICS resilience testing and simulation. This process provides valuable insights, enhances organizational preparedness for unforeseen events, and improves the overall security and reliability of critical industrial systems.

11.15 Cyber-Physical Attacks for Resilience Assessment

The purpose of cyber-physical attacks for resilience assessment is to evaluate the resilience and response capabilities of critical systems, particularly those that integrate digital and physical components. These attacks can be simulated or controlled. Understanding how cyber-physical systems, smart grids, and ICS resist and recover from advanced threats is crucial. By simulating real-world cyber-physical threats, organizations can identify vulnerabilities, assess the effectiveness of security protocols, and refine incident response strategies. This approach contributes to a proactive and adaptive cybersecurity posture, enhancing system resilience and ensuring the security and functionality of critical infrastructure despite evolving cyber threats.

a. **Recognizing the Interaction:** Cyber-physical attacks exploit weaknesses in both the physical and digital (cyber) components of an ICS. Simulating such attacks is essential to understanding how these components interact during a breach.
b. **Evaluating Response Mechanisms:** Organizations can assess how effectively their systems respond to complex breaches by modeling cyber-physical attacks. This evaluation facilitates improvements in incident response procedures and damage mitigation strategies.
c. **Enhancing Preparedness:** Conducting exercises helps educate personnel and improve their readiness to handle sophisticated cyberattacks. This ensures a swift and accurate response while minimizing disruptions to critical operations.

11.16 Advancements in Testing Methodologies for Industrial Control Systems

Testing methodologies for ICS have advanced significantly, playing a crucial role in evaluating and improving critical infrastructure resilience. These methodologies include cyber-physical attack simulations, penetration testing, and resilience testing. Penetration testing is essential for identifying ICS vulnerabilities and ensuring robust cybersecurity defenses. Resilience testing evaluates the system's ability to recover from various disruptions, while cyber-physical attack simulations assess system responses by considering the intricate interactions between digital and physical components during a breach. These advancements support continuous improvements in incident response procedures and personnel readiness, strengthening ICS defenses against evolving cyber threats. The integration of cutting-edge testing methodologies highlights a commitment to ensuring the security, reliability, and robustness of critical infrastructure in the ever-evolving cybersecurity landscape.

a. **Realistic Test Environments:** Modern testing methodologies provide realistic test environments that simulate ICS systems. These environments include various devices, protocols, and configurations, offering comprehensive testing scenarios.
b. **Continuous Monitoring and Feedback:** Contemporary testing methods incorporate continuous monitoring and feedback loops. This approach helps identify vulnerabilities, track remediation efforts, and validate fixes through re-evaluation.
c. **Threat Intelligence Integration:** Testing methodologies increasingly incorporate threat intelligence to simulate both existing and emerging threats. This integration enables ICS resilience assessments against contemporary cyber threats.

11.17 Blockchain for Industrial Control Systems Resilience

ICS can be made more resilient through blockchain technology, a decentralized and immutable ledger. Its key features—decentralization, immutability, transparency, and consensus mechanisms—make it well-suited for enhancing the security, integrity, and availability of critical ICS data and processes.

Blockchain technology enhances ICS resilience by providing a secure and decentralized ledger that ensures data integrity. Decentralization reduces the risk of system-wide failures, while smart contracts automate and enforce predefined rules, minimizing manual errors. Transparent and traceable transactions create an audit trail that helps detect unauthorized activities. Blockchain's resistance to cyberattacks and its role in supply chain security further improve ICS reliability. By reducing reliance on centralized third parties, blockchain offers greater operational control. Ultimately, blockchain provides a secure, transparent, and resilient foundation for critical industrial processes, playing a vital role in safeguarding ICS against cyber threats [123]. Some use cases of blockchain in enhancing ICS resilience include:

a. **Supply Chain Security:** Blockchain ensures authenticity and traceability, reducing counterfeit and tampered ICS component supply chains.
b. **Identity and Access Management:** Blockchain secures identities and access credentials, preventing unauthorized access to critical ICS systems [123].
c. **Smart Contracts for Automation:** Smart contracts in ICS can automate processes and trigger predefined actions, optimizing operations and improving efficiency.
d. **Data Integrity and Provenance:** Blockchain securely records sensor and device data, ensuring an immutable record that prevents unauthorized modifications.

11.18 Security and Privacy Aspects of Blockchain

ICS leverages blockchain to enhance critical infrastructure security by integrating essential security and privacy features. Blockchain's immutability prevents unauthorized modifications, ensuring data integrity. Decentralization strengthens system resilience by minimizing single points of failure. Cryptographic mechanisms and consensus protocols guarantee transaction trustworthiness. Privacy concerns are addressed by balancing transparency with confidentiality measures. Data availability and redundancy further bolster ICS reliability. Secure smart contract implementation prevents unintended consequences.

a. **Privacy-Preserving Mechanisms:** Blockchain technologies with privacy-enhancing mechanisms, such as zero-knowledge proofs, protect sensitive data, improving ICS confidentiality [123].
b. **Encryption and Hashing Methods:** Strong encryption and hashing techniques safeguard blockchain data, particularly critical ICS operational data.
c. **Access Control:** Blockchain-based access control ensures that only authorized personnel can access ICS applications, maintaining security and privacy.
d. **Compliance with Data Privacy and Security Standards:** Ensuring compliance with regulatory requirements when integrating blockchain into ICS is vital for avoiding legal and security risks [46].

11.18.1 Step 1: Set Up Ethereum Development Environment

First, you'll need to set up your Ethereum development environment. You can use tools like Ganache for a local blockchain and Truffle Suite for smart contract development [123].

11.18.2 Step 2: Create a Smart Contract

The core of this example is the Solidity smart contract, which will store sensor data on the blockchain. The contract includes a structure for the sensor data (timestamp, temperature, and pressure) and a function to log the data.

```solidity
// SPDX-License-Identifier: MIT
pragma solidity ^0.8.0;

contract SensorData {
    struct Record {
        uint timestamp;
        uint temperature;
        uint pressure;
```

```
 9          }
10
11          mapping(uint => Record) public records;
12          uint public recordCount;
13
14          function logSensorData(uint _temperature, uint
      _pressure) public {
15              records[recordCount] = Record(block.timestamp,
      _temperature, _pressure);
16              recordCount++;
17          }
18      }
```

Listing 11.1 Sensor Data Logging Smart Contract

11.18.3   Step 3: Deploy the Smart Contract

Using Truffle, the next step is to deploy the smart contract to a local Ethereum blockchain (Ganache). The deployment script defines how the contract is deployed to the network.

```
1  // File: migrations/2_deploy_sensor_data.js
2
3  const SensorData = artifacts.require("SensorData");
4
5  module.exports = async function (deployer) {
6      await deployer.deploy(SensorData);
7  };
```

11.18.4   Step 4: Interact with the Smart Contract

Create a JavaScript script to interact with the deployed smart contract and log sensor data. The code below is referred from: https://github.com/Sunzidasiddique1/ICS.

```
1  // File: scripts/log_sensor_data.js
2
3  const SensorData = require('./build/contracts/SensorData.json')
      ;
4  const Web3 = require('web3');
5  const web3 = new Web3(new Web3.providers.HttpProvider("http://
      localhost:8545"));
6
```

```javascript
const sensorData = new web3.eth.Contract(SensorData.abi,
    SensorData.address);

async function logSensorData(temperature, pressure) {
    await sensorData.methods.logSensorData(temperature,
    pressure).send({ from: "0xYourAccountAddress" });
}


// Example usage
(async () => {
    await logSensorData(25, 100); // Log temperature: 25\,^\
    circ C, pressure: 100 kPa
    await logSensorData(30, 110); // Log temperature: 30\,^\
    circ C, pressure: 110 kPa
})();
```

Explanation

- **Smart Contract:** The Solidity smart contract `SensorData` defines a structure for sensor records and a function `logSensorData` to log temperature and pressure data.
- **Deployment:** The Truffle deployment script deploys the `SensorData` contract to the local blockchain.
- **Interaction:** The JavaScript script uses `Web3.js` to interact with the deployed contract, logging sensor data by calling the `logSensorData` function.

11.18.5   Example Input and Output

Input: Logging sensor data (temperature: 25°C, pressure: 100 kPa)

```javascript
await logSensorData(25, 100);
```

Output: The sensor data is logged to the blockchain, and the `records` mapping in the smart contract is updated.

Input: Logging sensor data (temperature: 30°C, pressure: 110 kPa)

```javascript
await logSensorData(30, 110);
```

Output: The sensor data is logged to the blockchain, and the `records` mapping in the smart contract is updated.

This example demonstrates how blockchain technology can be used to securely log sensor data from an ICS environment, ensuring data integrity and resilience against tampering.

Table 11.1 IT security aspects

Aspect	IT security
Focus	Data and information protection
Objective	Protecting confidentiality, integrity, and availability of data
Environment	Office networks, servers, end-user devices
Threat landscape	Malware, phishing, unauthorized access
Downtime tolerance	Moderate to high
Patch management	Frequent updates and patching
Connectivity	Internet-Centric
Security implementation approach	More standardized
Regulatory compliance	GDPR, HIPAA
Incident response	Focuses on data breach mitigation
Connectivity	Internet-centric
Security implementation approach	More standardized

11.19 Differences Between IT security and ICS Security

While IT Security protects technological systems, Information Security preserves all information. A large discussion is provided in Table 11.1.

Both IT security and ICS security have the common goal of protecting digital assets and reducing cyber threats. However, they differ in terms of their scopes, priorities, risk profiles, technical architectures, regulatory requirements, and response strategies. Therefore, it is necessary to have separate approaches to ensure the security of information technology and industrial control environments. To ensure effective cybersecurity in critical infrastructure sectors, it is essential to have a comprehensive grasp of both IT and ICS security concepts and practices. This is necessary in order to effectively tackle the ever-changing problems presented by cyberattacks.

11.20 Modern Attacks on ICS with Impact

Industrial Control Systems (ICS) are vulnerable to a wide variety of sophisticated attacks, such as supply chain attacks, malware, phishing, and zero-day exploits [89]. These attacks have serious consequences, including the compromise of data integrity, disruption of vital processes, and monetary losses. Combating these modern ICS attacks is crucial for ensuring the resilience and security of critical infrastructure. Figure 11.5 from [78] illustrates this concept.

Figure 11.5 illustrates the evolution of cyber threats and risks over time, from 1990 to the present and beyond.

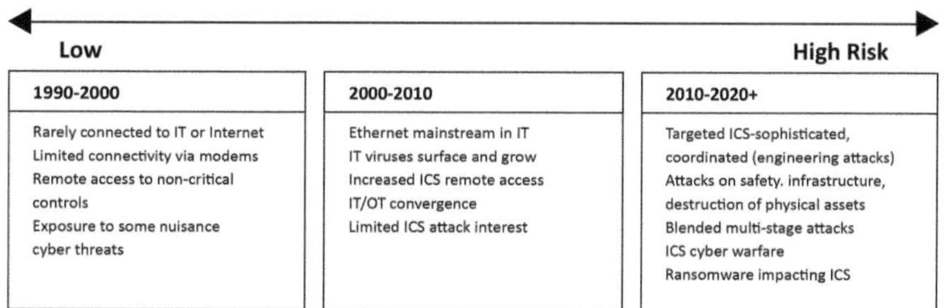

Fig. 11.5 Modern attack on ICS with impact

Beyond securing specific systems, it also includes safeguarding vital infrastructure, preserving data integrity and privacy, upholding public safety, sustaining national security, and ensuring economic stability. In a world where everything is connected, strong defenses against evolving threats are essential to mitigate their far-reaching impacts.

11.20.1   Ransomware Impact on Critical Infrastructure

Ransomware attacks on key infrastructure threaten vital services and businesses increasingly. Today's civilization relies on energy, transportation, healthcare, and other infrastructure. Attacks using ransomware exploit security flaws in these areas. Such attacks disrupt operations, compromise data, and endanger public safety. Ransomware damages essential infrastructure, causing financial losses, operational disruptions, and safety risks. Targeted firms face high ransom payments, compounding recovery expenses, and reputational harm. Additionally, ransomware incidents may delay important services and have economic and societal ramifications. To mitigate ransomware threats, critical infrastructure operators and governments must prioritize cybersecurity. Implementing strong security measures, system upgrades, personnel training, and incident response procedures is essential. To defend against ransomware and protect critical infrastructure from emerging cyber threats, public-private collaboration, threat information sharing, and fostering a cybersecurity culture are crucial.

11.21   Security Control Placement

Security control placement involves strategically positioning various security measures within an organization's IT infrastructure and operational environment to effectively minimize risks and safeguard critical assets. Organizations can enhance

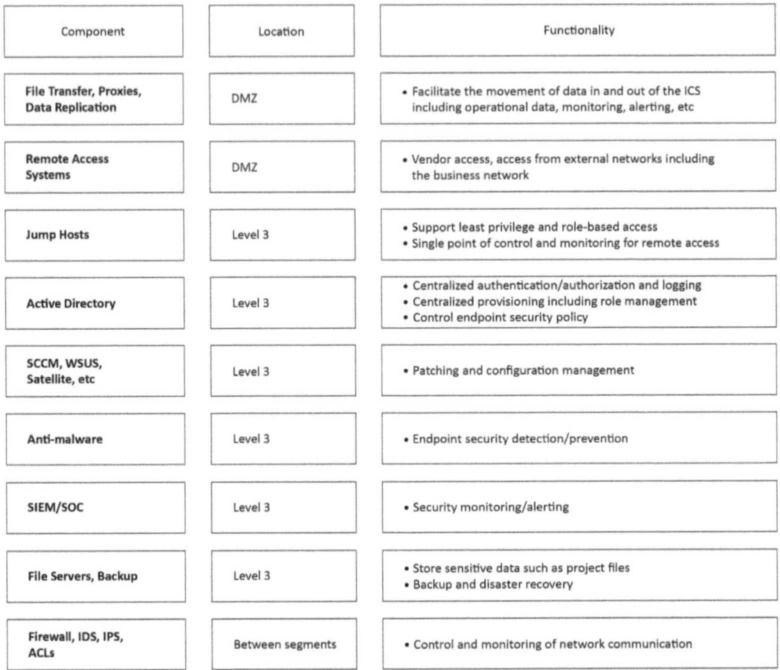

Fig. 11.6 Security control placement

their cybersecurity posture by implementing multiple layers of security controls, such as perimeter defenses, endpoint protection, network segmentation, application security, and data protection measures. These overlapping layers of defense collectively strengthen overall security. Furthermore, integrating security measures tailored to cloud environments, protocols for incident response and recovery, and initiatives for user awareness and training bolster defenses against evolving cyber threats. By strategically deploying and incorporating security measures, organizations can mitigate vulnerabilities, promptly detect and respond to security breaches, and ensure the confidentiality, integrity, and availability of their systems and data. Figure 11.6 is taken from [78].

Figure 11.6 illustrates the outlined components and functionalities.

In conclusion, strategic security control placement and integration within an organization's IT infrastructure are essential for risk mitigation and critical asset protection against emerging cyber threats. Organizations can enhance cybersecurity and safeguard their systems and data by implementing multiple layers of protection and promoting security awareness.

11.22 Overview of Active Directory

Microsoft's Active Directory is a robust directory service that serves as a central hub for managing resources, administering networks, maintaining security, and enabling communication within organizations. Active Directory relies on multiple protocols, such as the Lightweight Directory Access Protocol, to efficiently manage and retrieve directory information.

Active Directory functions as a centralized database where information about computers, devices, user accounts, groups, and network resources is stored and organized. Its hierarchical structure facilitates the logical representation of an organization's network environment, simplifying resource access management for administrators.

Resource management is a fundamental feature of Active Directory. Administrators can use this feature to establish and organize various resources, including printers, files, applications, and other network assets, in a way that allows authorized users to access them conveniently. Additionally, Active Directory offers capabilities for delegating administrative tasks, enabling administrators to grant specific permissions to users or groups based on their roles and responsibilities within the organization.

Active Directory is essential for network administration as it streamlines user account and system settings management. It consolidates authentication and authorization processes, allowing users to securely access resources from any network-connected device. Moreover, Active Directory facilitates Group Policy usage, enabling administrators to enforce security measures, control user desktop configurations, and automate software deployments across the network.

From a security perspective, Active Directory offers strong authentication and authorization capabilities to protect sensitive data and resources. The system supports multiple authentication methods, such as username/password, smart card, and biometric authentication, ensuring that only authorized users can access network resources.

Additionally, Active Directory serves as a framework for enforcing security measures, such as password complexity rules, account lockout policies, and access control lists (ACLs). These measures help organizations comply with industry standards and follow best security practices.

Active Directory is a highly adaptable and robust directory service crucial for efficiently managing and securing network infrastructure in Windows-based environments. With its extensive features and capabilities, Active Directory streamlines network management, enhances security, and fosters effective communication and collaboration within organizations.

11.23 Active Directory Working Procedure

Active Directory Domain Services (AD DS), a component of the Windows Server operating system, is the primary Active Directory service. Domain Controllers (DCs) are servers that run AD DS. Typically, an organization will have multiple directory copies, with each DC maintaining a copy of the entire domain. To keep all instances up to date, changes made to one Domain Controller's directory, such as password updates or user account deletions, are replicated to the other DCs.

Users and applications can locate objects in any domain within their forest using a Global Catalog server, which is a DC that maintains a partial copy of all objects from other domains in the forest and a complete copy of all objects within its domain. Although they do not run AD DS, desktops, laptops, and other Windows-based (non-Windows Server) devices can still be part of an Active Directory environment.

AD DS relies on multiple well-established protocols and standards, such as the Domain Name System (DNS), Kerberos, and the Lightweight Directory Access Protocol (LDAP).

11.24 Key Components of Active Directory

Active Directory is a sophisticated system consisting of several essential components that work together to provide directory services, authentication, and authorization within a Windows-based network environment. The following are the key elements of Active Directory:

1. **Domain Services:**

   a. **Purpose:** assists machines and domain users with authorization and authentication.
   b. **Function:** protects interactions between machines and users inside a domain.

2. **Certificate Services:**

   a. **Purpose:** Allows users and machines to connect securely by providing digital certificates.
   b. **Function:** Makes certain that digital conversations are secure and authentic.

3. **Federation Services:**

   a. **Purpose:** permits the use of Single Sign-On (SSO) in a variety of systems and apps.
   b. **Function:** simplifies the process of authenticating users, enabling smooth access to various services.

4. **Lightweight Directory Services (LDS):**

   a. **Purpose:**provides a flexible and lightweight directory service for apps that support directories.
   b. **Function:** supports features linked to directories without requiring the overhead of a large-scale directory server.

The core components of Active Directorycollaborate to provide centralized administration of network resources, secure authentication and authorization, and efficient directory services in Windows-based network settings.

11.25   Benefits of Active Directory

Active Directory (AD) offers several advantages that enhance corporate security and effectiveness. Modern businesses can benefit from Active Directory's adaptable capabilities, which help to create a secure and well-organized IT infrastructure.

1. **Centralized Management:**

   a. AD simplifies administration and boosts output by centralizing computer, resource, and user management.

2. **Single Sign-On (SSO):**

   a. When a user uses single sign-on (SSO), they just need to log in once to access multiple resources.

3. **Security and Access Control:**

   a. It restricts authorized users' access to resources through security policies and processes.

4. **Scalability:**

   a. When an organization grows, Active Directory expands with it seamlessly and doesn't compromise performance.

5. **Group Policies:**

   a. To simplify network security and configuration, administrators can establish standards for user accounts and workstations.

6. **Reduction in Redundancy:**

   a. It maintains network consistency and gets rid of redundant data.

7. **Efficient Resource Sharing:**

   a. Increases productivity by facilitating access to files, printers, and programs.

8. **Backup and Recovery:**

   a. Helps fix errors by restoring important settings and data.

9. **Integration with Other Services:**

   a. Enhances network performance overall by seamlessly integrating with Microsoft services.

Active Directory is a powerful and flexible framework that allows for efficient management of directory services, improved security, and optimized network performance in Windows-based systems. The many features and capabilities of this technology make it a crucial element of contemporary IT infrastructures, allowing firms to effectively fulfill their changing business requirements and quickly accomplish their strategic goals.

11.26   Backups and Baselines

Industrial Control System (ICS) security necessitates a deliberate focus on baselines and backups. Creating baselines tailored to ICS, keeping track of known-good images for comparison, and addressing ransomware's effects on controllers and project files are all crucial steps. Using specialized ICS file servers with Active Directory rights, utilizing appropriate configuration analysis tools, and adopting virtualization are some ways to fortify overall resilience. Geographic redundancy and offline backups via USB or tape strengthen ICS defenses. By using these procedures, enterprises may ensure the resilience of their ICS infrastructure and gain the ability to assess and respond to cyber threats efficiently.

a. **Develop ICS-specific baseline configurations for all devices:** Ensure customized baseline configurations tailored for Industrial Control Systems (ICS) devices.
b. **Keep a known-good image "on the shelf" for comparison:** Maintain a reference image, including OS, firmware, and project/configuration files, for easy comparison.
c. **Ransomware impact on ICS:** Recognize the potential impact of ransomware on ICS . Even if controllers are unaffected, project files are essential for updating programming.
d. **Use dedicated ICS file servers with permissions managed using Active Directory:** Enhance security by employing dedicated ICS file servers with permissions controlled through Active Directory.
e. **Use vendor tools or home-grown scripts for configuration and firmware comparison/validation:** Employ tools or scripts to copy and compare/validate configurations and firmware images, either provided by vendors or developed in-house.

f. **Leverage virtualization where it makes sense:** Explore virtualization opportunities for ICS components where applicable to improve efficiency and flexibility.

g. **Offline backups for ICS environments:** Implement offline backups for ICS environments using options such as tape, USB, or dedicated servers.

h. **Take advantage of sites in multiple geographies for redundancy:** Enhance system reliability by utilizing multiple geographical sites for redundancy in ICS operations.

5. **Rights Management Services (RMS):**

   a. **Purpose:** Protects sensitive data and controls access and use.

   b. **Function:** Safeguards confidential information, ensuring only authorized users have access while managing usage permissions.

Overall, baselines and backups are essential for ICS security and resilience. Fortifying ICS defenses requires tailoring baseline settings for ICS devices, preserving known-good images for comparison, and mitigating ransomware's possible effect on controllers and project files. Using dedicated ICS file servers with Active Directory rights, configuration analysis tools for validation, and virtualization where appropriate may boost resilience. Offline backups utilizing tape or USB and geographic redundancy for important systems enhance ICS infrastructure against cyberattacks. Organizations may strengthen their ICS settings and identify and react to cyber risks by taking these proactive steps.

11.27 Security Control system

Control system Cybersecurity, also known as Industrial Control System (ICS) security, is the process of guarding against deliberate or inadvertent disruptions to the normal operation of vital industrial automation and control systems. These systems, which rely on several components vulnerable to security flaws, oversee essential services like manufacturing, communications, energy, water, petroleum production, and transportation. The 2010 Stuxnet worm discovery brought to light how vulnerable these systems are to cyberattacks. Figure 11.7 is from [78].

Figure 11.7 illustrates an Internet enforcement boundary within a business network. It shows different levels (Level 4/5, Level 4-3, Level 3) with servers like AV Server and Patch Server (e.g., WSUS) placed at various points.

To effectively address the ever-changing environment of cybersecurity threats, companies must prioritize the implementation of a strong security control system [125]. This system is crucial for proactively managing risks, safeguarding important assets, and upholding trust and confidence in their operations, goods, and services.

Fig. 11.7 Security control system

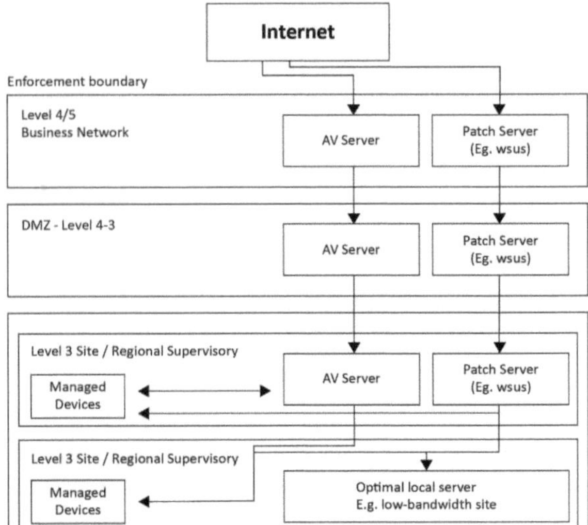

11.28   Industry Control System SIEM

The term "Industrial Control System Security Information and Event Management" (ICS SIEM) describes the combination of SIEM (Security Information and Event Management) solutions designed especially for Industrial Control Systems. SIEM is a holistic approach to security management that includes real-time information gathering, analysis, and monitoring of security. It becomes an ICS SIEM when used in conjunction with Industrial Control Systems. Figure 11.8 is from [78].

Figure 11.8 shows different levels and boundaries within a network infrastructure, including Enterprise Networks, Control Centers, ICS Data Centers, Business Networks, and Local Controllers. It also includes elements like DMZs, WAN, SCADA, SIEM, and Safety Systems.

Overall, ICS SIEM improves security monitoring and threat detection by combining ICS with specialist SIEM solutions. This comprehensive strategy enables critical infrastructure businesses to obtain real-time data, assess security incidents, and increase cyber resilience [16].

11.29   SIEM/SOC Recommendation

The Security Operations Center (SOC) and Security Information and Event Management (SIEM) guidelines cover the best ways to set up and maintain these essential cybersecurity infrastructure elements. Figure 11.9 is from [44].

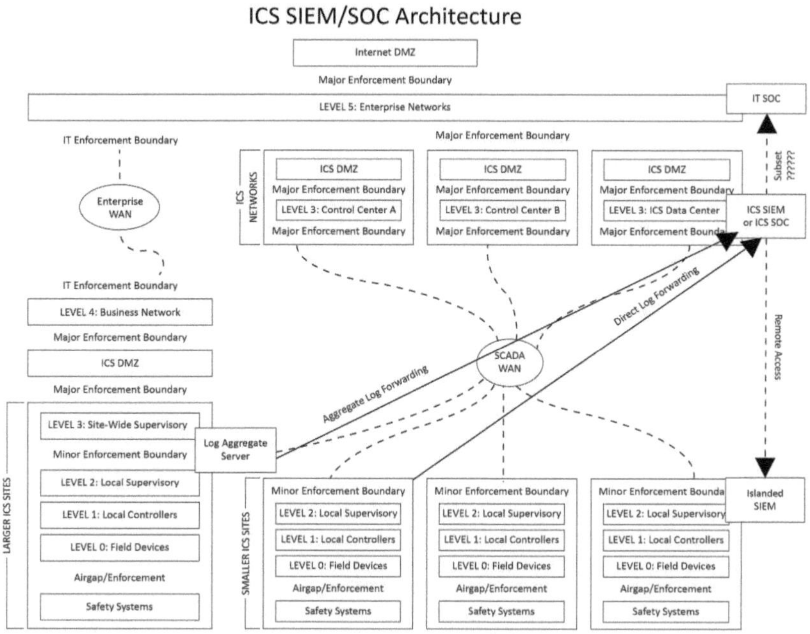

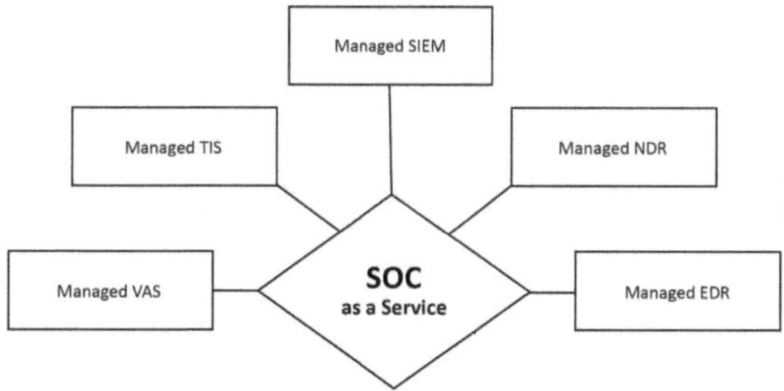

Fig. 11.8 ICS SIEM

Fig. 11.9 SOC management

Figure 11.9 shows different services such as Managed TIS, SIEM, Q, SOC, VAS, NDR, and EDR.

a. **SIEM Complexity:** SIEM solutions are complex. If an organization already has one, it should be leveraged effectively.
b. **Data Sending Strategy:** If there is no ICS SIEM, send all data. Even if an ICS SIEM is available, sending a subset of data and/or alerts may still be beneficial.
c. **Data Forwarding Through DMZs:** Forward data through DMZs, similar to historian practices

d. **Capacity for Smaller Organizations:** Smaller organizations may lack the resources to build a dedicated ICS SIEM. Start with a simple approach and scale gradually.

e. **Starting Point:** Begin with Purdue Level 3 devices. However, Level 1/2 logs can be highly valuable if collected.

f. **Leveraging Windows Event Log Collection:** Utilize Windows Event Log Collection to minimize the need for additional software installations.

g. **Caution with Automated Actions:** Be cautious with automated actions, as their impacts can be significant.

In summary, the guidelines for the Security Operations Center (SOC) and Security Information and Event Management (SIEM) emphasize the importance of strategic planning, data prioritization, and the careful implementation of automated processes. By following these principles, organizations can enhance their cybersecurity defenses and effectively counter emerging threats in today's evolving threat landscape.

11.30 Sliding Scale of Cybersecurity

The phrase *Sliding Scale of Cybersecurity* describes the concept that cybersecurity measures should be flexible and scalable based on an organization's unique requirements and risks. It emphasizes that cybersecurity is a dynamic, adaptable strategy rather than a one-size-fits-all approach.

According to this model, security measures can be adjusted based on various factors, including organizational size, industry, data sensitivity, budget constraints, and the evolving threat landscape.

Figure 11.10 is adapted from [77].

Figure 11.10 illustrates the security and defense sliding scale. It includes categories such as architecture, passive defense, active defense, intelligence, offense, and systems. The content covers aspects like planning, data collection, analysis, countermeasures, and self-defense systems.

In conclusion, the sliding scale of cybersecurity promotes flexible and scalable security solutions that meet enterprises' demands and threats. By embracing cybersecurity's dynamic nature and adapting methods, companies can safeguard their digital assets and become more resilient to emerging threats.

Fig. 11.10 Sliding scale of cyber security

11.31 Summary

Part 1's analysis of cyber reliance highlights the complex relationship between technology and human dependency. Several key insights have emerged, shedding light on the implications and intricacies of our dependence on digital technology.

First, it is evident that cyberspace is more than just a convenience; it is embedded in every aspect of modern life, from business and communication to politics and healthcare. The integration of technology into daily activities has transformed human interactions and societal structures.

Furthermore, vulnerabilities in our interconnected digital infrastructure pose significant risks to security and resilience. As discussed, cyberattacks and breaches can have severe consequences, including the disruption of critical services, the compromise of sensitive information, and threats to national security.

Moreover, cyber reliance extends beyond individuals and organizations, encompassing broader systemic dependencies that underpin global networks and economies. The interconnected nature of cyberspace means that disruptions in one area can trigger cascading effects across industries and nations, amplifying the impact and complicating mitigation efforts.

Chapter 12
Strengthening ICS Attack Resiliency: Advanced Incident Response, Threat Intelligence, and Cyber-Physical Systems Monitoring

Abstract This chapter provides a comprehensive examination of the active cyber defense cycle, encompassing threat information collection, detection techniques, and incident response processes. It offers insights into advanced persistent threats (APTs) and equips readers with skills in network traffic analysis and endpoint monitoring to identify suspicious activities. The chapter emphasizes the importance of cyber resilience and adherence to regulatory frameworks, such as the NIST Cybersecurity Framework. It explores the application of machine learning and deep learning techniques for threat identification and the protection of cyber-physical systems (CPS) against cyberattacks. Additionally, the chapter discusses strategies for cyber threat hunting, task prioritization procedures, and the deployment of honeypots and decoy systems. The following chapter will address resilience testing, red teaming exercises, and recovery strategies for Industrial Control Systems (ICS) in the aftermath of a cyberattack.

Keywords Active cyber defense · Threat information collection · Threat detection · Incident response · Advanced persistent threats (APTs) · Network traffic analysis · Endpoint monitoring · Cyber resilience · NIST cybersecurity framework · Machine learning · Deep learning · Cyber-physical systems (CPS) · Cyber threat hunting · Honeypots · Decoy systems · Resilience testing · Red teaming · ICS recovery strategies

12.1 Introduction

Industrial Control Systems (ICSes) are very important for running key services in areas like energy, transportation, and manufacturing. As these systems become more connected and automated, they also become more vulnerable to cyberattacks. These attacks can cause serious problems, such as service interruptions, financial loss, or even safety issues.

To keep these systems safe, it is important to test and improve how well they can handle cyber threats. This second part of our study looks at different ways to prevent and respond to attacks on ICSes.

  311
M. A. Rahman et al., *Securing Industrial Control Systems*,
https://doi.org/10.1007/978-3-032-03018-4_12

We focus on key areas that help protect ICSes, such as analyzing network traffic, using decoy systems, using threat intelligence, and applying active defense methods. By studying these areas, we want to help organizations better understand how to detect threats, respond to incidents, and reduce the risks of cyberattacks.

12.2   Active Cyber Defense Cycle

The Active Cyber Defense Cycle is a continuous and proactive way to protect digital systems. It includes several steps to help organizations find, prepare for, and react to cyber threats before or as they happen.

A key part of this approach is root cause analysis. In this step, security teams examine system logs to understand serious alerts from detection systems. This includes checking how malware may have entered a computer or device. Most organizations now use layered security systems to improve protection. Figure 12.1 is based on [131].

12.3   Active Cyber Defense Cycle

Active Cyber Defense includes five main parts: using threat intelligence, improving visibility, detecting threats, responding to incidents, and controlling the threat environment [63]. The focus is on acting before a cyberattack happens by using

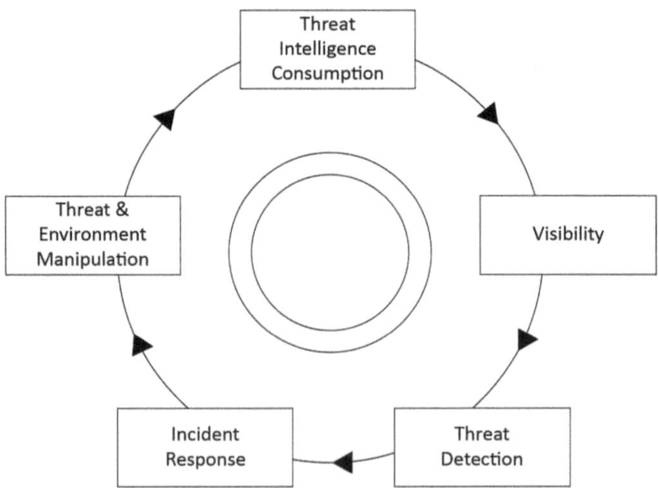

Fig. 12.1 Active cyber defense cycle

useful information to strengthen defenses and reduce risks. The cycle includes the following steps:

1. **Threat Intelligence Consumption**
2. **Visibility**
3. **Threat Detection**
4. **Incident Response**
5. **Threat Environment Manipulation**

12.4 Threat Intelligence Consumption

Threat intelligence is the process of gathering and understanding information about possible or real cyber threats. This is a key part of staying ahead in cybersecurity.

This information can come from many sources—forums, security companies, government agencies, and past incident reports. It includes useful data like malware signs, suspicious IP addresses, and new types of attacks. By using this information, organizations can better understand how attackers work, what methods they use, and what risks are rising. This helps them prepare, strengthen their systems, and respond quickly to threats. The use of threat intelligence improves detection, incident handling, and overall security. It also helps decision-makers choose the right tools, spend wisely, and reduce risk.

Visibility

Visibility means knowing what is happening across digital systems—networks, devices, apps, and data. This is important because it allows security teams to spot unusual behavior and react quickly. Tools like log analyzers, network monitors, and endpoint detection systems help organizations watch their systems in real time and stay alert to any problems.

12.5 Threat Detection

With good visibility, threat detection becomes more effective. The goal is to find and stop cyberattacks before they cause harm.

Modern tools use many techniques together—like artificial intelligence (AI), machine learning (ML), behavior analysis, and known threat patterns. These help detect unusual activities, hidden threats, and malware. By detecting threats early, organizations can protect sensitive data, keep services running, and build trust with users.

Incident Response

Incident response is the plan for what to do after detecting a cyberattack. It helps reduce damage and recover normal operations quickly.

This includes finding out what happened, stopping the attack, removing the threat, and restoring systems. Each team member—IT, legal, management, and communication—has a clear role to play. Good communication during an incident is also important. Clear channels help teams work together and respond fast. In some cases, organizations also need to coordinate with law enforcement and other partners. Having a solid plan helps avoid mistakes, speeds up recovery, and prepares the organization for future attacks.

Threat and Environment Manipulation
This step involves taking active steps to test and improve defenses. One method is penetration testing—trying to break into own system like a hacker would, to find weak points.

Another method is using decoys, or honeypots, to attract attackers and study their behavior. Red teaming is a similar technique where experts act like attackers to test how well the organization can defend itself. These activities help improve security by finding weaknesses before real attackers do. They also prepare the team to respond to new types of cyber threats.

12.6 Cyber Threat Intelligence

Cyber Threat Intelligence is the understanding of threats that can harm computer systems, computer networks, digital data, and other digital resources. To protect against cyberattacks, Cyber Threat Intelligence involves collecting, studying, and understanding information related to digital security.

Cyber Threat Intelligence includes several important parts. These include identifying threat actors, understanding their methods and behavior, recognizing warning signs of attacks called indicators of compromise, knowing about known system weaknesses, analyzing harmful software and malicious code, reviewing past and current cyber incidents, and being aware of trends in the digital threat environment.

By using Cyber Threat Intelligence, organizations can improve their digital security through early prevention, quick responses during attacks, smarter planning for security, and cooperation with other digital security teams. Sharing this information also helps reduce risk for others.

As cyber threats continue to grow, Cyber Threat Intelligence is an important part of protecting digital systems, reducing damage, and helping organizations continue to operate without interruptions.

12.6.1 Cyber Threat Detection Examples

The following is a simple Python program that detects cyber threats using log file data. It includes two methods: one for finding known attacks (misuse detection) and one for finding strange or unusual behavior (anomaly detection). Example input and output are provided to help understand the process.

```python
import re
from collections import defaultdict

# Sample log data
log_data = [
    "2024-12-25 12:00:00 - User login from IP 192.168.1.100
        Success",
    "2024-12-25 12:05:00 - User login from IP 192.168.1.101
        Failed",
    "2024-12-25 12:10:00 - Unauthorized access attempt on /
        admin - IP 192.168.1.102",
    "2024-12-25 12:15:00 - User login from IP 192.168.1.100
        Success",
    "2024-12-25 12:20:00 - User login from IP 192.168.1.101
        Success",
    "2024-12-25 12:25:00 - Unusual high traffic from IP
        192.168.1.103"
]

# Define known patterns for suspicious activities
suspicious_patterns = [
    re.compile(r"Failed"),
    re.compile(r"Unauthorized access attempt"),
    re.compile(r"Unusual high traffic")
]

# Detect misuse and anomalies
def detect_threats(log_data, suspicious_patterns):
    detected_threats = defaultdict(list)
    for log_entry in log_data:
        for pattern in suspicious_patterns:
            if pattern.search(log_entry):
                detected_threats[pattern.pattern].append(
    log_entry)
    return detected_threats

# Main process
detected_threats = detect_threats(log_data, suspicious_patterns
    )

# Output the detected threats
if detected_threats:
    print("Detected Threats:")
    for pattern, logs in detected_threats.items():
        print(f"\nPattern: {pattern}")
        for log in logs:
            print(f"- {log}")
else:
    print("No threats detected.")
```

Explanation

- **Sample Log Data:** The script begins with example log entries that are similar to real logs produced by software and systems. Each log includes a date and time, a short description of the event, and an Internet Protocol address.
- **Suspicious Patterns:** The script uses search patterns to describe known suspicious activities. These include failed login attempts, attempts to access systems without permission, and unusually high network traffic.
- **Threat Detection Function:** The `detect_threats` function takes two inputs: the log data and the suspicious patterns. It checks each log entry to see if it matches any suspicious pattern. If a match is found, the entry is saved as a detected threat.
- **Main Process:** The script checks the log data against the suspicious patterns. If it finds any threats, it prints them. If there are no matches, it prints a message saying no threats were found.

This script provides a basic example of detecting cyber threats using log analysis. It can be improved later by adding more advanced detection methods and custom patterns based on specific security needs.

Figure 12.2 outlines the process of threat intelligence, including planning, feedback collection, intelligence dissemination and deployment, structure and enrichment, and analysis.

Crucial stages of the threat intelligence lifecycle:

1. **Planning, Feedback:**

    a. During this early stage, organizations set objectives and identify the threat intelligence they need.

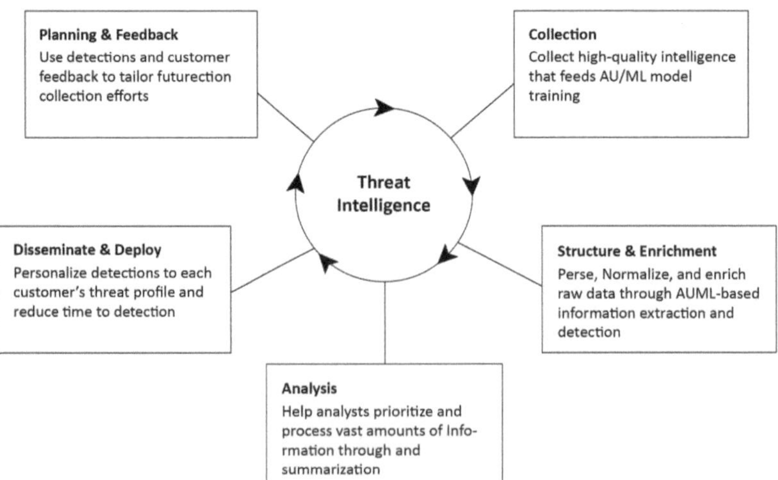

Fig. 12.2 Threat intelligence lifecycle

b. Connect threat intelligence to business and security needs.

c. Insight cycles past help strengthen future intelligence cycle planning.

2. **Collection:**

a. Internal and external sources provide data at this time.

b. Information may contain IoCs, threat actor TTPs, vulnerabilities, and impending dangers.

c. Utilize information sharing platforms, open-source intelligence, automated technology, and subscription services.

3. **Analysis:**

a. Data is processed to gain insights.

b. Analysts assess the context, relevance, and impact of a threat.

c. Involves data analysis, trend identification, and threat and organizational risk classification.

4. **Build and Enhance:**

a. Organize and enhance data to produce a comprehensive threat picture.

b. Use information standards such as STIX (Structured Threat Information eXpression) frameworks.

c. Add context through threat actor profiles, geographical information, or historical attack patterns.

5. **Analysis (Continuing):**

a. Structure and enrichment lead to deeper examination.

b. Consider the organization's resources, weaknesses, and operational environment.

c. Contextualize threat intelligence and rank hazards based on their impact on the organization.

6. **Share and Deploy:**

a. Share analyzed intelligence with the organization's stakeholders.

b. Stakeholders may include decision-makers, security teams, or incident response teams.

c. Implement security measures such as hacker rules, incident response plans, and vulnerability patches based on intelligence.

7. **Repeated Feedback:**

a. Gather feedback from intelligence utilization and deployment.

b. Analysts and security teams examine adopted procedures to assess risk reduction.

c. An ongoing feedback loop in the threat intelligence lifecycle improves planning and data collection for the future.

12.7 Sources of Threat Intelligence

Sources of threat intelligence are different places where we can find information about possible cyber threats. These sources can be public websites, government offices, security companies, or an organization's own records. Some common sources are:

1. **Open Source Intelligence (OSINT):** Information from free and public sources like news, social media, and websites.
2. **Commercial Intelligence:** Information bought from companies that collect and study cyber threat data.
3. **Government and Law Enforcement Agencies:** Information from government offices that monitor cybercrimes and threats.
4. **Information Sharing Communities:** Groups or forums where security teams and companies share threat-related information.
5. **Internal Incident Data:** Information from past or current security problems within the organization.

12.8 Categories of Threat Intelligence

Threat intelligence can be grouped by its use and the kind of information it gives. Each type helps in a different way to improve cybersecurity. By using these different kinds of intelligence, organizations can better understand threats, prepare defenses, and respond faster during an attack. Threat intelligence can be divided into the following types:

1. **Strategic Intelligence:** Gives high-level information about the long-term plans, goals, and trends of attackers. It is mainly used by senior leaders and decision-makers to guide overall security policies and investments. Strategic intelligence helps answer questions like "Who is targeting us and why?" or "What are the risks over the next year?"
2. **Tactical Intelligence:** Focuses on how attackers work, including their tools, tricks, and warning signs. It is useful for security teams who set up defenses like firewalls or antivirus rules. Tactical intelligence shows patterns and behaviors that help block attacks before they happen.
3. **Technical Intelligence:** Gives detailed data about the software, tools, and methods used in attacks. This includes IP addresses, malware hashes, URLs, and other indicators of compromise (IOCs). Technical intelligence is very helpful for quickly detecting and stopping threats in real time.
4. **Operational Intelligence:** Helps understand short-term risks and how well a system can respond to attacks. It deals with real-time information about current threats and active campaigns. This type of intelligence is useful for incident

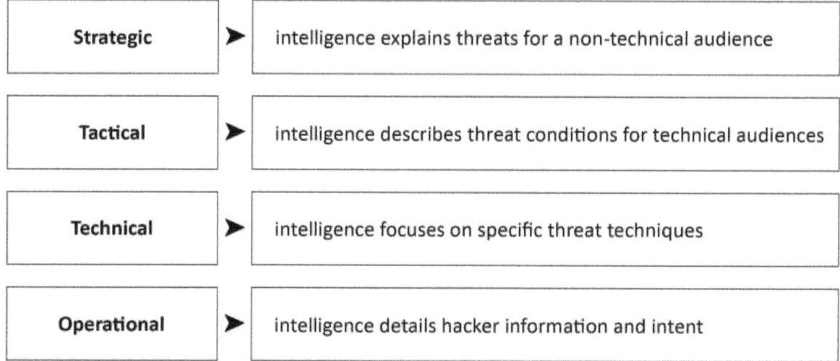

Fig. 12.3 Types of threat intelligence

response teams and helps improve reaction speed and coordination during an ongoing attack.

Each type of intelligence plays an important role in building a strong cybersecurity system. When combined, they provide a full view of both current and future risks, allowing better planning, protection, and response. Figure 12.3 shows some of the main types of threat intelligence, based on [139].

Figure 12.3 shows different types of threat intelligence: strategic, tactical, technical, and operational. It mainly gives information about hackers and how they carry out attacks. Each type explains the kind of threats, tools, and situations it deals with.

12.9 Threat Intelligence Tools

Threat Intelligence Tools are software systems that help find, study, and share information about cyber threats. They collect data from many sources like public sites, threat feeds, and private research. These tools give useful and timely information to help protect computers and networks. As online threats become more complex, these tools are important to stop malware, phishing, ransomware, and advanced attacks. Below is a list of the top 10 cyber threat intelligence tools [90]:

a. **Cisco Secure Malware Analytics:** Checks how malware behaves and rates how dangerous it is. It also supports APIs for easy use and automation.
b. **SpectralOps:** Protects source code using AI tools and scans to help developers.
c. **SIRP:** A risk-based tool that helps teams make smarter security choices.
d. **Echosec:** Collects threat data from social media, forums, and the dark web to help keep businesses safe.
e. **ETP Suite by IntSights:** Gives real-time threat information from the dark web and automates threat responses.
f. **Jit.io:** Adds security features directly into code using open-source methods.

g. Security by OX: Helps manage software risks with automatic tracking and risk scores.

h. Security Connection: Uses intelligence, risk levels, and automation to give useful actions.

i. Zero-Focus: Focuses on threats on social media and protects users from scams and cybercrimes.

j. Recorded Future: Uses machines to collect and study real-time data to give a full picture of cyber threats.

12.10 Threat Intelligence Platform

A Threat Intelligence Platform (TIP) is a full system that collects, studies, and manages information about cyber threats. It helps companies understand and deal with threats more effectively. A TIP brings together data from many sources, such as firewalls, antivirus tools, and public threat feeds, to create a clearer picture of the risks facing an organization. One of the main benefits of a TIP is that it automates the process of sorting and analyzing large amounts of threat data. This saves time and reduces human error. TIPs also allow security teams to share intelligence with other organizations, improving overall defenses. In addition, TIPs can be used to send alerts, update defense systems automatically, and support fast responses during attacks. By using a TIP, companies can stay ahead of attackers and protect their systems more efficiently. Figure 12.4 shows how a TIP works.

Figure 12.4 shows different cybersecurity tools such as Security Information and Event Management (SIEM), Firewall, Intrusion Prevention System (IPS), and Security Orchestration, Automation, and Response (SOAR).

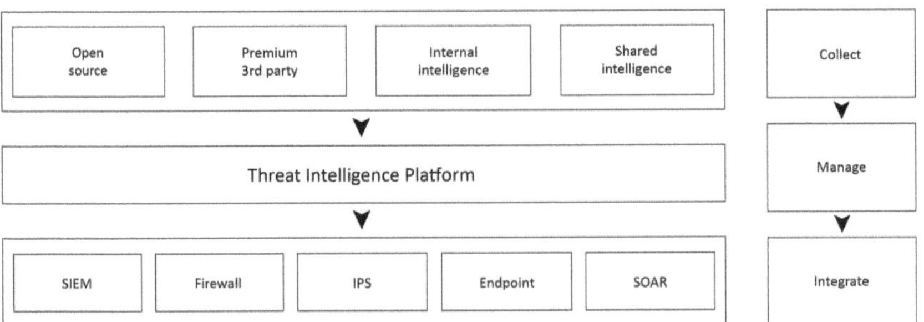

Fig. 12.4 Threat intelligence platform

12.11   Advanced Persistent Threats

An Advanced Persistent Threat (APT) is a type of cyberattack where a person or a group of people secretly break into a network and stay there for a long time to steal important data.

1. APTs usually focus on important targets like government systems or large companies. The attacker studies the target carefully before the attack. The damage caused by APTs may include:

   a. Stealing secret information such as trade secrets or patents.
   b. Leaking personal data, like customer or employee records.
   c. Disrupting important systems like databases.
   d. Taking full control of websites.

2. APT attacks need more skill, time, and money than normal cyberattacks. These attacks are often done by expert hacker groups and sometimes supported by governments. APTs are different from other threats because:

   a. They are more advanced and harder to detect.
   b. The attacker stays hidden in the system for a long time to collect more information.
   c. The attack is done step-by-step by people, not automatically.
   d. The goal is to take over the whole network, not just a part of it.

Attackers usually start with basic tricks like SQL injection, cross-site scripting (XSS), or remote file inclusion (RFI) to get into the network. Once inside, they use harmful software like Trojans or hidden backdoors to stay and move through the system.

12.11.1   Advanced Persistent Threat Lifecycle

An APT attack usually follows a series of steps, called the APT lifecycle.

1. APT attacks last longer and are more advanced than normal cyberattacks.
2. **Set the Goal:** Decide who to attack, what to get, and why are you doing the attack.
3. **Find Helpers:** Choose skilled team members and try to get inside help from the target system.
4. **Build or Collect Tools:** Use existing tools or create new ones needed for the attack.
5. **Study the Target:** Learn about the target's systems, people, software, and hardware.
6. **Test Detection:** Do a small test attack to see if the system detects it. Watch how the system reacts.

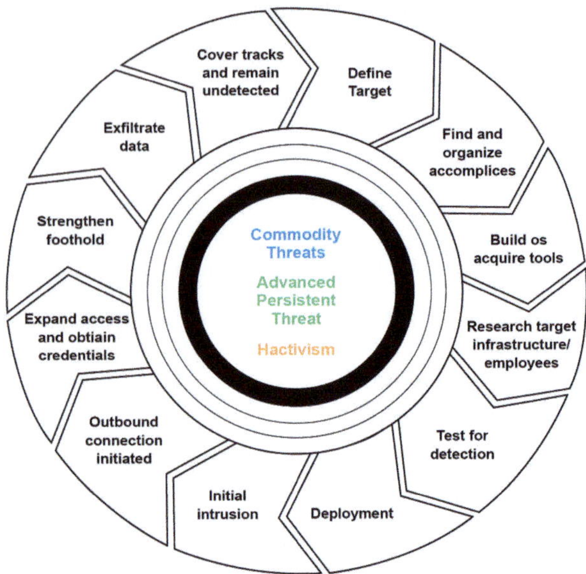

Fig. 12.5 Advanced persistent threat lifecycle

7. **Start the Attack:** Begin the main attack and try to get into the system.
8. **First Entry:** After entering, explore the system to locate important targets.
9. **Create Hidden Connection:** Send information out of the system using a secret method.
10. **Increase Access:** Create a fake network inside and use it to get deeper into the system.
11. **Secure Access:** Use system weaknesses to take more control or access other parts of the system.
12. **Steal Data:** Copy important data and send it to a safe place.
13. **Hide Activities:** Stay hidden, delete signs of the attack, and avoid being discovered.

The steps can vary, but a common model includes the phases shown in Fig. 12.5, based on [103].

Figure 12.5 shows the steps in the life cycle of an Advanced Persistent Threat (APT). These steps include hiding tracks, building a strong presence, gaining more access, making outside connections, entering the system, choosing targets, finding helpers, making or stealing tools, learning about the system and people, and testing detection.

12.11.1.1 Important Facts About Advanced Persistent Threats

Advanced Persistent Threats (APTs) are secret and skilled cyberattacks done by expert hackers. These hackers often work for activist groups, criminals, or governments. Key facts about APTs:

1. **Goals and Targets:** APTs usually attack governments, military systems, banks, and other important industries for political or financial reasons.
2. **Attack Methods:** APTs use tricks like social engineering, hidden access, and harmful software to break into systems quietly.
3. **Attack Steps:** APTs break in, stay hidden, gain more access, learn the system, move around, collect data, and complete their goal without being noticed.
4. **Hard to Detect:** Normal security tools often cannot find APTs. Watching network traffic and checking system logs is very important.

12.12 Network Traffic Analysis

Network Traffic Analysis (NTA) is the process of studying the data that moves through a network. It helps improve security, speed, and system performance, and makes sure the network follows rules. NTA tools collect and check network traffic. They learn what normal activity looks like and find suspicious behavior by spotting unusual patterns.

Figure 12.6 shows a layout of a Security Operations Center. It includes parts such as tools for analyzing network traffic, basic data about other data (called metadata),

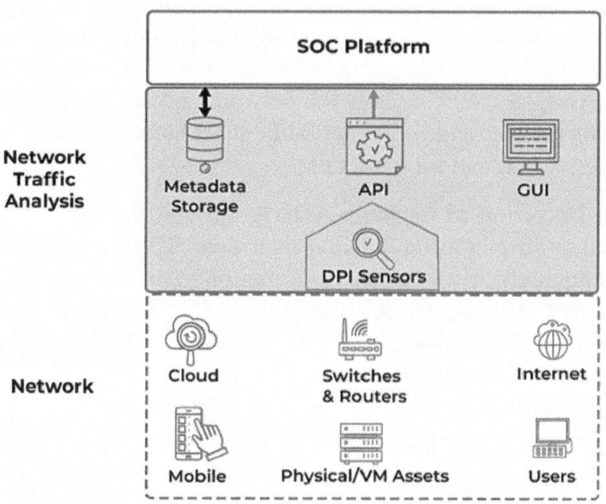

Fig. 12.6 Network traffic analysis

interfaces for connecting software programs (application programming interfaces), user-friendly display systems (graphical user interfaces), data storage units, tools for checking every detail in network packets (deep packet inspection tools), network switches hosted in cloud systems, Internet routers, mobile phones and tablets, both physical and virtual computers, and the people using the system. The figure shows how all these parts are connected and work together in a computer network.

This setup helps security teams respond quickly to problems and investigate after something suspicious happens. Analyzing network traffic helps find slow areas, test how well the network works, and check if the system follows privacy and security rules. It also compares known attack signs with new traffic to detect possible cyberattacks. Checking network traffic is a very important part of staying safe in the digital world. It helps protect the system and make it work better.

12.12.1 Importance of Network Traffic Analysis

Analyzing network traffic is very important for any business that wants to keep its computer network safe and working well. It means watching and recording the data that moves through the network to find unusual behavior, unsafe activity, or weaknesses in the system.

As networks grow more complex, traffic analysis helps find issues like slow Internet paths or overused resources before they become serious problems. By always checking the traffic, businesses can learn how their network is doing, what users are doing, and whether there are any signs of a security problem. This helps prevent digital attacks, reduce time when systems are down, and make the system work more smoothly. In the modern digital world, where cyberattacks are getting more advanced, watching network traffic is important for protecting personal data and following safety rules. It looks out for signs like harmful software or someone trying to break into the system without permission. This helps the security team react quickly. Also, by checking how the network and its data flow behave, it helps make the network faster and more stable. In short, network traffic analysis makes the system safer, faster, and more reliable.

1. **Automatic Detection of Strange Activity:** Network traffic analysis tools can find unusual or suspicious actions on their own. This helps security teams save time and act quickly, using helpful information to understand and fix the issue.
2. **Stable Network Access:** These tools can quickly find what is causing a network issue. This helps the technical team fix the problem faster, which means fewer complaints and less system downtime.
3. **Full View of the Network:** With more use of cloud systems, smart devices, and remote work, it is harder to see everything that is happening on a network. But network traffic analysis tools give a full view of network activity by watching many types of network data. They help find information that other tools may miss and help draw a clear picture of the network.

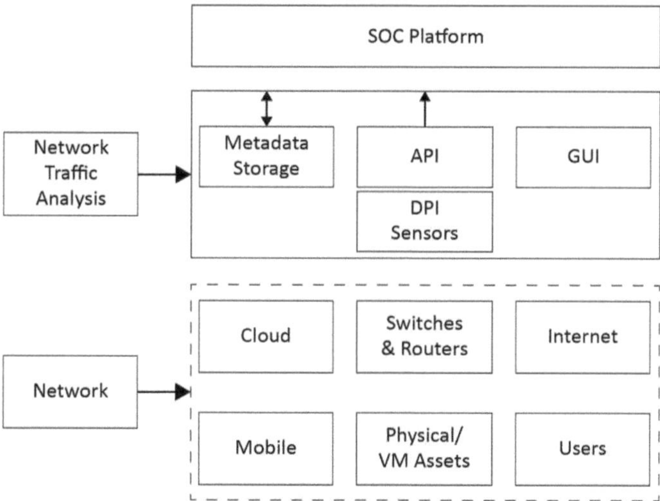

Fig. 12.7 Importance of network traffic analysis

4. **Better Security:** These tools help find unusual data and possible attacks early. By finding threats quickly and stopping attackers before they cause harm, the system becomes more secure. Automated tools also allow the security team to focus on removing threats while simple tasks are done automatically.

Figure 12.7 shows the parts and features of an automated system that finds strange or risky behavior in a computer network. The key parts include fast network performance, easy and stable network access, full visibility of network activity, and strong protection from threats.

12.12.1.1   How to Start Using Network Traffic Analysis

Setting up network traffic analysis means putting in place software and tools to watch and understand how data moves in the network. This helps find strange patterns, security issues, or slow parts of the system. The setup usually includes tools for watching traffic, collecting data, and analyzing that data. All these parts work together to show how well the network is working and how safe it is.

1. **Step 1: Find Data Sources** Look for data sources such as software programs, servers, and network devices within the organization. Using automatic tools can help make this easier.
2. **Step 2: Choose How to Collect Data** Decide whether to use tools with or without agents to collect data. Collecting data without agents gives a basic view but less detail. Using agents gives more detail but may cause storage and processing issues.

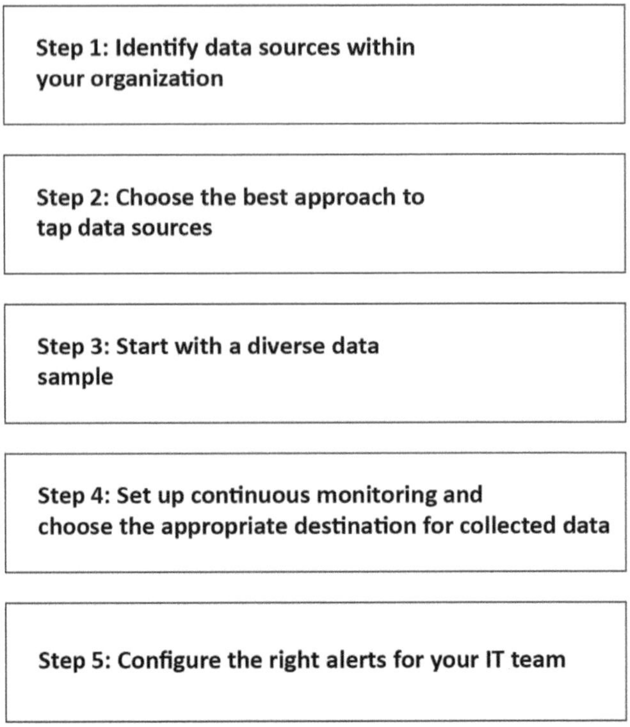

**Step 1: Identify data sources within
your organization**

**Step 2: Choose the best approach to
tap data sources**

**Step 3: Start with a diverse data
sample**

**Step 4: Set up continuous monitoring and
choose the appropriate destination for collected data**

Step 5: Configure the right alerts for your IT team

Fig. 12.8 Network traffic analysis implementation steps

3. **Step 3: Start with a Varied Set of Data** Collect data from different parts of the organization before starting the network traffic analysis. This will help to find problems that affect the whole system.
4. **Step 4: Keep Monitoring All the Time** Watch network traffic regularly to find security problems, slow performance, or useful patterns. Continuous monitoring helps respond quickly to issues.
5. **Step 5: Set Up Alerts for the IT Team** Create alerts based on the size and skill level of the IT team. This helps make sure no major issues are missed. Set the right alert levels to avoid too many notifications and help focus on real problems.

Figure 12.8 shows the steps for managing data correctly in an organization. These steps include finding where the data comes from, choosing how to collect it, using different types of data, always watching the network, setting up a place to store the data, and creating alerts.

12.13   Endpoint Monitoring

Endpoint monitoring means watching and managing every device connected to a network, like computers, servers, and phones. It helps find and stop harmful activity. The system collects and checks the actions of each device. If something strange happens, it can quickly find it and alert the team.

It also helps companies manage mobile devices using special software called Mobile Device Management [136]. This monitoring helps find security risks, like open ports that could be attacked. In today's world, keeping devices safe is a key part of protecting sensitive data and responding to attacks.

Key Features of Endpoint Monitoring

Endpoint monitoring includes watching and managing how well devices such as laptops, desktop computers, servers, smartphones, and tablets are working and how safe they are on a network. The security and health of a company's network depend on these main features.

Figure 12.9 shows an example from [135].

Figure 12.9 illustrates key capabilities related to Keypoint Monitoring. The capabilities include Intelligent-Driven Detection, Data Encryption, Behavioral Analysis, and Regular Software Updates.

The significance of the following essential skills is listed:

1. **Intelligence-Driven Detection:**
   Cyberattackers' methods help security companies identify attack trends and create effective repair solutions.
2. **Behavioral Analysis:**
   Endpoint monitoring automatically identifies anomalous endpoint actions using AI and machine learning to avoid further assaults [135].

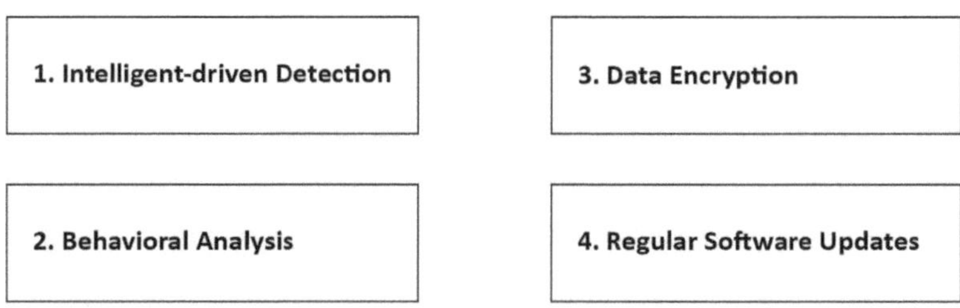

Fig. 12.9 Key capabilities of endpoint monitoring

3. **Data Encryption:**
   Protects client records and source codes by detecting unprotected devices and encrypting viewable data.
4. **Regular Software Updates:**
   Alerts obsolete devices to update software to reduce vulnerabilities and improve cybersecurity against sophisticated attacks.

12.14 Cyber Resilience and Cyber Security Regulations

Cyber resilience refers to an organization's ability to foresee, endure, and bounce back from cyberattacks or interruptions, while guaranteeing the uninterrupted functioning of activities and essential services [16]. Cybersecurity regulations include a set of legally required rules, policies, or standards that are established by governmental or regulatory entities to enforce cybersecurity measures and safeguard sensitive information from cyber threats. The integration of these components is crucial for many businesses as they strengthen essential resources, reduce the impact of cyber threats, guarantee adherence to regulatory requirements, cultivate customer confidence, and improve the ability to withstand interruptions in linked supply networks. Organizations may successfully protect themselves against emerging cyber risks, reduce downtime, and maintain brand reputation in today's digital environment by following cyber resilience principles and regulatory standards [16]. Digital landscape protection is based on two key components:

1. Cyber resilience
2. Cyber security legislation

With an emphasis on readiness and continuity, cyber resilience denotes an organization's ability to foresee, react, respond, and recover from cyberattacks.

Cybersecurity regulations, on the other hand, are codified rules and specifications that are imposed by authorities to protect the privacy of data, harmonize security procedures, and assign responsibility. As a unit, they provide a strong defensive plan that helps firms stay ahead of the curve in terms of cyber threats and meet industry and regulatory standards for a safe online space.

12.14.1 NIST Risk Management Framework

The National Institute of Standards and Technology (NIST) created the structured and all-inclusive NIST Risk Management Framework (RMF) as a means of controlling information security risk. The framework is intended to assist enterprises in managing and reducing cybersecurity risks efficiently. Figure 12.10 is from [138].

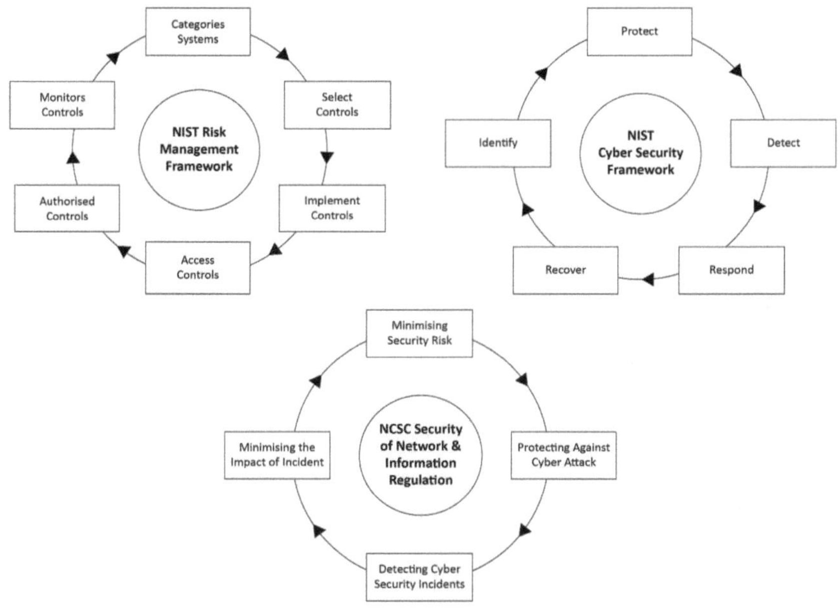

Fig. 12.10   Core security functions, principles, and activities of NIST risk management framework

Figure 12.10 shows different security-related topics like systems, controls, the NIST risk management framework, security controls, cyberattacks, incident detection, and ways to reduce security threats.

12.14.2   The NIST RMF's Key Core Security Functions

The main goal of the NIST Risk Management Framework (RMF) is to give organizations a clear and organized way to handle information security risks. It helps them find, measure, and reduce the risks of using information systems. The main steps of the NIST RMF are:

1. **Prepare:** *Idea:* Understand the background, goals, and resources needed to manage risk.

   Tasks: Set risk limits, plan resource use, and create a risk management plan.

2. **Authorize:** *Idea:* Officially approve the use of information systems after checking how well the security controls work.

   Tasks: Prepare an authorization package with risk assessments, control tests, and documents for review.

3. **Monitor:** *Idea:* Keep checking security controls to make sure they still work and to see if any risks have changed.

Tasks: Set up regular checks, review reports, and share findings with decision-makers.

4. **Assess:** *Idea:* Test and check security controls to make sure they are working correctly.

    Tasks: Run tests using set rules, review results, and decide what to fix first.

5. **Respond:** *Idea:* Build an incident response plan to handle security problems properly.

    Tasks: Make a response plan, manage incidents, and train teams to handle them.

6. **Categorize:** *Idea:* Use impact levels to classify systems and choose suitable security steps.

    Tasks: Decide how systems affect people, assets, and operations, then assign impact levels.

12.15 NIST RMF's Fundamental Security Principles

The NIST RMF is based on a "risk-based approach." This means choosing security steps based on business goals, how much risk the organization can accept, and the possible effects of threats.

1. **Risk-Based Method:** *Idea:* Focus on the most important risks based on goals and possible impact.
2. **Continuous Monitoring:** *Idea:* Regularly check the system to see if the security setup still works well.
3. **Life Cycle Security:** *Idea:* Think about security from the start of system design until it is no longer used.
4. **Link to Enterprise Risk:** *Idea:* Match information security steps with the larger organization's risk plan.

12.15.1 NIST RMF's Core Security Activities

The NIST RMF includes common practices that help manage information security risks in a structured way. These are:

1. **Risk Assessment:** *Task:* Find, measure, and sort risks to people, assets, and operations.
2. **Select and Apply Controls:** *Task:* After sorting systems, pick and apply the right controls to reduce risk.
3. **Evaluate Controls:** *Task:* Test controls to see if they work properly.
4. **Authorize Systems:** *Task:* After checking risks and applying controls, officially approve system use.

5. **Ongoing Monitoring:** *Task:* Keep watching systems and controls to stay up to date with security.
6. **Incident Response:** *Task:* Build and use a plan to deal with security events.
7. **Security Training:** *Task:* Train staff and raise awareness about security risks and safe practices.

12.16 Detection of Cybersecurity Threats in ML and Deep Learning

Machine Learning (ML) can use past malware examples to find new ones, making it helpful for detecting harmful software. Even if malware tries to hide, it still behaves in ways that can be detected. Artificial Intelligence (AI) helps in cybersecurity by finding, stopping, and managing online threats. Machine learning looks for patterns in data to detect unusual behavior, predict attacks, and improve protection.

ML can also use known malware samples to spot new ones, making it useful for malware detection. Even if malicious code is hidden to look like safe code, its behavior can still be caught. This is because ML models can learn the behavior of normal software and then detect anything that acts differently. Deep Learning, which is a part of ML, goes a step further by working with large sets of data and learning more complex patterns.

AI-based systems can also monitor network activities, spot unusual behavior, and take steps to respond. These systems can work in real-time, allowing for faster threat detection and quicker actions to stop attacks before they cause damage. They can alert security teams or automatically block suspicious actions.

One common use of ML in cybersecurity is in spam and phishing detection. Email filters trained with large datasets can recognize suspicious emails, even if the attacker changes the message slightly. Another use is in intrusion detection systems (IDS), where ML models learn what normal network traffic looks like and alert administrators if something unusual happens.

Deep learning models, such as neural networks, are especially powerful because they can analyze huge amounts of data, including logs, files, and system behavior, to identify threats that traditional tools might miss. These models are now being used in advanced antivirus software, endpoint detection and response (EDR) systems, and cloud security platforms [24].

In modern cybersecurity systems, ML and deep learning are used together with other tools like firewalls and antivirus programs. This creates a stronger defense system that can adapt to new types of threats. AI-based models continue to improve over time as they learn from new data, helping organizations stay prepared against both known and unknown attacks. An example is shown in Fig. 12.11, as seen in [144].

Figure 12.11 shows how machine learning (ML) and deep learning (DL) are used in cybersecurity to detect malware, botnets, intrusions, and fileless attacks [7].

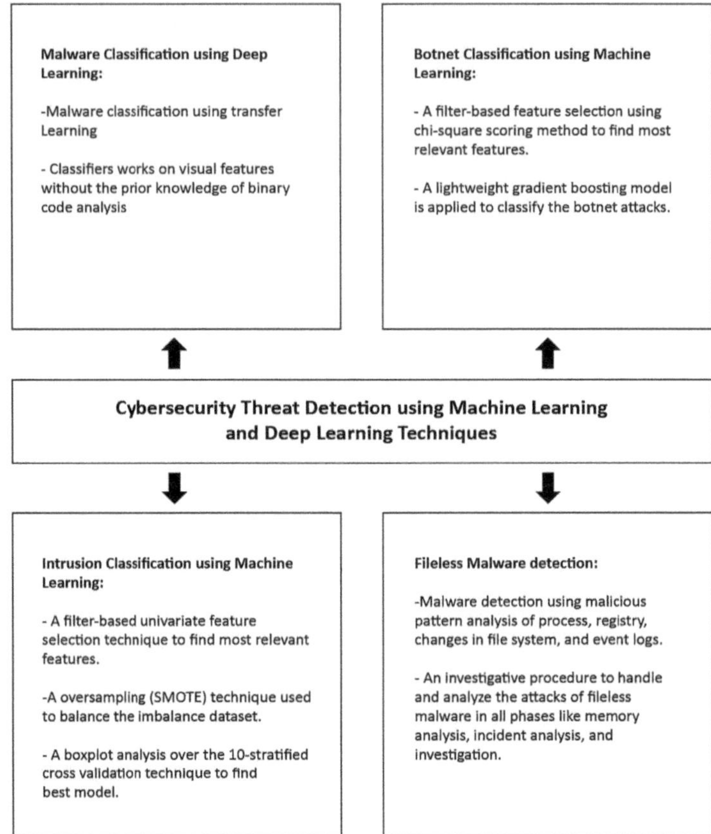

Fig. 12.11 Cybersecurity threat detection in ML and deep learning

It also includes methods like feature selection, oversampling, and gradient boosting to improve how well threats are detected.

12.17 ML-DL Synergy for Cyber Resilience

Machine learning and deep learning have greatly improved to detect cyber threats. These methods scan large sets of data to find patterns and unusual activity using smart models and algorithms.

ML can learn from past data and improve over time, helping it find both known and new threats. DL uses neural networks to look deeper into data and find complex patterns, improving detection accuracy.

By training on labeled data, these models learn to tell good behavior from harmful activity. After training, they can spot malware, phishing, intrusions, and other threats.

Real-time monitoring lets systems react quickly to threats. In short, ML and DL help systems detect and respond to cyberattacks faster, making security better and reducing damage.

12.18   Cyber-Physical System

Being ready in advance is very important to protect cyber-physical systems [98]. These systems combine both digital and physical parts, so they need both cyber-security and physical protection. To keep these systems running smoothly during attacks or disruptions, organizations need detailed plans like a Business Impact Analysis (BIA), Disaster Recovery Plan (DRP), Business Continuity Plan (BCP), and Incident Response Plan (IRP).

The "Defense-in-Depth" (DiD) approach uses layers of protection such as restrictions, variety in controls, hidden system details, and simplified security settings. This makes it harder for attackers to reach important systems. If one layer fails, others still protect the system. The aim is to stop, detect, and respond to attacks before they can do serious harm.

Preventing Cyberattacks on CPS

Cyber-Physical Systems (CPSes) must be protected to keep important operations and assets safe. These systems should be separated into secure zones, and strong access controls like multi-factor authentication and role-based access must be used.

Regular patching and secure system setup help reduce known weaknesses. Continuous monitoring and intrusion detection systems help detect abnormal activity as it happens. Machine learning and anomaly detection tools can find changes in normal patterns that suggest an attack. Besides that, employees must be trained to understand security threats, and every organization should have a response plan ready to act when an attack happens.

By following these steps, businesses can protect CPSes from cyber threats and make them strong and reliable. Figure 12.12 is taken from [98].

Figure 12.12 demonstrates a security strategy to protect against attacks. It includes a Threat Actor that generates an Attack, which is directed towards Controls that aim to protect against the attack. The Controls consist of Deter, Detect, Prevent, and components, each connected to Exploit, Vulnerability, Impact, and Ameliorate future attacks.

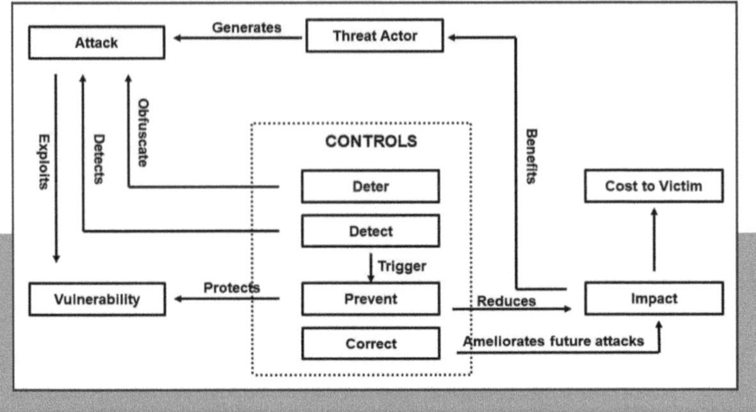

Fig. 12.12 Holistic view to protect CPSes from cyber intrusions

12.19   Attack Types in a Cyber-Physical System

A variety of different types of attacks can jeopardize the security and integrity of a CPS, which combines digital computing and communication with physical processes. These attacks may target either the physical or digital aspects of a system, exploiting vulnerabilities to disrupt or compromise system functionality. In a CPS, the following attack types are frequently seen: DoS attacks, data manipulation, unauthorized access, and physical tampering. These attacks can lead to system malfunctions, damage to critical infrastructure, or unauthorized control over operational processes, posing significant risks to both safety and business continuity. Furthermore, attackers can also exploit communication vulnerabilities within CPS to alter data or inject malicious signals into the system, causing widespread damage. As a result, safeguarding both the physical and digital components of CPS is essential to ensure their secure and reliable operation. Even small disruptions in one part of the system can cause larger failures across connected devices. Therefore, early detection and response to any suspicious activity are key to minimizing the impact of such attacks.

1. **Integer Flow:**

    a. **Occurs:** When arithmetic procedures result in values of integers that are unduly large.
    b. **Description:** The phenomenon that results from arithmetic operations producing integer numbers that are too big to be shown.
    c. **Impact:** Could lead to unanticipated events or security holes.

2. **Overflow of the Buffer:**

    a. **Occurs:** When a buffer is written with more data than it can manage.

    b. **Description:** Frequently occurs when too much data is written to a buffer, which can cause system crashes or illegal access.

    c. **Impact:** Could lead to unauthorized access, code execution, or system crashes.

3. **Worms:**

    a. **Occurs:** Malware that replicates itself and spreads throughout networks without human assistance.

    b. **Description:** Software that propagates via networks and replicates on its own is malicious.

    c. **Impact:** High resource use, quick dissemination, and possible system damage.

4. **Structure Attack with a String:**

    a. **Occurs:** Memory manipulation due to errors in string formatting techniques.

    b. **Description:** Memory manipulation through the use of flaws in string formatting techniques.

    c. **Impact:** Information disclosure, illegal access, or code execution.

5. **Hijacking Control:**

    a. **Occurs:** Malicious code is executed by manipulating a program's control flow.

    b. **Description:** The control flow of a program is redirected to allow for the execution of harmful or unauthorized code.

    c. **Impact:** Unauthorized control, code execution, or compromise of the system.

6. **The Middle Man:**

    a. **Occurs:** Covertly listens in on two parties' conversations and maybe modifies them.

    b. **Description:** Unauthorized communication between two parties intercepted to change the content of the conversation.

    c. **Impact:** Unauthorized access, communication eavesdropping, or data manipulation.

7. **Rootkits:**

    a. **Occurs:** Malicious software that grants persistent, undetectable access to a network or machine.

    b. **Description:** Software created to provide constant, hidden access to a machine or network .

    c. **Impact:** Unauthorized access, command, or stealthy living.

8. **Trojan Horses:**

    a. **Occurs:** Malicious software poses as reliable applications.

   b. **Description:** Malicious software that poses as trustworthy but actually performs harmful operations.
   c. **Impact:** Unauthorized access, data theft, or malicious conduct.

9. **Malware:**

   a. **Occurs:** A general word for software (such as viruses and worms) intended to damage or take advantage of systems.
   b. **Description:** A catch-all word describing a variety of applications designed to compromise or harm computer systems.
   c. **Impact:** Vary according to the particular kind of malware (e.g., viruses, worms).

10. **Viruses:**

   a. **Occurs:** When trustworthy programs are run, malware gets infected and spreads.
   b. **Description:** Malicious software that affixes itself to trustworthy programs in order to infect and proliferate.
   c. **Impact:** Corrupted data, unauthorized access, or system outage.

11. **ARP Spoofing:**

   a. **Occurs:** False MAC address is linked to an IP address by manipulating Address Resolution Protocol (ARP).
   b. **Description:** ARP manipulation is a deceptive technique used to link a fictitious MAC address to a real IP address.
   c. **Impact:** Makes it possible for eavesdropping, network traffic interception, and man-in-the-middle attacks.

12. **IP Spoofing:**

   a. **Occurs:** A network packet's source IP address is forged in order to conceal the sender's identity.
   b. **Description:** The fraudulent act of falsifying the sender's IP address to hide the actual sender.
   c. **Impact:** Makes it possible to carry out attacks surreptitiously, avoid security measures, and conceal one's identity.

13. **Traffic Sniffing:**

   a. **Occurs:** Network traffic is captured and examined to extract private information or prevent unwanted access.
   b. **Description:** Unauthorized interception of network traffic with the intention of gathering private information or gaining illegal access.
   c. **Impact:** Leads to unapproved access to personal data and data interception.

14. **Spoofing:**

   a. **Occurs:** Acts in an attempt to trick and obtain unauthorized access by impersonating a user, device, or system.

b. **Description:** The act of deceiving someone by pretending to be a reputable organization in order to deceive and obtain unauthorized access.

c. **Impact:** May result in data alteration, illegal access, or system compromise.

15. **Eavesdropping:**

a. **Occurs:** Unauthorized monitoring and interception of two parties' communications occurs.

b. **Description:** Listening in on conversations between two people covertly.

c. **Impact:** Violates privacy and exposes private information.

16. **GPS Spoofing:**

a. **Occurs:** Fakes GPS signals to trick navigation systems.

b. **Description:** This is a deceptive technique that involves tricking navigation systems by sending out fake GPS signals.

c. **Impact:** Resulting in misleading guidance and location manipulation is the impact.

17. **Replay Attack:**

a. **Occurs:** Captures and retransmits valid data to gain illegal access.

b. **Description:** The act of illegally recording and retransmitting legally captured material.

c. **Impact:** Causes information to be disclosed and permits illegal access.

18. **Injection of SQL:**

a. **Occurs:** Makes use of SQL query flaws to gain access to or alter a database.

b. **Description:** The act of injecting malicious SQL code to take advantage of weaknesses in database queries.

c. **Impact:** May lead to unapproved access, data manipulation, or exfiltration.

19. **Injections of Code:**

a. **Occurs:** When malicious code is inserted into software or programs.

b. **Description:** Inserting malicious code into software or scripts to change their behavior.

c. **Impact:** A system compromise, data tampering, or illegal access could result from this.

20. **Remote File Injection:**

a. **Occurs:** Puts malicious files onto a server that is far away.

b. **Description:** Intrusion and execution of malicious files on a remote server is a malicious practice.

c. **Impact:** Potential consequences include code execution, data compromise, and unwanted access.

21. **DoS (Denial of Service):**

    a. **Occurs:** Overloads a system, preventing authorized users from accessing it or allowing regular operation.
    b. **Description:** The deliberate act of flooding a system with traffic in order to interfere with regular operation and prevent authorized users from accessing it.
    c. **Impact:** Interrupts services and momentarily disables the system.

22. **Distributed Denial of Service (DDoS):**

    a. **Occurs:** Coordinates the use of several devices to launch a denial-of-service assault.
    b. **Description:** An organized attempt to overload a system or network with traffic from multiple sources, making its mitigation difficult.
    c. **Impact:** Causes a great deal of disruption and puts an inordinate amount of strain on system resources.

23. **Permanent DoS:**

    a. **Occurs:** Causes irreversible damage to a system, making it useless.
    b. **Description:** Intentional acts that render a system permanently damaged and unable to function normally.
    c. **Impact:** Leads to irreversible injury and lasting system collapse.

24. **Cross-Site Scripting:**

    a. **Occurs:** Unauthorized access is gained by inserting malicious scripts into websites that are accessible.
    b. **Description:** Injects malicious scripts to take advantage of vulnerabilities in online applications, which are subsequently executed by people who visit the hacked websites.
    c. **Impact:** Makes it possible for illegal access, data theft, or session hijacking.

25. **Shell Injection:**

    a. **Occurs:** Provides instructions to execute any code in a system shell.
    b. **Description:** Inserts instructions into system shells to enable the execution of arbitrary code with the compromised system's privileges.
    c. **Impact:** Causes code execution, illegal access, or system breaches.

26. **Reflected Attacks:**

    a. **Occurs:** Takes use of holes in web servers when malicious code is mirrored off of them.
    b. **Description:** Exploits weaknesses wherein injected code is mirrored from a web server and runs within the user's browser.
    c. **Impact:** Causes unapproved access, data modification, or code execution.

Figure 12.13 is from [98].

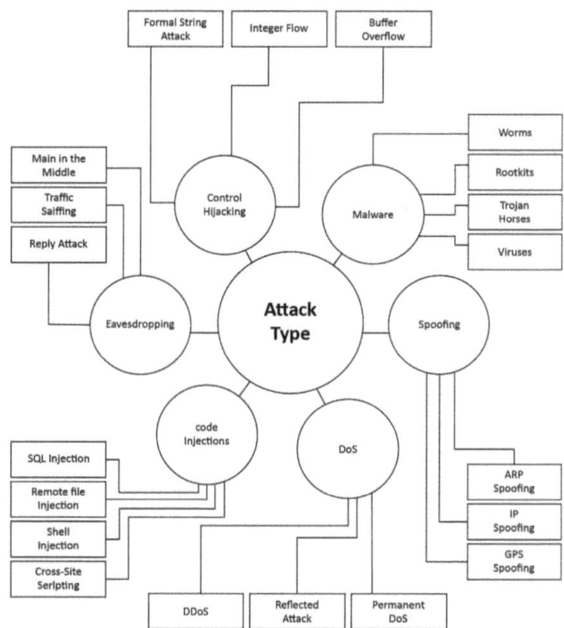

Fig. 12.13 Attack types in a cyber-physical system

Figure 12.13 illustrates various cybersecurity attacks and threats, including Buffer Overflow, Worms, Rootkits, Malware, Viruses, DoS attacks, SQL Injection, and others.

12.20 Cyber Threat Hunting

The proactive cybersecurity approach known as "cyber threat hunting" entails cybersecurity specialists, sometimes referred to as "threat hunters," consistently and continuously scanning the network or systems of a company for security flaws and potential threats. This approach transcends conventional threat detection methods, which frequently rely on recognized patterns and indicators of compromise. Cyber threat hunting seeks to identify threats that might have evaded automated security measures, such as advanced persistent threats (APTs), zero-day vulnerabilities, and cunning insider attacks. Threat hunters use a range of state-of-the-art tools, techniques, and human analysis to locate these threats. Unlike more passive cybersecurity tactics like automated threat detection systems, cyber threat hunting is the active search for threats that have gone unidentified, undiscovered, or unresolved and that could circumvent an organization's automated defensive measures. In order to ensure Internet security, a corporation must take a proactive and forward-thinking strategy by actively searching for hidden security vulnerabilities inside its network . Threat hunters do this in an effort to strengthen the organization's security posture by seeing threats early and taking the necessary precautions to avert potential harm or breaches.

12.21 Importance of Cyber Hunting

A vital component of security operations center services that concentrates on proactive threat detection and/or remediation is cyber threat hunting. Threat-hunting is a highly valuable technical resource that can be utilized to detect warning indicators of a potential cyberattack or to locate malware that utilizes sophisticated evasion methods. Figure 12.14 is from [3].

Figure 12.14 illustrates the Threat Hunting Core Methodology, which includes aspects such as effort, scope, and prioritization, iterative approach, attack framework, user interaction, pursuit of attacks, mapping, visualization, and various hunting techniques.

It also covers components like adversarial attack, techniques, data sources, models, TTPs, tools, data analysis techniques, detection capabilities, hunting scope, intelligence input, threat actors profiles, SOC monitoring, threat intelligence platform, and threat hunting process. The diagram emphasizes proactive searching, transformation, automation, and the continuous feedback loop in threat hunting operations.

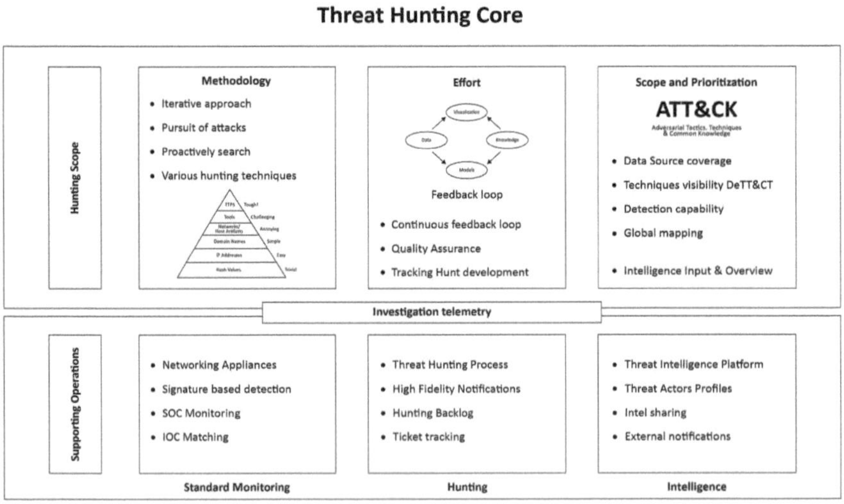

Fig. 12.14 Cyber threat hunting

12.22   Threat Hunting Intelligence Sources

It's important to realize that threat hunting calls for knowledge and a methodical approach before getting into the specifics. Threat hunters need to know where to search, what to look for, how to assess evidence, foresee possible outcomes, and how to respond appropriately. Security teams use the following primary intelligence sources before starting a threat hunt. Figure 12.15 is from [3].

Figure 12.15 illustrates a process that involves creating hypotheses, investigating using tools and techniques, threat-hunting loop analytics, informing and enriching, uncovering new patterns, and dealing with TTPs.

1. **Compromise Indicators (IOCs):**

   a. **Definition:** Signs pointing to a compromised network or endpoint.
   b. **Examples include:** Strange DNS queries, unusual patterns of network traffic, DDoS attacks, abnormal login attempts, atypical activities related to accessing databases, file manipulation, callbacks to the same file, etc. [23].

2. **Attack Indicators (IOAs):**

   a. **Definition:** Indiscriminate signals reflecting the purpose and motive of an ongoing network or endpoint attack.
   b. **Examples include:** Chatter between hosts after an infection, communication over nonstandard ports, surges in SMTP traffic, covert communication between public servers facing the Internet and individual hosts, connection attempts from various geographic locations, and a lower-than-ideal malware clearance rate.

3. **Procedures, Techniques, and Tactics (TTPs):**

   a. **Definition:** Gaining insight into the tactics and behaviors of threat actors to improve threat-hunting skills.
   b. **Suggestions:** For thorough TTP research, visit the official MITRE page. Use the NIST framework to support this and offer countermeasures or defensive tactics.

Fig. 12.15 Threat hunting life cycle

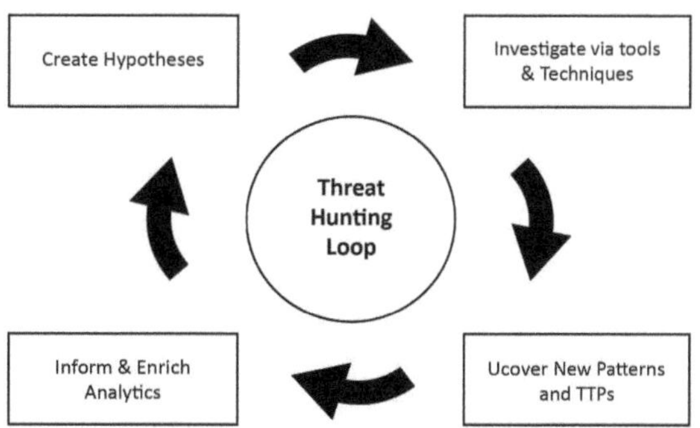

These intelligence sources provide danger hunters with the knowledge and skills they need to recognize and neutralize any threats. Comprehending IOCs, IOAs, and TTPs is crucial for an all-encompassing threat-hunting approach.

12.23 Threat Hunting Life Cycle

Cybersecurity experts use the Threat Hunting Life Cycle, a methodical process, to proactively find and eliminate possible security risks on an organization's network. In order to find hidden threats, irregularities, and malevolent activity that might elude conventional security procedures, it entails a number of phases or actions. The Threat Hunting Life Cycle comprises the following crucial stages:

1. **Create Hypotheses:**

   a. Start by understanding the organization's network, systems, applications, and attack surfaces.
   b. Form theories about undiscovered risks based on knowledge of the surroundings and prospective threat landscape.

2. **Inform and Enrich Analytics:**

   a. Collect relevant data from the organization's logs, network traffic, endpoint data, and threat intelligence feeds.
   b. Normalize and correlate the data to create a baseline and find abnormalities or trends indicating malicious behavior.
   c. Utilize threat intelligence to detect Indicators of Compromise (IoCs) and attack trends, improving analytics.

3. **Discover New TTPs and Patterns:**

   a. Advanced Analytics: Find odd patterns, anomalies, and departures from norms using advanced analytics, machine learning, and data mining.
   b. Pattern Recognition: Identify threats that may not match attack signatures or IoCs using statistical analysis and pattern recognition.

4. **Investigate Tools and Techniques:**

   a. Use automatic and manual technologies to investigate abnormalities and dangers.
   b. Employ specialized threat-hunting tools and platforms to analyze and investigate possible threats.
   c. Analyze network traffic and endpoint activity to identify threats and determine compromises.
   d. Collaborate with incident response teams and exchange investigative findings to improve cybersecurity.

THREAT HUNTING STEPS

1. THE TRIGGER
Points threat hunters to a specific area of the network for investigation. When potential malicious activity is detected

2. INVESTIGATION
Threar hunter uses EDR (Endpoint Detection and Response) technology to deep dive into potential malicious compromise of a system

3. RESOLUTION
Communicating malicious activity intelligence to operations and security teams so they can respond to the incident and mitigate threats

Fig. 12.16 Threat hunting steps

Threat Hunting Step

Threat hunting is a proactive way to deal with cybersecurity problems that aims to find and reduce possible threats that might get past regular security measures. The steps of threat hunting involve looking for signs of bad behavior in a company's network, systems, and apps in a planned way. Several crucial actions are included in the Threat Hunting Life Cycle to proactively detect and manage network threats, ensuring timely identification, investigation, and resolution of emerging security issues. Usually, these actions consist of. Figure 12.16 is from [40].

Figure 12.16 represents the steps involved in threat hunting, including Trigger, Investigation, and Resolution.

1. **The Trigger:**

   a. Started by powerful detection techniques detecting suspicious behavior.
   b. Hypotheses regarding emerging risks might potentially trigger targeted system or network analysis.
   c. Example: Finding sophisticated threats utilizing fileless malware.

2. **Investigation:**

   a. Threat hunters use EDR to examine unusual activity and possible breaches.
   b. Continues until the activity is benign or malevolent conduct is understood.

3. **Resolution:**

   a. Operations and security teams get harmful activity intelligence to react and mitigate risks.

Threat Hunting Capability Maturity Model	Level 1 INITIAL	Level 2 MANAGED	Level 3 DEFINED	Level 4 QUANTITATIVELY MANAGED	Level 5 OPTIMISING
People	• Existing SOC analyst • Resourcing needs not known • Training needs not known • Performance not managed • Lack of career development plan • Normal systems behaviour not sufficiently understood	• Threat Hunting lead • Informal view of resourcing • Informal view of training • Performance is qualitatively managed • Career development informally managed • Normal systems behaviour is moderately understood	• Dedicated threat hunters • Formal recruitment plan • Formal training plan • Performance expectations defined with role profiles • Formalised career development plan • Normal systems behaviour is fully understood	• SOC analysts rotated for L&D • Succession plans in place • Training completion tracked • Metrics utilised for team performance • Mission critical systems identified	• Teams integrated across SOC • Resourcing needs integrated • Training needs integrated • Improvement plans to address underperformance • Situational awareness
Process	• Hypothesis generation is unstructured • Hunts occur od-hoc, if at all • Little or no data collected • Little understanding of anomalies indicative of malicious activity • Abnormalities not routinely searched for	• CTI and Domain Expertise used to generate hypotheses and prioritisation by lead • Hunts occur occasionally • Moderate data collection from key areas • Basic threat feeds with IOCs utilised • Targeting of IOCs at bottom of POP	• CTI and Domain Expertise used to generate hypotheses and prioritisation by lead • Hunts occur occasionally • Moderate data collection from key areas • Basic threat feeds with IOCs utilised • Targeting of IOCs at bottom of POP	• Manual risk scoring e.g.Crown Jewels • Hunts occur frequently • Moderate data collection from most of estate • CTI tailored to organisation • Targeting of IOCs at top of POP	• Automated risk scoring e.g machine learning • Hunts occur continuously • High data collection from full estate • Hunt analytics and IOCS shared across community • Automated TTP and campaign tracking
Tools	• Reactive SOC tools • Little or no automation • Little or no documentation produced	• Basic searching via text or SQL-like queries • Automatic matching of IOSc • Documentation using basic office suites	• Statistical analysis techniques • Library of hunt procedures automated on regular schedule • Central workflow and knowledge repository tools • Lab environments used to aid hypothesis generation and testing	• Visualisation tools utilised,and analytics tested for effectiveness • Library of hunt procedures automated on frequent schedule • Dashboards utilised	• Machine learning is leveraged, with horizon scanning maintained • Library of hunt procedures automated continuously • Central workflow and knowledge repository are integrated and shared

Fig. 12.17 Threat hunting capability maturity model

b. Malicious and benign data may improve automated technologies without human interaction.

Threat Hunting Capability Maturity Model

The Threat Hunting Maturity Model outlines how well-equipped a company is to detect and neutralize threats online. The Hunting Maturity Model level, with HMM0 being the least capable and HMM4 being the most efficient, rises with the business's level of ability. Figure 12.17 is from [40].

Figure 12.17 illustrates different levels of Threat Hunting capabilities.

12.24   Triage in Cyber Security

Cybersecurity triage is a key incident response approach for managing today's various digital threats and vulnerabilities. Security teams may efficiently deploy resources by quickly reviewing and ranking security incidents by severity. The goal is to quickly address the most significant risks to speed up IT incident response and avoid escalation. Triage prioritizes urgent and severe occurrences above less crucial ones. It simplifies incident response, protects analysts from alert overload, and guides resource allocation for efficient issue management and resolution. Triage is the key to expedient incident containment, elimination, and recovery.

12.25 Triage Analysis Process Steps

The triage analysis process plays a critical role in cybersecurity by enabling security teams to efficiently detect, assess, and respond to security incidents. It ensures a structured approach to handling potential threats and minimizing their impact. The process begins with an **incident being declared**, which is triggered by security alerts, anomaly detection, or suspicious activity reports. This phase marks the official recognition of a security event that requires immediate attention.

Once an incident is declared, the **investigation phase** commences, where security analysts gather information, assess the nature of the incident, and determine its scope and severity. This involves reviewing system logs, monitoring network traffic, and analyzing forensic evidence to identify the root cause and affected systems. Proper classification of the incident is crucial at this stage to prioritize response efforts effectively.

Following the investigation, the **fixing phase** involves mitigation efforts to contain and remediate the issue. This may include applying patches, revoking compromised credentials, isolating affected systems, or deploying countermeasures to prevent further exploitation. Security teams work swiftly to eliminate the vulnerability or threat while ensuring that normal business operations are minimally disrupted.

Once the immediate threat is addressed, the **monitoring phase** begins. This phase involves continuous observation of the affected systems to confirm that the issue has been resolved and that no residual threats persist. Security teams analyze system performance, monitor logs for any suspicious activity, and validate the effectiveness of the applied fixes. If additional anomalies are detected, further actions may be taken to strengthen security defenses.

After thorough monitoring and verification, the incident moves to the **closure phase**, where the incident is officially marked as resolved. A post-incident review is conducted to analyze the root cause, document lessons learned, and recommend improvements to security policies and response protocols. Organizations use these insights to refine their cybersecurity strategies and enhance preparedness for future incidents.

A structured triage analysis process is essential for minimizing damage, reducing response time, and ensuring a proactive cybersecurity approach. Figure 12.18 visually represents the triage analysis process, showcasing the different stages involved in handling a security incident [40]. By following these systematic steps, organizations can enhance their cybersecurity resilience, prevent recurring threats, and maintain a robust security posture [30].

Figure 12.18 depicts different stages of an incident resolution process. It includes statuses such as Incident Declared, Investigating, Fixing, Monitoring, and Closed.

1. **Detection:** Verify security alert or event not a false positive.
2. **Scoping:** Quickly examine event to find attack information, impacted assets, and other signs.

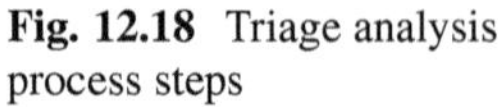

Fig. 12.18 Triage analysis process steps

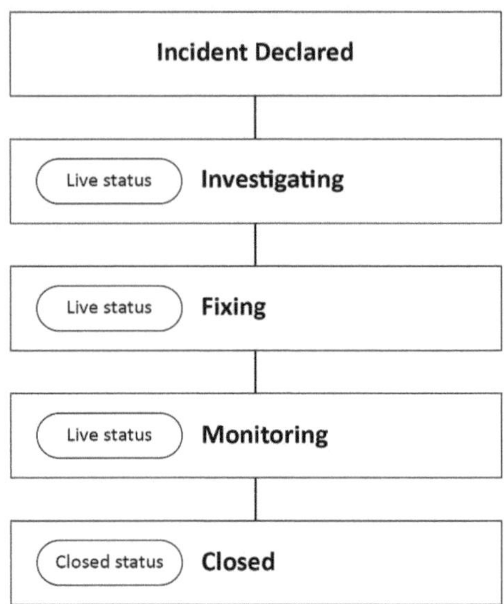

3. **Severity Classification:** Based on probable effect and harm, classify severity (low/medium/high).
4. **Escalation:** Report the event to relevant people depending on severity.
5. **Containment:** Initiate containment of high/critical incidents to isolate and limit damage.
6. **Queuing:** Add lower-severity situations for resource-based response.
7. **Eradication:** For major events, remove environmental risks.
8. **Recovery:** Restore systems and data after catastrophic incidents.
9. **Circle back:** Analyze and triage fresh security warnings.

12.26 Difference Between Triage and Threat Intelligence

Triage and threat intelligence are two critical components of a cybersecurity strategy, but they serve different purposes and are applied at distinct stages of incident response and threat management.

Triage refers to the initial process of assessing and prioritizing security incidents or alerts to determine their severity and potential impact. It involves filtering out false positives and categorizing the events based on their urgency, so that resources can be allocated effectively to the most critical issues. The primary goal of triage is to ensure that the most pressing threats are addressed first, allowing security teams to respond promptly and prevent further damage.

Threat intelligence, on the other hand, involves the collection, analysis, and dissemination of information regarding potential or active cyber threats. This

Table 12.1 Comparison of triage and threat intelligence

Aspect	Triage	Threat intelligence
Purpose and Focus	Incident response and prioritization of incidents	Proactive collection, analysis, and understanding of potential threats and vulnerabilities
Timing	Reactive, triggered by incidents	Proactive, ongoing effort unrelated to specific incidents
Action vs. Analysis	Immediate actions for incident resolution	Analysis and understanding of potential threats for strategic decision-making
Scope	Focuses on managing and prioritizing existing incidents	Broad analysis of various sources for emerging threats, vulnerabilities, and attacker behaviors
Outcome	Immediate incident resolution and containment	Enhanced situational awareness, informed decision-making, and improved security strategies

information typically includes data about attack methods, tactics, techniques, and procedures (TTPs), as well as indicators of compromise (IOCs) and threat actor profiles. Threat intelligence helps organizations anticipate and prepare for future attacks by providing insights into emerging threats and helping to strengthen defensive measures. It plays a vital role in proactive cybersecurity, helping to identify vulnerabilities before they can be exploited.

While triage is more focused on immediate response to security incidents, threat intelligence is a proactive and ongoing process that helps organizations stay ahead of evolving threats. Together, these two components provide a comprehensive approach to managing cybersecurity risks effectively (Table 12.1).

12.27 Honeypots in Cybersecurity

A security tool called a honeypot sets up a virtual trap to entice intruders. Attackers can take advantage of flaws in purposefully compromised computer systems, which can examine to strengthen security protocols.

An organization's intrusion detection system (IDS) and threat response should be improved with the help of a honeypot to put it in a better position to manage and stop attacks. Production and research honeypots are the two main types. To attract and watch cyberattackers, honeypots are set up on vulnerable systems or networks. The system traps attackers, exposing their methods and objectives. This method informs companies about emerging threats and strengthens their defenses. Figure 12.19 is from [82].

Figure 12.19 shows a network set up with a Scanning Attack coming from the Internet, going through a DMZ, and ending at a HoneyPOT. The attack is logged and triggers actions on the Service Router, HoneyPOT, and Firewall in the network hub, which is linked to the internal network.

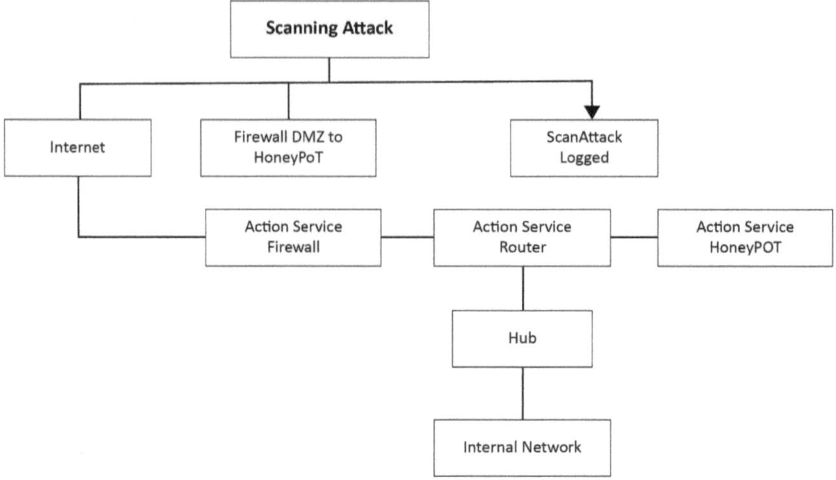

Fig. 12.19 Architecture of a network with a honeypot

12.28   Types of Honeypots in Cybersecurity

Honeypots are fake networks or systems that are meant to attract attackers and find out their strategies, techniques, and procedures (TTPs). Inside the field of hacking, honeypots come in different types, each with its own use and amount of contact. These are the main kinds:

1. **Low-Interaction Honeypots:** Low-interaction honeypots replicate a restricted set of services and have little attacker engagement.
2. **Medium-Interaction Honeypots:** Medium-interaction honeypots replicate more services and interactions than low-interaction ones, but are not entire systems.
3. **High-Interaction Honeypots:** High-interaction honeypots simulate a full computing environment for attackers.
4. **Pure Honeypots:** Pure honeypots are systems or networks used only to entice and investigate harmful behavior.

12.29   Cyber Incident

A cyber incident is an occurrence that may compromise the availability, confidentiality, or integrity of digital data or information systems. The Federal Government is particularly concerned about cyber events that cause substantial harm.

Since malware includes a wide range of attack types, including ransomware, trojans, spyware, viruses, worms, keyloggers, bots, cryptojacking, and other attacks

Fig. 12.20 Cyber incident
life cycle

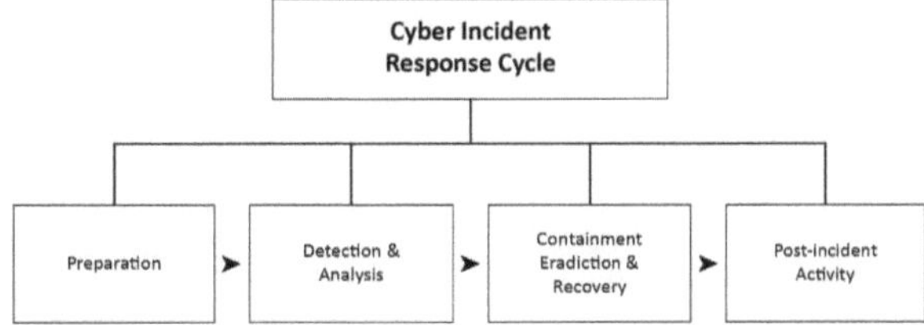

that exploit software for nefarious purposes, malware is the most prevalent kind of cyberattack. To identify and stop cybersecurity attacks, it relies on the use of preventive measures like firewalls and software updates.

Cyber Incident Life Cycle

The process of responding to and handling a cybersecurity problem is referred to as the "Cyber Incident Life Cycle." These phases usually involve containment, eradication, recovery, detection and analysis, preparedness, and post-event actions. To properly handle and mitigate the effects of a cybersecurity event, each stage is essential.

1. **Preparation:** Companies prepare by assessing risks, scanning for vulnerabilities, and creating an incident response plan.
2. **Detection and Analysis:** This phase detects suspicious activity by continuously monitoring systems and networks. To determine the severity of an abnormality, analysis and forensic investigation are done.
3. **Containment, Eradication, and Recovery:** The goal is to isolate compromised systems to avoid additional harm. To maintain business continuity, malicious components are eliminated, and systems are restored.
4. **Post-Incident Activity:** Post-mortems follow event resolution. The event, root cause analysis, and improvement areas are reported.

Figure 12.20 is from [14].

Figure 12.20 represents the Cyber Incident Response Cycle. It includes stages like Containment, Preparation, Detection and Eradication, Post-incident Analysis, and Recovery Activity.

12.30   Decoy Systems in Cybersecurity

In cybersecurity, decoy systems are deliberately placed elements or entities within a network that imitate genuine resources, programs, or weaknesses. Their main objective is to deceive would-be assailants by directing their focus and interactions to these mimicked components. These spoofs are designed to seem just like real systems or services, which makes it easier for attackers to interact with fictitious, non-critical targets.

Example Consider a corporate network with various servers and workstations. To deceive potential attackers, decoy systems could be deployed, resembling high-value targets like a finance server. If an attacker engages with this decoy server, security teams are alerted, allowing them to analyze the attack and fortify the actual finance server's defenses.

Decoy systems are crucial in cybersecurity since they improve threat detection, provide early alerts for possible security breaches, and collect knowledge on attackers' strategies and objectives. When implemented and overseen efficiently, they may greatly enhance an organization's capacity to safeguard against cyber threats and secure its vital assets.

12.30.1   Key Purposes of Decoy Systems in Cybersecurity

Decoy systems, commonly referred to as honeypots or deception technology, provide many important functions in cybersecurity. These include improving the detection of threats, collecting valuable information about attacker strategies, and providing timely alerts about possible security breaches. Decoy systems function as strategic elements inside a network, imitating authentic assets with the purpose of luring and misleading attackers. They have significance for several reasons:

1. **Misleading Attackers:** Decoy systems create a false environment, fooling attackers into thinking they have discovered targets.
2. **Realistic Simulation:** The decoys simulate production vulnerabilities and setups to seem legitimate, inviting attackers.
3. **Early Threat Detection:** Security teams may spot suspicious activity and threats early by using decoy systems.
4. **Gathering Threat Intelligence:** Attacker interactions with decoy systems reveal attack methods, tools, and intentions.
5. **Enhancing Incident Response:** Security teams need decoy systems to watch and analyze attacker activity for incident response.
6. **Buy Time and Delay Attacks:** Decoy system attackers spend time and resources postponing their assault.

12.31   Relevance of Decoy Systems in ICS Cyberattacks

Honeypots or decoy networks are a strategic way to shield Industrial Control Systems (ICS) from cyberattacks. These considerations demonstrate the importance of decoy systems in ICS security in Table 12.2.

Table 12.2 Relevance of decoy systems in ICS cyberattacks

Aspect	Relevance to ICS cyberattacks
Identifying targeted attacks	Decoy systems imitate ICS parts to draw in and detect attackers trying to take advantage of weaknesses
Understanding attack tactics	Decoy systems facilitate threat actor analysis of attack strategies and tactics in ICS environments
Diverting and misleading attackers	Attackers' focus is diverted from real ICS components by decoy systems, which also reveal information about their tactics
Enhanced incident response	In ICS contexts, interaction with a decoy system initiates alarms and notifications, facilitating quick incident response and reducing possible harm

12.32 Resilience Testing and Red Teaming for Industry Control System

The cybersecurity technique known as resilience testing evaluates a system's capacity to resist and bounce back from a variety of cyber threats and delays. To assess an ICS's ability to sustain essential operations and services in the face of unfavorable circumstances, such as cyberattacks, natural disasters, or other occurrences, controlled and methodical testing is required.

Red teaming is a cybersecurity technique that tests the efficacy of an organization's security defenses by mimicking actual attacks. Within the framework of information and communication security, Red Teaming assesses an organization's ability to defend against external and internal threat actors that impersonate crafty and adaptable adversaries. Comparison tables are shown in Table 12.3.

12.33 Industry Control System Resilience Assessment for Post-Cyberattack

ICS resilience evaluation refers to the process of systematically assessing the ability of an ICS to withstand, recover from, and adapt to cyber threats following an attack. Post-cyberattack mitigation involves analyzing how well the system can continue its operations while addressing vulnerabilities and weaknesses that were exposed during the incident. This evaluation is essential for identifying areas where security

Table 12.3 Comparison of resilience testing and red teaming in cybersecurity

Categories	Resilience testing	Red teaming
Purpose	Measure system's ability to withstand attacks	Simulate real-world attacks to assess defenses
Execution	Controlled and systematic	Mimics sophisticated, adaptive threat actors
Focus	Internal system behavior	External and internal threat actor behaviors
Scope	Typically broader system analysis	Focused on specific target and attack vectors
Testing frequency	Periodic and regular	Occasional or as needed
Team expertise	Strong understanding of ICS	Diverse expertise covering ICS and threat actor tactics

measures need to be reinforced to prevent future attacks and ensure operational continuity.

The assessment process consists of multiple phases, starting with analyzing the extent of the damage inflicted on the ICS. This includes identifying which components were compromised, whether data integrity was affected, and the impact on overall system functionality [123]. Security analysts perform forensic investigations to trace the origin of the attack, understand the techniques used, and uncover any lingering threats that could lead to future security breaches.

Another critical aspect of resilience assessment is identifying vulnerabilities that were exploited during the attack. Understanding these weak points allows organizations to implement necessary security patches, update configurations, and deploy stronger access controls to mitigate the risk of recurrence. Additionally, evaluating the effectiveness of incident response and recovery protocols is crucial. This involves assessing how quickly the organization detected, contained, and remediated the attack, as well as how efficiently operations were restored to normal.

Organizations use the insights gained from this assessment to enhance their cybersecurity posture and develop robust resilience strategies. Implementing proactive security measures, such as network segmentation, threat intelligence sharing, and continuous monitoring, strengthens ICS against future threats. Moreover, conducting regular resilience testing and training security personnel ensures that teams are well-prepared to respond swiftly to emerging cyber risks.

By systematically evaluating resilience post-attack, businesses can fortify their critical infrastructure, minimize operational disruptions, and establish a stronger defense against evolving cyber threats. A structured resilience assessment not only improves immediate recovery efforts but also lays the foundation for a long-term cybersecurity strategy that enhances overall system robustness.

12.34   Conducting Impact Analysis on ICS Cyberattacks

A systematic methodology is required for a thorough investigation of the effect of cyber-attacks on Industrial Control Systems (ICS). The process begins with identifying the specific components that were compromised during the attack, determining whether these were hardware elements, network segments, or software applications critical to ICS functionality. Establishing the level of the compromise is crucial, as it helps assess whether the attack resulted in partial system failures, complete shutdowns, or unauthorized alterations to system operations.

An essential aspect of impact analysis is evaluating the functional damage caused by the attack. This includes analyzing disruptions to automated processes, sensor manipulations, control logic modifications, and unauthorized data access. Understanding the attack vector, including whether it originated from phishing, malware, insider threats, or supply chain vulnerabilities, is crucial in pinpointing weaknesses in security protocols. Moreover, determining the time required for detection and response is vital in assessing the extent of the harm. Delayed identification of threats often leads to prolonged exposure, enabling attackers to cause severe disruptions. The ability of an organization to respond quickly and effectively to a cyber-attack significantly influences the recovery process and minimizes operational downtime. Beyond technical implications, the financial, operational, and reputational consequences must be carefully evaluated. Financial losses may stem from production downtime, regulatory fines, or litigation costs. Operational disruptions can lead to inefficiencies in industrial processes, supply chain breakdowns, or delays in service delivery. Reputational damage, especially in industries where reliability and security are paramount, can impact customer trust and investor confidence.

Lastly, it is crucial to determine the extent to which the attack affected key services and whether it posed risks to human safety. In critical infrastructures such as power grids, water treatment plants, and manufacturing systems, cyberattacks can have severe real-world consequences, including hazardous chemical releases, power outages, or compromised public safety. A comprehensive impact analysis informs mitigation strategies, strengthens security frameworks, and enhances the overall resilience of ICS against future threats.

12.35   Recovery Plans for ICS Resilience

Strategic and systematic resilience methods are essential for rebuilding and strengthening Industrial Control Systems (ICS) after a cyberattack or disruptive incident. These strategies aim to minimize downtime, restore critical functions, and enhance the security posture of ICS to prevent future threats. Recovery plans should be comprehensive, addressing both immediate response measures and long-term improvements in system resilience.

A well-structured recovery plan includes incident response escalation procedures and communication protocols to ensure coordinated action across teams. Clearly defined roles and responsibilities help facilitate efficient decision-making, allowing organizations to contain and remediate threats swiftly. Enhancing communication among security teams, operational staff, and external cybersecurity experts ensures that all stakeholders are informed and aligned in their response efforts.

To protect vital data and ensure integrity, security measures such as encryption, automated backups, and validation mechanisms play a crucial role. Secure backup systems should be regularly updated and tested to enable seamless recovery of compromised data and configurations. Additionally, redundancy activation procedures allow ICS to transition to backup systems or alternative operational modes, minimizing service disruptions.

Infrastructure restoration efforts involve identifying compromised assets, repairing affected components, and deploying system patches. Patch management and vulnerability mitigation are essential for addressing security gaps exploited by attackers, ensuring that weaknesses are promptly eliminated. Regular penetration testing and system audits help validate the effectiveness of applied security controls.

Comprehensive incident recording and forensic analysis are critical for extracting lessons from past attacks. By documenting attack patterns, security teams can refine incident response strategies, strengthen defenses, and prevent recurrence. Furthermore, continuous security monitoring and threat intelligence integration enhance early threat detection and proactive mitigation.

A robust cybersecurity culture within the organization is key to maintaining long-term resilience. Regular cybersecurity training for employees, awareness programs, and simulated attack exercises improve preparedness and response capabilities. Encouraging a proactive security mindset ensures that personnel remain vigilant and adhere to best practices, reducing the likelihood of human errors that could expose ICS to future cyber risks.

By implementing a well-defined recovery plan, organizations can effectively limit damage, restore normal operations swiftly, and fortify their Industrial Control Systems against evolving cyber threats. These strategies help build a resilient ICS infrastructure capable of withstanding and recovering from even the most sophisticated cyberattacks.

12.36 Conducting Impact Analysis on ICS Cyberattacks

A systematic methodology is required for a thorough investigation of the effect of cyberattacks on Industrial Control Systems (ICS). It starts with identifying the compromised components, establishing the level of the compromise, and assessing the functional damage produced. It is critical to understand the attack vector and how it entered the system. Furthermore, determining the time required for discovery and reaction is critical in determining the extent of the harm. It is also necessary to assess the financial, operational, and reputational consequences. Finally, determining the extent to which the assault affected key services and possibly jeopardized human safety is crucial.

12.37 Summary

Conclusively, the analysis of ICS attack resilience in Part 2 highlights the need of a proactive and comprehensive strategy to cybersecurity in critical infrastructure environments. A comprehensive and flexible attitude is necessary for resistance against cyber threats after investigating a variety of tactics and procedures, from the use of decoy systems to the consumption of threat information.

Organizations may strengthen their defensive capabilities and more accurately predict, identify, and mitigate cyberattacks on ICS settings by adopting ideas like threat hunting, resilience testing, and the integration of machine learning and deep learning technology. Furthermore, the focus placed on post-attack evaluations and regulatory compliance highlights the need of ongoing development and attention to detail to protect vital infrastructure assets.

Investing in strong cybersecurity measures for ICS systems is becoming more and more important as the cyber threat environment keeps changing. The successful mitigation of threats presented by cyber adversaries and the maintenance of critical infrastructure systems' dependability and security in an increasingly interconnected world can only be achieved via coordinated efforts, teamwork, and a commitment to resilience.

Chapter 13
ICS Security Requirements: Cybersecurity Frameworks, Incident Response Strategies, and IoT Device Compliance

Abstract This chapter offers a comprehensive understanding of the security requirements for Industrial Control Systems (ICS). It discusses the importance of risk assessments, security architectures, and boundary defense mechanisms. The chapter covers a variety of security solutions, including vulnerability management, configuration control, and multi-point malware protection. Readers will examine the role of cybersecurity audits and the CIA triad model (Confidentiality, Integrity, Availability) in assessing ICS security. The chapter highlights the differences between ICS and conventional IT security, emphasizing the need for robust physical and digital security protocols. It also addresses the impact of Internet of Things (IoT) devices on ICS security and compliance and explores mitigation strategies. By understanding these elements, readers will be equipped to develop resilient security measures to safeguard their Industrial Control Systems.

Keywords Industrial Control Systems (ICS) · Security Requirements · Risk Assessments · Security Architectures · Boundary Defense · Vulnerability Management · Configuration Control · Malware Protection · Cybersecurity Audits · CIA Triad · IT vs. ICS Security · Physical Security · Digital Security · IoT Security · Compliance · Mitigation Strategies

13.1 Introduction

Industrial Control System (ICS) security is crucial in today's technology-driven world. ICS, which rely more on digital controls, support industries like manufacturing, electricity, water, and transportation. While connectivity improves efficiency, it also exposes ICS to cyber threats that can impact public safety and economic stability.

A well-rounded approach is needed to address the unique challenges of ICS security, including protecting against cyberattacks, insider threats, and system failures. This section outlines the key security measures needed to protect ICS from evolving threats and maintain essential services.

ICS stands for Industrial Control Systems, referring to hardware and software used to monitor, control, and automate critical processes in industries like manufacturing, transportation, and energy. These systems are vital for infrastructure and have a direct impact on daily life.

ICS security focuses on protecting these systems to ensure they run safely and efficiently. As ICS manage critical services, protecting them from cyber threats is essential.

The goal of ICS security is to maintain the integrity and function of systems that control essential infrastructure like water, electricity, and transportation. As ICS rely more on software and networks, securing these systems is critical to minimize cyber threats and protect public safety.

During emergencies, organizations can contact an ICS security hotline for immediate support. Key challenges include securing system components and addressing protocol weaknesses. Effective security measures help ensure ICS continue to operate safely while minimizing risks.

13.2 ICS Security Measures: Protecting Critical Infrastructure

1. **Hardware Protection:** Protect physical components like sensors, actuators, and programmable controllers to prevent unauthorized access and maintain system security.
2. **Software Protection:** Strong cybersecurity measures are needed to protect control software and operating systems, ensuring the integrity and reliability of critical operations.
3. **Network Defense:** Securing the ICS network is crucial. This includes protecting communication protocols and network devices from unauthorized access to ensure safe data transmission.
4. **Endpoint Protection:** Implement security measures on all ICS endpoints, including computer systems and human-machine interfaces (HMIs), to reduce the risk of cyber threats and unauthorized access.

13.3 Importance of ICS Security

ICS support vital infrastructure like water supply, electricity, and transportation. As these systems become more digital, protecting them becomes even more important. Challenges include the risk of system failures and the use of outdated protocols. Solutions involve applying security best practices and, where necessary, upgrading systems. Best practices include access control and abuse prevention. Reasons to prioritize ICS security include:

1. **Safety of Operations and Personnel:**Protects systems to prevent accidents that can harm people.
2. **Reliable Infrastructure:** Ensures stability and continued operation of critical services.
3. **Protection from Cyber Threats:** Guards ICS against unauthorized access, data breaches, and cyberattacks.
4. **Resilience and Business Continuity:** Improves system resilience for fast recovery after an attack.

13.4   Benefits of ICS Cybersecurity Risk Assessment

Organizations using ICS for critical infrastructure benefit from a cybersecurity risk assessment, which helps protect systems and assets. It reduces downtime by identifying and addressing vulnerabilities early. The main benefits include:

1. **Improved Safety, Availability, and Reliability:**

   a. Detecting and fixing vulnerabilities enhances operational safety.
   b. Cybersecurity controls reduce downtime, improving availability.
   c. Improved reliability reduces interruptions and ensures stable operations.

2. **Building Investor Confidence:**

   a. Investors value organizations that actively manage cybersecurity risks.
   b. Strong cybersecurity measures demonstrate commitment to asset protection.

3. **Compliance with Regulations:**

   a. Many industries must follow strict cybersecurity regulations.
   b. Risk assessments help identify compliance gaps and meet regulatory requirements.

4. **Improved Corporate Reputation:**

   a. A focus on cybersecurity shows ethical business practices.
   b. Proactive risk management strengthens the organization's reputation.

5. **Reducing Legal Risks:**

   a. Effective cybersecurity risk management lowers legal liability for data breaches or system failures.
   b. Evidence of due diligence can be used as a legal defense.

6. **Supporting Long-Term Business Success:**

   a. Integrating cybersecurity into the business strategy ensures long-term sustainability.
   b. Protecting systems improves resilience against cyber threats and supports business continuity.

13.5   ICS Security Architecture

The ICS Security Architecture involves planning and designing security measures to protect ICS networks, data, and components. The main goals are to ensure system availability, confidentiality, and integrity. Key components include:

1. **Network Firewalls:** Firewalls control traffic between secure networks and protect ICS from unauthorized access.
2. **Dual Network Interface Cards (NICs):** Dual-homed PCs with firewalls safeguard communication between corporate and ICS networks.
3. **Firewalls Between Corporate and ICS Networks:** A firewall improves security by controlling traffic between corporate and ICS networks.
4. **Firewalls and Routers Between Corporate and ICS Networks:** Using both firewalls and routers enhances security and reduces the load on firewalls.

These components work together to create an effective defense strategy for protecting critical infrastructure from cyberattacks [162]. ICS security architecture is a framework for managing security across different network levels, including corporate services, business networks, and operational systems.

13.5.1   ICS Security Assessment

ICS security assessment is a formal review of the security controls and vulnerabilities in an organization's ICS. It identifies risks and weaknesses that could affect the integrity and availability of critical systems. Key aspects of the assessment include:

1. **Monitoring Connected Devices:** Ensure that only authorized devices have access to the ICS network.
2. **Multi-layered Defense:** Use multiple security measures like firewalls and intrusion detection systems to protect the network.
3. **Vulnerability Management:** Regularly scan for vulnerabilities and fix them to maintain system security.
4. **Login Activity Monitoring:** Monitor login attempts to detect unauthorized access.
5. **Secure Configurations:** Ensure devices and software are configured according to security best practices.
6. **Malware Defense:** Protect against malware with antivirus solutions and endpoint protection.
7. **Finding Security Vulnerabilities:** Continuously monitor and assess security to find and fix vulnerabilities.

13.5.1.1 ICS Security Solutions

ICS security solutions use tools, rules, and technologies to protect control systems and important infrastructure from cyber threats. As these systems become more digital and connected, strong security is needed to keep operations safe and stable. Common threats include malware, unauthorized access, insider actions, and advanced attacks.

A strong ICS security plan uses many layers of protection. Network segmentation helps by keeping key ICS systems separate from other networks. Firewalls, intrusion detection systems (IDS), and intrusion prevention systems (IPS) monitor and block bad traffic. Using multi-factor authentication (MFA) ensures that only trusted users can access important systems. Protecting devices directly is also important. Antivirus, host-based intrusion detection (HIDS), and application whitelisting stop harmful software. Keeping software updated with patches helps fix known problems. Security information and event management (SIEM) tools give alerts and track threats in real time. Threat intelligence platforms [6] give useful data about new risks, helping teams update defenses. Training workers on cybersecurity best practices helps prevent human errors. Regular checks, like audits and penetration tests, improve readiness and recovery. Following standards like IEC 62443, NIST 800-82, and ISO 27001 helps ensure proper protection. By combining these methods, organizations can better protect their ICS systems from cyberattacks. Figure 13.1 is an excerpt from [60].

Figure 13.1 shows ICS cybersecurity solutions, focusing on vendor access, collaboration controls, sensor security, threat authentication, detection, and risk mitigation.

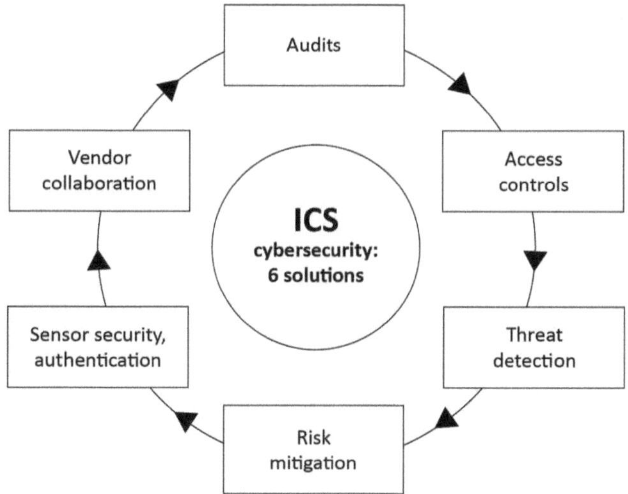

Fig. 13.1 ICS security solution

Tracking Connected Devices

An ICS security assessment thoroughly examines the security controls and vulnerabilities within an organization's industrial control systems. The goal is to identify any threats or weaknesses that could compromise the availability, integrity, and confidentiality of critical infrastructure and processes managed by these systems.

13.6   Boundary Defense

Boundary defense, which is the 12th control in the CIS Critical Controls, falls under the network category. It includes 11 subsections covering areas such as DMZs, firewalls, proxies, IDS/IPS, NetFlow, and remote access. Boundary defense acts as the first line of defense, protecting an organization from external threats. It involves monitoring and controlling external communications to prevent unauthorized access and attacks using boundary protection devices like gateways, routers, firewalls, and encrypted tunnels. To implement CIS Control 12, a multi-layered approach is recommended, including network segmentation and data flow management.

To start, implement a demilitarized zone (DMZ) to separate the internal network from the Internet. Use application layer proxies to protect communication and outbound proxies to block malicious websites. Block malicious IP addresses using blacklists, or only allow traffic from trusted sources using whitelists. Ensure proxies can decode and log network traffic to prevent unauthorized data exfiltration. Firewalls should be configured to block all non-business-related outbound traffic and inspect all network traffic. Deploy IDS to alert security teams about attacks and IPS to block malicious network activity. Place IDS sensors carefully and fine-tune IDS/IPS devices to minimize disruption. Tools like Snort, Suricata, BroIDS, and commercial firewall software with IPS capabilities can help with this.

13.7   Multi-layered Boundary Defense

Multi-layered perimeter defense in ICS uses several security layers to protect critical infrastructure such as power plants, water treatment facilities, and industrial plants. ICS are connected to external networks, making them vulnerable to cyber threats. A multi-layered security approach helps reduce the risks of unauthorized access, malware, and cyberattacks.

Network segmentation and firewalls help minimize exposure to external threats. By adding layers of defense, even if one layer is compromised, others still provide protection. Industries must follow standards like NERC-CIP and IEC 62443 to maintain robust security controls. While one security layer may be breached, additional layers lower the likelihood of a successful attack, enhancing resilience against evolving threats.

Monitoring traffic at multiple points enhances the ability to detect unusual or suspicious activities. Multi-layered security improves incident response and strengthens overall protection. A strong boundary protection strategy must be in place for ICS, and effective network design is critical for successful deployment.

Segmenting the network based on the function and sensitivity of the data is a key strategy. Firewalls, intrusion prevention, and access controls should be implemented at segment boundaries to create multiple layers of protection. Network monitoring technologies should be used to detect abnormal activities. Regular maintenance and updates are essential to keep up with emerging threats and evolving organizational needs. A multi-layered defense approach ensures robust protection for critical infrastructure, helps prevent attacks, and supports smooth industrial operations.

13.8 Complete Vulnerability Management

Cybersecurity vulnerabilities are weaknesses in processes, networks, hardware, or software that attackers can exploit for unauthorized access, disruption of operations, or stealing sensitive information. These weaknesses are like open doors or windows in virtual systems, offering entry points for attackers. Identifying and fixing these vulnerabilities is key to improving defenses and reducing the chances of security breaches. Vulnerabilities are an inevitable part of cybersecurity because they are the access points for attackers, they can damage data integrity, and they can lead to financial and reputational harm. Addressing these vulnerabilities is essential for meeting regulatory requirements, as many standards require organizations to identify and fix these risks. Moreover, as technology and cybercrime evolve, it is crucial to actively manage vulnerabilities to maintain strong security. Some common types of vulnerabilities include:

1. **Software Vulnerabilities:** Flaws in software applications, operating systems, or libraries, such as coding bugs, misconfigurations, and unpatched security weaknesses.
2. **Network Vulnerabilities:** Weaknesses in network protocols, devices, or configurations, such as open ports, weak encryption, and misconfigured firewalls.
3. **Human-Related Vulnerabilities:** Risks related to human behavior or mistakes, such as using weak passwords, falling for social engineering attacks, or insider threats.
4. **Zero-Day Vulnerabilities:** These are vulnerabilities that attackers exploit before the defenders are aware of them, and a fix is available.

Unpatched vulnerabilities are significant threats to the security of an organization. They can lead to data breaches, illegal access, or disruptions in operations. Data loss or theft is a major concern, as these breaches can cause financial damage and harm the organization's reputation. Not addressing vulnerabilities can also result in legal penalties and lawsuits. Moreover, vulnerabilities make systems susceptible to cyberattacks, leading to downtime and productivity loss. Finally, unmanaged vulnerabilities can damage customer, partner, and stakeholder trust, which can harm the organization's reputation.

Vulnerability management is a systematic and proactive process that is essential for cybersecurity. It involves identifying, evaluating, prioritizing, and fixing vulnerabilities to minimize the chances of security breaches. By addressing vulnerabilities before attackers can exploit them, organizations can focus on the most critical issues. Effective vulnerability management shows compliance with regulations and demonstrates a commitment to cybersecurity. Proper management prevents costly downtime, ensures smooth operations, and protects the organization's reputation from data breaches and security attacks.

In ICS, vulnerability management is crucial to ensuring system integrity. The first step is to conduct regular vulnerability scans using automated tools. Detected vulnerabilities should be ranked based on risk and potential impact on operations. A plan should be developed to fix and mitigate identified vulnerabilities. It's recommended to create a centralized system to track, prioritize, and address vulnerabilities effectively, forming a solid vulnerability management plan for ICS security.

13.9   Monitoring Login Activity

Monitoring login activity is an important security practice for ICS. Centralized logging systems should be set up to collect and analyze logon data efficiently from all ICS components. The login activity data should be analyzed using Security Information and Event Management (SIEM) software to detect unusual patterns or signs of unauthorized access. Clear policies and automatic alerts for suspicious login attempts help ensure quick action in case of security breaches. Regular auditing and reviewing of these logs helps establish access restrictions and maintain security in the ICS environment.

13.10   Secured Configuration in ICS

It is essential to ensure that ICS configurations are secure to maintain safety, integrity, and resilience against cyberattacks. Proper configuration management ensures that all system components, such as controllers, sensors, actuators, and network devices, operate in a secure and controlled state. From deployment, all devices should follow a baseline configuration that acts as the foundation for secure operation. These configurations must prevent unauthorized changes, incorrect configurations, and vulnerabilities that attackers can exploit. Configuration management tools are vital for enforcing, auditing, and maintaining compliance with security baselines. These tools can quickly detect deviations from security policies and fix security risks. Securing ICS configurations also involves setting up strong authentication, disabling unnecessary services, and using encryption for communication to protect against data interception and tampering [109].

A key concept in ICS configuration security is the principle of least privilege (PoLP). This principle ensures that users, applications, and processes only have the minimum access required to perform their tasks. By restricting access, organizations can significantly reduce the risks of unauthorized entry, privilege escalation, and insider threats. Role-based access control (RBAC) and multi-factor authentication (MFA) add extra layers of security, ensuring that only authorized individuals can modify critical configurations. Periodic audits and configuration reviews are necessary to stay aligned with emerging cyber risks. Regular security evaluations help identify outdated configurations, weak access controls, and insecure communication protocols that could be vulnerabilities. Any new configuration changes must be tested in a controlled environment before being deployed to minimize disruption and avoid unforeseen issues. Organizations should enforce policies to automatically log and monitor all configuration changes. Automated configuration management software can track system settings continuously, alert administrators to unauthorized changes, and restore systems to their last secure state in case of anomalies.

Compliance with industry best practices and standards, such as NIST SP 800-82, IEC 62443, and ISO 27001, provides a structured approach to configuration security. These standards offer guidelines for ICS configuration security, defining access controls and change management processes.

By following these configuration security practices, organizations can maintain a strong and proactive security posture in their ICS environments. A good configuration management strategy not only improves system reliability and performance but also reduces the risk of cyber incidents that could disrupt critical industrial processes.

13.11   Malware Defense at Multiple Checkpoints

ICS require a strong and comprehensive strategy to defend against malware.

Using powerful antivirus and anti-malware software, specifically designed for business environments, is one of the most important practices. Anti-spam filters also help block malicious links and attachments from being delivered. Deploying endpoint protection across all devices significantly strengthens overall security. Network-based malware detection and protection software is essential for ICS. Staff should also receive regular security training to identify and report potential malware issues. This proactive approach helps maintain a strong defense against malware threats, protecting critical infrastructure and operations.

Malware defense should be implemented at multiple points within the ICS to minimize risk. It helps reduce the impact of malware-caused disruptions, ensuring continuous operations. A solid defense not only protects sensitive data but also maintains privacy, enhances security, and prevents unauthorized use or harmful code execution. A layered defense strategy, combining technical controls, monitoring, and user awareness, provides comprehensive protection.

Endpoint security is crucial, meaning systems should have up-to-date operating systems, firewalls, and application whitelisting. Protecting industrial controllers, such as PLCs and RTUs, involves installing security protocols, improving detection mechanisms, and monitoring for suspicious activity. Antivirus and anti-malware software play a critical role in maintaining robust ICS security.

13.12   Cybersecurity Audits in the Industry Sector

Cybersecurity audits are detailed reviews of a company's IT systems, policies, procedures, and security controls. The goal is to ensure that the company follows standard cybersecurity rules, guidelines, and best practices. These audits help find any gaps in compliance or potential weaknesses in the company's cybersecurity environment. By conducting thorough evaluations, organizations can spot areas that need improvement and strengthen their overall security to better resist cyberattacks. Figure 13.2 is from [42].

Fig. 13.2 Security audit consulting

13.13   Cybersecurity Audits in the Industry Sector

Figure 13.2 shows a security program with awareness, training, policies, controls, managed services, design, and implementation in a corporate setting. There are two types of security audits:

1. Internal and External
2. Focus Areas for Audits

a. Internal and External Audits

a. **Internal Audits:** These are conducted by the company's IT department using internal information to evaluate the success of current cybersecurity policies.
b. **External Audits:** These are performed by outside experts who can identify risks and weaknesses that internal staff may miss.

b. Areas of Focus for Audits

a. **Data Security:** Examines how well the company protects sensitive information, including encryption, data flow, and access controls.
b. **Operational Security:** Reviews the company's policies, procedures, and controls to ensure they are adequate and compliant.
c. **Network Security:** Looks at antivirus settings, network restrictions, and monitoring systems to ensure the network is protected and can detect threats [35].

Regular audits are essential for improving cybersecurity in industries. They should be done at least once a year, more often for highly regulated sectors or those handling sensitive data. Cybersecurity audits help organizations identify vulnerabilities, assess risks, and defend against attacks (Table 13.1).

Table 13.1 Comparison of internal versus external security audits

Internal security audits	External security audits
Limited to the organization's systems	Third-party auditors look beyond the company's systems
Conducted regularly	Done based on regulations and standards
Less expensive	Special audits can be costly
Produces a basic report	Produces a detailed report

13.13.1   Primary Goals of a Cybersecurity Audit

The main goals of a cybersecurity audit are to assess the effectiveness of an organization's cybersecurity controls, identify weaknesses, and ensure compliance with legal and industry standards. Key objectives include:

1. Identify and assess risks related to the IT infrastructure, policies, and data.
2. Ensure compliance with relevant laws, regulations, and standards.
3. Evaluate the adequacy of security controls and procedures.
4. Assess the organization's ability to handle cybersecurity challenges.

13.13.1.1   Key Elements of a Cybersecurity Audit

1. **Scope Definition:** Define the systems, procedures, and areas to be audited.
2. **Policy and Procedure Review:** Ensure compliance with existing cybersecurity policies and regulations.
3. **Risk Assessment:** Identify potential risks, vulnerabilities, and threats to security.
4. **Compliance Verification:** Check compliance with industry standards and regulations.
5. **Technical Testing:** Conduct penetration tests and vulnerability assessments to identify system weaknesses.
6. **Security Controls Assessment:** Evaluate security measures like network segmentation, encryption, and access controls.
7. **Incident Response Simulation:** Test incident response protocols to ensure readiness for security breaches.
8. **Employee Awareness and Training:** Measure staff understanding of cybersecurity to reduce human-caused risks.
9. **Documentation Review:** Check incident logs, access logs, and security policies for completeness and accuracy.

13.14   CIA Triad: The ICS Security Assessment Model

Information security is built on the CIA triad: Confidentiality, Integrity, and Availability. This model is essential for improving the security of Industrial Control Systems (ICS).

1. **Confidentiality:** Prevent unauthorized access to sensitive information by ensuring only authorized users can view it.
2. **ICS Security Assessment for Confidentiality:**

   a. **Access Controls:** Ensure ICS components are accessible only to authorized users based on roles.

b. **Encryption Protocols:** Use encryption to protect ICS communication from unauthorized access.

3. **Integrity:** Ensure data accuracy, reliability, and trustworthiness by preventing unauthorized changes.
4. **ICS Security Analysis for Integrity:**

   a. **Data Validation Mechanisms:** Verify that ICS systems maintain data accuracy and consistency.
   b. **Change Control Procedures:** Manage changes in the ICS environment to maintain data integrity.

5. **Availability:** Ensure ICS systems are always available and minimize downtime.
6. **ICS Security Assessment for Availability:**

   a. **Cyber Resilience:** Test ICS infrastructure's ability to withstand cyberattacks and reduce downtime.
   b. **Disaster Recovery and Backup Systems:** Ensure backup systems and disaster recovery plans are in place for quick recovery.

13.14.1   Security Standards for Industrial Control Systems

ICS security standards offer guidelines and best practices for managing and securing critical infrastructure. Two main standards are ISA/IEC 62443 and NERC CIP.

1. **ISA/IEC 62443:**

   a. Developed by the International Society of Automation (ISA) and the International Electrotechnical Commission (IEC), it focuses on securing industrial automation systems.
   b. **Key Objectives:**

      I. **Defense-in-Depth:** Use access control, network segmentation, and incident response to protect ICS.
      II. **Risk Management:** Provides guidance for assessing and reducing ICS-specific risks.
      III. **Security Lifecycle:** Covers system design, implementation, operation, and maintenance.

   c. **Components of ISA/IEC 62443:**

      I. **Policies and Procedures:** Establish security policies for ICS.
      II. **Asset Inventory and Classification:** Identify and classify assets based on criticality.
      III. **Access Control:** Limit access to critical ICS components.
      IV. **Network Segmentation:** Divide ICS networks to reduce threats.

V. **Incident Response:** Plan for incident response to manage security breaches.

2. **NERC CIP:**

a. Developed by the North American Electric Reliability Corporation to secure the electric grid.
b. **Critical Infrastructure Protection:** Focuses on minimizing cyber threats to the grid.
c. **Compliance and Enforcement:** Requires compliance with cybersecurity standards.
d. **Elements of NERC CIP:**

   a. **CIP-002 to CIP-011:** These standards cover critical asset identification, security controls, and recovery planning.
   b. **Security Controls:** Includes access management, incident response, and disaster recovery to protect infrastructure.

Both ISA/IEC 62443 and NERC CIP provide structured approaches to ICS security, offering essential guidelines for improving security in critical infrastructure.

13.14.1.1   Layered Access Control for ICS Security

In an ICS network, managing data flow and access is crucial, as highlighted in "Layered Access Management in ICS Security." Separating the network into different layers and regulating data flow and access between them are key for security. Specifically, controlling access to hardware at the control network level is essential. This improves security and reduces the chances of unauthorized or malicious use. It ensures that critical hardware is only accessible to those with the right permissions, safeguarding the network's integrity and creating a safer environment for operations. Figure 13.3 is taken from [41].

Figure 13.3 shows the network design of an industrial system. It includes workstations, servers, routers, PLCs, HMIs, and control servers. The figure also shows how the corporate network, control network, and various devices are connected.

In industrial systems, using layered access control is an important security method. It helps protect important infrastructure and reduce the risk of cyberattacks.

13.15   Network Firewall

A network firewall is either software or hardware that checks and controls all incoming and outgoing network traffic based on set security rules. It acts as a barrier between untrusted external networks like the Internet and trusted internal networks.

Fig. 13.3 Access control for ICS security

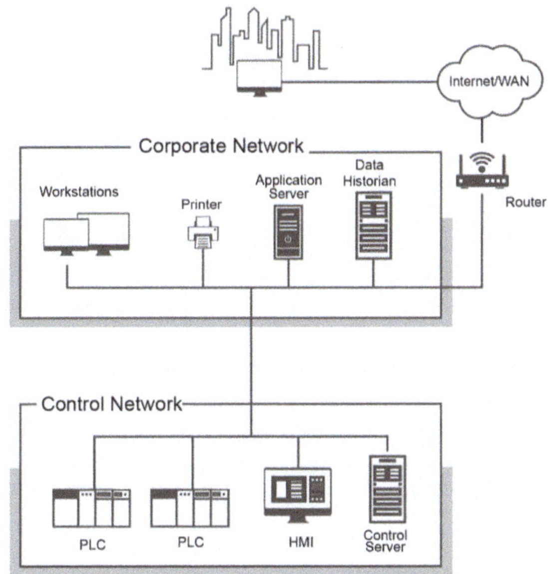

The main purposes of a firewall are to stop unauthorized access, manage data traffic, and keep the internal network secure.

13.15.1   Code Example: Basic Firewall Configuration (Pseudo-Code)

The following is a simple pseudo-code for a firewall configuration in Python. This example demonstrates how a basic firewall checks traffic against predefined rules.

```python
class Firewall:
    def __init__(self):
        self.rules = []

    def add_rule(self, rule: str) -> None:
        """Adds a new rule to the firewall."""
        self.rules.append(rule)
        print(f"Rule added: {rule}")

    def check_traffic(self, traffic: str) -> str:
        """Checks if the given traffic matches any blocking rules."""
        for rule in self.rules:
            if rule.split(":")[1] in traffic:
                return "Traffic blocked"
        return "Traffic allowed"
```

```
16
17  # Example usage
18  firewall = Firewall()
19  firewall.add_rule("BLOCK:192.168.0.1")
20  traffic = "192.168.0.1 -> 192.168.0.2"
21  result = firewall.check_traffic(traffic)
22  print(result)
```

Listing 13.1 Basic Firewall Configuration

This pseudo-code, referred from: https://github.com/Sunzidasiddique1/ICS, defines a *Firewall* class with two main functions: adding rules and checking network traffic. The *add_rule* function allows rules to be added, while the *check_traffic* function checks if the traffic matches any rule. If it matches a rule, the traffic is blocked. If not, the traffic is allowed.

For example, a rule is added to block traffic from `192.168.0.1`. When traffic from that IP is checked, the firewall blocks it and shows the message `"Traffic blocked"`. This is a simple way to show how firewalls filter network traffic using rules.

13.16 Security Objectives of ICS in Network Firewall

In Industrial Control Systems (ICS), a network firewall protects important systems and data from unauthorized access and cyberattacks. It acts as a barrier between trusted internal networks and untrusted external ones. Firewalls help stop data theft, system disruptions, and changes made without permission by using strict security rules. The key goals of ICS firewalls include:

1. **Access Control and Segmentation:**

   a. Block unauthorized users and devices from reaching sensitive parts of the ICS network.
   b. Divide the network into smaller parts to protect critical systems from unwanted access.

2. **Threat Prevention and Detection:**

   a. Detect and block cyber threats targeting the ICS network.
   b. Use tools like IDS (Intrusion Detection Systems) and IPS (Intrusion Prevention Systems) to spot and stop malware and suspicious activity.

3. **Traffic Monitoring and Filtering:**

   a. Watch and control network traffic to follow security rules.
   b. Allow only safe data to pass while blocking dangerous or unauthorized traffic.

4. **Protecting Critical Systems and Data:**

   a. Secure vital control systems and private data from cyber threats.
   b. Stop unauthorized access to key hardware and control systems.

5. **Alerting and Reporting:**

   a. Send real-time alerts and reports about network activity and security issues.
   b. Help administrators respond quickly and track patterns to improve security.

6. **Compliance with Rules and Standards:**

   a. Meet industry and legal cybersecurity requirements.
   b. Help create a secure system to avoid legal problems and follow regulations.

7. **System Resilience and Continuity:**

   a. Strengthen the ICS network against cyber threats and ensure critical operations continue without interruption.

13.17 Common ICS Network Segregation Architectures

In ICS cybersecurity, separating networks (called segregation) helps stop cyberattacks from spreading. Some common ways to do this include:

1. **Dual-Homed Computers:** Computers with two network connections that manage traffic between different network sections.
2. **Dual-Homed Server with Personal Firewall:** A server with two network connections and a personal firewall to filter and separate traffic.
3. **Packet Filtering Router or Layer-3 Switch:** A device that filters traffic between the Process Control Network (PCN) and Enterprise Network (EN) using rules.
4. **Two-Port Firewall:** A firewall with two ports that controls traffic between PCN and EN.
5. **Router/Firewall Combination:** A setup that uses both router and firewall to manage and protect traffic between PCN and EN.
6. **Firewall with Demilitarized Zones (DMZ):** A firewall that uses DMZs to isolate and secure communication between PCN and EN.
7. **Paired Firewalls:** Two firewalls placed one after another to make a more secure connection between PCN and EN.

13.18 Network Firewalls with Their Functions

Network firewalls are tools used to block or allow network traffic. They protect systems from hacking, malware, and unauthorized access by checking data packets.

Functions:

1. **Packet Inspection:** Firewalls scan each data packet and check details like source and destination IP, port, and protocol. They use deep inspection to detect threats and stop dangerous traffic.
2. **Access Control:** Firewalls use access control lists (ACLs) to decide which traffic is allowed or denied. These rules can be based on address, service type, or port number.
3. **Traffic Monitoring:** Firewalls constantly check network traffic for anything unusual, like denial-of-service (DoS) attacks. They can block harmful traffic or send alerts to system administrators.

13.19 Segregation of Network with Benefits

Network segregation means dividing a network into smaller parts (segments or subnetworks) and controlling the access between them. This helps reduce security risks and stops attackers from moving freely across the network.

Benefits:

1. **Smaller Attack Area:** Reduces the number of places attackers can target.
2. **Protect Important Systems:** Keeps sensitive parts of the system safe from other, less secure areas.
3. **Better Performance and Security:** Organizes traffic more efficiently and improves both speed and safety.

Dual Network Interface Card

A dual NIC (Network Interface Card) has two network ports. This allows a device to connect to two different networks at the same time.

Features:

1. **Traffic Control:** Manages and routes data between the two networks.
2. **Better Security:** Can apply specific security settings for each network.

Firewall Between Corporate and ICS Network

A firewall placed between the corporate network and the Industrial Control System (ICS) network helps control and protect the flow of data between them.

Advantages:

1. **Secure Separation:** Blocks unwanted access between corporate and ICS systems.
2. **Traffic Rules:** Controls what data can pass through based on security policies.

13.20   Basic Security Requirements for ICS Protection

Protecting Industrial Control Systems (ICS) is important to keep industrial processes safe and running smoothly. The following basic security steps help ensure this:

Access Control:

1. Only allow trusted users to access the system.
2. Use strong login methods like multi-factor authentication.
3. Set access rights based on user roles and responsibilities.

Network Segmentation:

1. Divide the network to limit the spread of attacks.
2. Use strict rules to control traffic between segments.

Endpoint Protection:

1. Add security to devices like sensors, PLCs, and HMIs.
2. Keep software updated to fix known problems.

Encryption:

1. Use encryption to protect data moving between ICS parts.
2. Apply strong encryption methods to avoid data leaks.

Patch Management and Risk Checks:

1. Regularly scan the system for risks and weaknesses.
2. Apply updates quickly to fix known security problems.

Incident Detection and Response:

1. Use monitoring tools to detect strange or harmful actions.
2. Prepare and test plans to respond quickly to attacks.

Security Awareness and Training:

1. Train staff and stakeholders on good security practices.
2. Provide regular sessions to improve security knowledge.

Data Backup and Disaster Recovery:

1. Make regular backups to protect important data.
2. Create and test recovery plans to restore systems fast.

Secure Configuration:

1. Set secure default settings for each ICS device.
2. Review and update settings to follow best practices.

Physical Security:

1. Use locks and physical controls to limit access to equipment.
2. Monitor entry points to stop unauthorized access.

Firewall and Perimeter Defense:

1. Use firewalls to manage and check network traffic.
2. Add extra layers of protection around the ICS network.

Auditing and Logging:

1. Keep records of system activity to track changes.
2. Check logs regularly to find any unusual behavior or rule violations.

13.21   ICS Security Versus Traditional IT Security

Information Technology (IT) and Operational Technology (OT) are two areas of modern systems. IT deals with storing, processing, and sharing data using computers and networks. OT focuses on monitoring and controlling industrial operations.

A comparison between ICS security and traditional IT security is shown in Table 13.2.

Table 13.2 Comparison of ICS and traditional IT security requirements

Aspect	ICS security requirements	Traditional IT security requirements
Focus	Protects physical processes and machinery	Protects digital data and information
Priority	Emphasizes availability to prevent downtime	Balances confidentiality, integrity, and availability
Lifecycle and Systems	Involves legacy systems and long lifecycles	Focuses on modern, regularly updated systems
Integration with the physical world	Primarily digital interactions	Primarily digital interactions
Real-time Requirements	Operates in real-time or near real-time	May not have strict real-time processing requirements
Threat Landscape	Unique threat landscape, including physical threats	Predominantly digital threats, malware, etc.
Interconnectedness and Complexity	Interconnected with limited bandwidth/resources	Highly interconnected, extensive use of bandwidth/resources
Human Safety and Environment	Emphasizes human safety and environmental protection	Focuses on data security and privacy
Regulatory and Compliance Differences	Subject to specific industry regulations and standards	Follows general frameworks and industry standards
Security Impact on Operations	Considers the impact of security on physical operations	Focuses on minimizing security risks to digital operations

13.21.1   Physical Security in ICS Environments

Physical security in Industrial Control Systems (ICS) is important to keep critical infrastructure, equipment, and processes safe. It helps protect the system from physical threats such as unauthorized access, damage, theft, and sabotage.

Main features of ICS physical security:

1. **Access Control:** Limit entry to important areas using ID cards, biometric systems, or locked doors.
2. **Video Surveillance:** Use cameras to monitor areas in real-time and review incidents when needed.
3. **Perimeter Security:** Use fences, gates, and barriers to protect the outer area of the facility.
4. **Security Staff:** Have trained security guards to handle emergencies and follow safety procedures.
5. **Environmental Controls:** Maintain proper temperature and humidity to protect sensitive equipment.

13.21.2   Digital Security in ICS Environments

Digital security helps keep ICS operations safe, reliable, and accurate. ICS security is more challenging because it combines Operational Technology (OT) and Information Technology (IT), and often needs high availability and real-time performance.

Key parts of ICS digital security:

1. **Firewalls:** Control network traffic between corporate and ICS networks to stop unauthorized access.
2. **Network Segmentation:** Divide the network into smaller sections to prevent cyber threats from spreading.
3. **Intrusion Detection and Prevention (IDS/IPS):** Watch network activity and respond to any suspicious or harmful actions.
4. **Encryption:** Protect data during transfer by making it unreadable to outsiders.
5. **Access Controls:** Allow only authorized people to use or control the system.
6. **Security Training:** Teach staff about cybersecurity and how to recognize and respond to threats.

13.22   Incident Detection and Response Strategy for ICS Security

An incident response strategy is necessary to protect ICS and OT systems from cyberattacks. The incident response process involves finding the affected systems, stopping the damage, reducing its impact, and finding out what caused it. Every organization must act quickly and effectively during a cyberattack. Compared to general IT security, incident response is more active, as it includes tasks like minimizing damage and checking the situation after the attack.

Incident response is especially important in industrial control system networks because these systems are often targets of cyberattacks. The United States Department of Homeland Security considers ICS a part of critical infrastructure, like the power grid and water supply. Because of their importance, protecting ICS is a major challenge for many companies. A strong strategy should include ways to detect threats, define who is responsible, and describe how to respond to different types of incidents. To improve ICS security, organizations should include a detailed incident response plan as part of their risk management strategy. This helps detect, prevent, and contain intrusions and damage quickly. Details of this strategy are shown in Table 13.3.

The National Institute of Standards and Technology provides an incident response model to help organizations in different industries. It includes four main steps:

1. **Preparation:** Monitoring and identifying critical information technology assets in case of an emergency.
2. **Detection and Analysis:** Gathering information to predict and analyze incidents.
3. **Containment and Eradication:** Stopping and containing attacks to prevent further damage.
4. **Recovery:** Planning for future prevention and learning from past incidents.

13.23 Incident Response Planning in ICS

The National Institute of Standards and Technology (NIST) offers a clear framework for building an incident response program. It includes training, assigning roles, setting success measures, and defining short- and long-term goals. To reduce risks effectively, organizations need a clear incident response plan that includes specific goals, assigned roles, and proper actions to follow.

13.24 Impact of IoT Devices on ICS Security and Compliance

Internet of Things (IoT) devices have a major effect on the security of Industrial Control Systems (ICS). In the past, ICS used physical separation for protection. But now, they are often connected to business networks. This connection improves operations and visibility but also brings new security risks. If an attacker enters the business network, they may access ICS as well.

Table 13.3 Incident response plan aspects

Aspect	Description
Define objectives and scope	Define event kinds, assets to safeguard, and geographical scope
Develop incident response plan	Create a detailed incident management plan with roles, responsibilities, and processes
Implement monitoring tools	Use IDS and SIEM to detect odd activities
Continuous network monitoring	To detect and address suspicious activity, monitor network traffic and systems 24/7
Establish incident categorization	Prioritize issues by severity to allocate resources efficiently
Automated alerts and notifications	Automate notifications for compromise thresholds or indicators
Conduct regular incident drills	Prepare for incidents with table-top and simulated drills
Post-incident investigation	Investigations should determine an incident's breadth, effect, and causes
Communication and reporting	Clear avenues for incident reporting to stakeholders and authorities
Legal and compliance considerations	Response to incidents must follow legal and industry standards
Documentation and learning	To develop and learn, keep complete incident and reaction records
Updating and improving strategies	Review and update incident response plans to address new threats and vulnerabilities

Managing privileged access is important, especially as new devices connect to the Internet, making them more open to attacks [111]. The merging of IT and OT systems adds complexity and increases cyber risks. Also, more IoT devices mean more chances for attackers to find weak spots.

In short, IoT devices help ICS but also increase security risks. To reduce these risks, strong protection and planning are needed.

Key Risks of IoT in ICS:

1. **More Attack Paths:**

   a. IoT devices collect data and automate tasks, but they also expand the attack surface.

2. **Added Complexity:**

   a. IoT makes ICS networks more complex.
   b. Managing many devices can be hard.

3. **Connected Systems:**

   a. IoT devices often connect to ICS, creating possible entry points for attackers.
   b. Weaknesses in IoT can lead to ICS being compromised.

4. **Data Privacy:**

   a. IoT devices often handle sensitive data.
   b. Rules like GDPR and HIPAA must be followed when using these devices.

13.25 Reducing IoT Compliance Issues in ICS

To protect IoT devices in ICS, follow these good practices:

1. **Device Inventory:**

   a. Keep a list of all IoT devices and rate them by importance and risk.

2. **Access Control:**

   a. Only allow trusted users to access devices.
   b. Use role-based access rules.

3. **Network Segmentation:**

   a. Keep IoT devices separate from the main ICS.
   b. This limits how far an attacker can move inside the network.

4. **Penetration Testing:**

   a. Regularly test device security.
   b. Find and fix any security weaknesses.

5. **Encryption:**

   a. Use encryption to protect data shared by IoT and ICS systems.

6. **Patch Management:**

   a. Build a plan to quickly fix security issues in IoT devices.

13.26   Compliance Challenges in Remote ICS Locations

Managing ICS security in different locations brings new challenges. These include:

1. **Different Laws:**

   a. Security rules may vary by country or region.

2. **Communication:**

   a. Good communication is key to coordination.

3. **Common Standards:**

   a. Use the same security rules at all locations to simplify compliance.

4. **Third-Party Providers:**

   a. Outside vendors are common and must be managed securely.

5. **Regular Checks:**

   a. Audit each site often to find and fix problems.

6. **Training:**

   a. Run awareness programs and training for staff.

7. **Incident Handling:**

   a. Use one incident response plan for all locations.

8. **Central Monitoring:**

   a. Use tools that watch over all sites from a central place.

13.27   Why ICS Security Is So Important

Industrial Control Systems (ICS) are foundational to the functioning of the national critical infrastructure. These systems are employed in the supervision and control of essential sectors such as energy, water treatment, manufacturing, transportation, and chemical processing. As ICS environments directly interact with physical processes, any compromise can result in significant safety risks, environmental damage, service disruptions, or financial losses. Therefore, the security of ICS is not only a technical requirement but also a matter of national safety and economic stability.

Unlike conventional Information Technology (IT) systems, ICS environments are designed for high availability, real-time performance, and long-term operation. Many control systems remain in service for decades and often lack modern security features found in IT environments. Furthermore, the operational disruption caused by system downtime or patching in ICS can be far more severe. Consequently, ICS security demands tailored solutions that account for both cyber and physical threats.

To safeguard these critical systems, a multi-layered defense strategy must be implemented [162]. Key security measures include:

- **Access Control:** Restricting system access to authorized personnel only.
- **Network Segmentation:** Isolating ICS networks from corporate or public networks to reduce attack surfaces.
- **Encryption:** Protecting sensitive communication and data to prevent interception or tampering.
- **Continuous Monitoring:** Actively observing system behavior to detect and respond to anomalies or threats in real-time.

In addition to technical safeguards, compliance with internationally recognized standards is essential. Standards such as **ISA/IEC 62443** and **NERC CIP** offer comprehensive guidelines for risk assessment, system design, incident response, and ongoing security maintenance. These frameworks are specifically developed to address the unique requirements of industrial environments.

While IT security principles provide a foundational understanding, ICS security must consider operational constraints such as 24/7 uptime, system legacy, vendor limitations, and the potential physical impact of cyber incidents. Therefore, an integrated approach that bridges the gap between IT and operational technology (OT) is essential for effective ICS protection.

In summary, the importance of ICS security lies in its direct impact on critical services and physical infrastructure. A well-designed and properly enforced ICS security framework ensures not only operational resilience but also the safety, reliability, and continuity of essential industrial processes. By combining robust technical measures, strict access controls, continuous monitoring, and adherence to industry standards, organizations can effectively mitigate risks and safeguard their critical systems against cyber threats.

13.28   Summary

Finally, in a world that is becoming more interconnected every day, the security specifications for industrial control systems are crucial to protecting critical infrastructure from cyberattacks. Organizations that oversee industrial control systems must prioritize security measures that comply with industry best practices and regulatory requirements as technology evolves and cyberattackers' capabilities improve.

Stakeholders can strengthen industrial control systems against various threats by implementing strong authentication procedures, encryption techniques, intrusion detection systems, and other security measures. This will enhance resilience and ensure that vital services continue to operate without interruption. Promoting cooperation between government offices, businesses, and cybersecurity experts is also essential to share threat information and coordinate responses to emerging cyber threats.

Investing in the security of critical infrastructure components protects public safety, national security, and economic prosperity. As the digital world continues to advance, maintaining the reliability, availability, and integrity of services society depends on will require a proactive and comprehensive approach to industrial control system security.

Chapter 14
Static Defense Strategies for ICS: Prioritizing Patch Management, Defense-in-Depth Approaches, and Regulatory Compliance

Abstract This chapter describes static defense strategies for securing industrial control systems, highlighting their critical importance. It details seven key strategies, including application whitelisting, configuration management, and attack surface reduction. The chapter provides real-world examples and evaluates the effectiveness of these strategies. It also explores dynamic defense measures, such as network segmentation and device hardening, emphasizing their role in reducing threat exposure. The chapter covers regulatory compliance with frameworks like NERC CIP and HIPAA, underlining their importance. Additionally, readers will gain insights into industrial control system communication technology migration and defense-in-depth strategies for proactive security. Risk management aspects, including risk identification, mitigation, and monitoring, are also discussed. The chapter further explores the technologies employed in static defense and potential implementation challenges. By understanding these concepts, readers will be equipped to implement layered security measures to protect their industrial control system environments.

Keywords Static defense strategies · Application whitelisting · Configuration management · Attack surface reduction · Network segmentation · Device hardening · Regulatory compliance · NERC CIP · HIPAA · ICS communication technology · Defense-in-depth · Risk management · Risk identification · Risk mitigation · Risk monitoring

14.1 Introduction

Static defense means using fixed and planned security methods to protect industrial control systems (ICSes) from cyberattacks, physical damage, and system weaknesses. These methods include firewalls, software updates (patch management), dividing networks into smaller parts (network segmentation), access control, keeping track of assets, checking third-party vendors, regular audits, security reviews, physical protection, and strengthening devices. They form the base of a complete cybersecurity plan by reducing the chance of attacks and helping keep ICS environments strong. ICSes are used in important infrastructure like factories,

M. A. Rahman et al., *Securing Industrial Control Systems*,
https://doi.org/10.1007/978-3-032-03018-4_14

transport systems, water plants, and power stations. These systems help provide essential services in modern life. However, they face many cyber risks because they connect with IT networks and the Internet.

14.2   Static Defense for Industrial Control Systems

To protect ICSes, it is important to understand the link between threats, system weaknesses, and the security actions taken [152]. This understanding helps build a strong, layered defense strategy where static defense is the first line of protection [162]. Static defense involves finding and fixing weak points in hardware, software, and processes before attackers can use them. This approach reduces the number of ways attackers can enter and lowers the chance of a successful attack. It also supports other digital security methods like intrusion detection and access control. These methods work better when the system is already secure through updates, assessments, and system hardening. Static defense is not a one-time task—it needs regular checks and updates. Key actions include regular security reviews, managing system settings, dividing networks, and improving security processes. By focusing on static defense and encouraging a culture of security, organizations can better protect their ICSes from new and changing threats.

14.3   Advantages of Static Defense Measures

Static defense, especially when using layers of protection, has many benefits [96]:

1. **Step-by-Step Attack Prevention:** Layered defense stops attacks that need to happen in a certain order by blocking them at different stages [4].
2. **Layered Security Design:** This approach uses a structured design with several layers of protection, as suggested by models like the one from Messnarz et al.
3. **Finding Risk Points:** The layered method helps find weak points and possible attack paths using models like STRIDE.
4. **Better Security Planning:** Using the AutoDL (Automotive Defense Layers) idea, defenders can better understand the steps attackers take, making the system more secure.
5. **Understanding Attack Patterns:** Instead of just focusing on single attacks, layered defense shows common attack behaviors, improving awareness of threats.
6. **Fewer Weak Spots:** Many analysis methods miss attack patterns. Layered defense gives a clear structure to improve system checks and reduce blind spots.
7. **Stopping Chain Reactions:** Understanding how attacks happen step by step helps stop one breach from causing more damage in the system.

8. **Network Segmentation:** Splitting the ICS network into smaller parts helps stop attackers from moving around easily.
9. **Smaller Attack Surface:** Closing unused ports, turning off unnecessary services, and limiting external access reduce the number of entry points.
10. **Hardening Devices and Systems:** Making hardware and software stronger by removing unneeded features and applying updates helps block attacks.
11. **Using Approved Apps Only:** Allowing only listed and approved programs to run helps stop harmful software, even if antivirus software misses it.
12. **One-Way Data Flow (Data Diode):** Limiting two-way data flow and using tools like data diodes improves security by stopping data from flowing back into critical parts of the system.

14.4 Significance of Static Defense Measures

Static defense is a proactive security method used to find and fix weaknesses in Industrial Control System (ICS) hardware, software, and processes before attackers can take advantage of them. This early action helps reduce possible entry points for attackers, lowering the chance of an attack.

Think of it like maintaining a castle—patching holes in the wall and locking the gates to stop intruders. The main benefits of static defense are described below:

1. **Preventing Threats:** Static defense blocks attacks before they can cause harm. These methods create strong barriers that reduce the chance of security failures.
2. **Reducing Weak Points:** By hardening systems and devices, limiting access, and shrinking the attack surface, static defense lowers the number of system weaknesses. This makes it harder for attackers to exploit anything.
3. **Meeting Legal Requirements:** Static defense helps organizations follow laws and regulations, especially in sensitive areas like banking and healthcare. These include standards like PCI-DSS and HIPAA.
4. **Improving Reliability and Strength:** Static defense helps critical systems run smoothly by stopping unwanted access and cyber threats. This keeps ICS operations secure and stable.

14.5 Seven Strategies to Defend ICSs

Protecting Industrial Control Systems is very important for the safety and performance of essential services like manufacturing, water treatment, and power supply. As ICSs become more connected with IT systems and start using new technologies like the Industrial Internet of Things (IIoT) and cloud-based monitoring, the risk of cyberattacks increases. While this connectivity improves efficiency and remote access, it also opens up new ways for attackers to strike—so strong cybersecurity is needed.

Without proper defense, ICSs can face serious problems such as halted operations, money loss, environmental damage, or even danger to human life. Well-known attacks like ransomware, data theft, and threats like Stuxnet show the need for strong protection. Unlike regular IT systems, ICSs run for longer periods and can't be updated as easily, which makes them more vulnerable. Attackers may use different methods, such as encrypting data (ransomware), flooding systems (DoS attacks), unauthorized access by insiders, or logging in remotely without permission. These can lead to system shutdowns, production delays, or even dangerous failures. To stop these threats, ICSs should use a layered security system that includes:

- Dividing networks (network segmentation)
- Strong user verification (authentication)
- Constant system watching (real-time monitoring)
- Regular system updates

Organizations should also:

- Train employees in cybersecurity
- Follow strict security rules
- Work with others in the industry to share knowledge and tools

Figure 14.1, based on [43], shows these practices in a clear format. Each method helps reduce risks, limits unwanted access, and increases ICS safety and business continuity.

Figure 14.1 shows the main steps for securing an application. These include allowing only trusted applications, using secure settings and updates, managing

Fig. 14.1 Incident mitigation by strategy (FY 2014–2015)

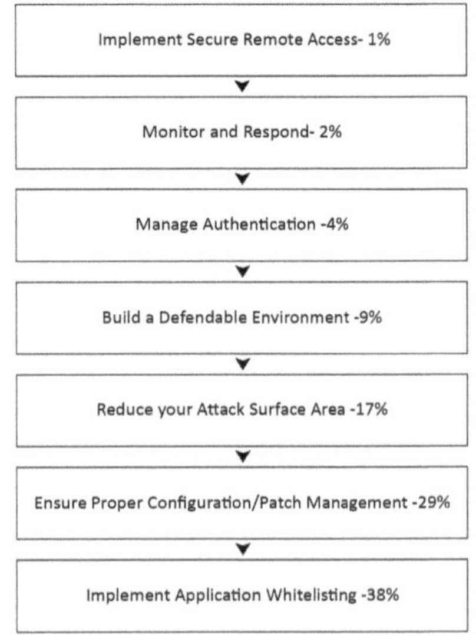

remote access, reducing weak points, monitoring for issues, creating a strong system setup, and managing who can log in.

The seven key ways to protect Industrial Control Systems (ICSs) are:

1. Allow Only Trusted Applications—38%
2. Use Secure Remote Access—1%
3. Keep Systems Updated and Well-Configured—29%
4. Limit Weak Points (Attack Surface)—17%
5. Watch Systems and Respond Quickly—2%
6. Manage User Access—4%
7. Build a Strong and Safe System—9%

(a) **Allow Only Trusted Applications:** Make sure only approved software runs in the ICS. This blocks harmful or unknown programs and helps prevent cyber threats.
(b) **Use Secure Remote Access:** Control who can access ICS systems from outside. This helps stop hackers from getting in remotely.
(c) **Keep Systems Updated and Well-Configured:** Update and adjust systems regularly to fix security holes. This makes the system stronger and safer.
(d) **Limit Weak Points:** Reduce the number of ways attackers can get in. This includes dividing the network, turning off unused features, and protecting devices.
(e) **Watch Systems and Respond Quickly:** Keep an eye on system activity and act fast if something strange happens. This helps stop attacks early.
(f) **Manage User Access:** Control who is allowed to use which part of the system. This keeps critical parts safe from misuse.
(g) **Build a Strong and Safe System:** Use good design, strong layers of protection, and clear security rules to make the system hard to attack.

14.6   Allow Only Trusted Applications (Application Whitelisting)

Application Whitelisting (AWL) is one of the best ways to stop unwanted software in ICS systems. It blocks everything except approved software, making it much harder for harmful programs to run, and it can use built-in tools or buy trusted software to manage the whitelist. This list can also include specific items like plugins, libraries, or settings files.

Instead of trying to spot bad programs (like antivirus does), AWL just blocks all unapproved ones. This approach gives a strong line of defense. It also helps avoid data leaks and system damage by stopping unknown software from running.

Since ICS systems usually do the same jobs in stable conditions, AWL fits well. Less critical tasks that don't need to respond quickly—like data analysis or machine learning—can be done by a separate helper system (AWL process). This helps the main system stay focused and efficient. Still, it's important to plan carefully so these helper systems don't cause problems with communication or security.

Devices like database servers or Human-Machine Interface (HMI) computers are good places to start using AWL. Since these tools usually run the same few programs, it's easier to control what's allowed and block everything else.

14.6.1 Real-World Example

A notable incident reported by the Industrial Control Systems Cyber Emergency Response Team (ICS-CERT) involved a manufacturing organization that experienced a severe malware infection, compromising more than 80% of its networked systems. The attack caused widespread operational disruption and forced the company to completely rebuild its network infrastructure—a process that required significant time, financial resources, and technical effort.

What made this incident particularly concerning was that the malware evaded detection by the organization's traditional antivirus solutions. This highlights a critical weakness in relying solely on signature-based defenses, which are often ineffective against new or modified threats.

Had the company implemented Application Whitelisting (AWL), the damage could likely have been prevented or significantly minimized. AWL works by allowing only pre-approved applications to run on a system, effectively blocking unauthorized or unknown software—even if it is malicious. In high-security environments like ICS, where system availability and integrity are paramount, AWL provides a strong additional layer of protection by reducing the risk of executing harmful code.

This case underscores the importance of adopting advanced and proactive security measures in ICS environments, where conventional tools may not provide sufficient defense against modern cyber threats.

14.7 Working with Vendors for Trusted Setups

Because ICS systems are stable and predictable, they work well with fixed settings and software lists. To make the system more flexible, tasks that don't need instant results can be moved to an external system (AWL).

Jobs like data analysis, machine learning, or tasks that need special knowledge are good to move outside the core ICS. This improves system speed and makes future changes easier. Still, the connection and security between the main system and the helper system must be carefully planned.

Setting up AWL works best when ICS vendors help. They can give the right list of safe programs and help tune the system. This way can get both strong security and smooth system operation.

14.8   Keeping Systems Updated (Patch and Configuration Management)

Effective patch and configuration management is essential for maintaining the security, stability, and reliability of Industrial Control Systems (ICS). This process involves systematically identifying, evaluating, and applying updates—commonly referred to as patches—to operating systems, applications, firmware, and device configurations. Timely patching not only addresses known vulnerabilities but also enhances system performance and ensures compliance with security standards.

Unpatched systems are a common target for cyberattackers, as they often contain well-documented weaknesses that can be exploited with minimal effort. For ICS environments—where safety, availability, and continuity are critical—the risks associated with outdated systems are particularly severe. As such, ICS operators must establish a well-defined patch management strategy that balances security with operational demands.

A comprehensive patch management process begins with the creation and maintenance of a detailed inventory of all hardware and software assets. This allows security teams to prioritize updates based on risk, focusing first on vulnerabilities that could cause the most harm if exploited. Once updates are selected, they should be tested in a controlled environment to ensure compatibility and avoid unintentional disruptions.

After successful testing, patches can be deployed to production systems during scheduled maintenance windows, minimizing the impact on critical operations. Following deployment, systems should be monitored and verified to confirm that updates were applied correctly and that normal functionality is maintained.

Vendors play a key role by releasing patches and configuration guidelines tailored to specific ICS products. These updates often address not only security concerns but also reliability issues, which is especially important in environments where downtime must be avoided. Configuration management, which includes securely managing system settings and access controls, is equally important to ensure that changes do not introduce new vulnerabilities.

Ultimately, patch and configuration management is a continuous process that requires coordination between cybersecurity teams, operational staff, and vendors. By implementing a structured and risk-based approach, organizations can reduce their exposure to cyber threats while maintaining the high availability demanded by industrial operations.

14.9   Why Asset Inventory Is Important in Patch Management

An accurate and up-to-date asset inventory forms the foundation of an effective patch management strategy in Industrial Control System (ICS) environments.

Before any security updates or configuration changes can be applied, it is essential to first identify and document all hardware and software components within the operational network. This includes workstations, servers, controllers, switches, Human-Machine Interfaces (HMIs), field devices, operating systems, applications, and firmware.

Without a comprehensive asset inventory, organizations risk overlooking critical systems, leaving them exposed to vulnerabilities that could be exploited by malicious actors. Unidentified assets may continue to operate with outdated software or insecure configurations, creating weak points in the infrastructure that compromise the overall security posture of the environment.

A well-maintained inventory enables security and operations teams to:

- Track the current patch level of each asset.
- Identify which systems are affected by newly disclosed vulnerabilities.
- Prioritize patch deployment based on asset criticality and threat exposure [152].
- Ensure regulatory compliance by demonstrating due diligence in managing system components.

Furthermore, an asset inventory supports broader cybersecurity initiatives, such as risk assessments, incident response, and network segmentation. It also helps in coordinating with vendors for targeted patch releases and compatibility assessments.

In the context of ICS, where legacy equipment, vendor-specific technologies, and limited downtime windows are common, having visibility into all system components is not just useful but essential. Asset inventory tools that offer automated discovery, real-time updates, and integration with patch management platforms can significantly streamline the update process while reducing the risk of human error.

Overall, maintaining a detailed and accurate asset inventory is a critical prerequisite for secure and efficient patch management. It allows or

14.10 Four Practical Patch Management Techniques

1. **Install Patches on Time**

   - **Act fast:** Apply patches as soon as they are available. Delaying updates can leave systems exposed to known threats.
   - **Use automation:** Set up automated tools to install patches quickly and reduce human errors or delays.

2. **Check That Patches Work**

   - **Test carefully:** After installing a patch, make sure everything works correctly. Focus on critical functions to avoid problems.
   - **Watch the system:** Keep monitoring systems to catch any issues caused by the patch.

3. **Patch Before Problems Happen**

   - **Stay updated:** Follow security news and vendor alerts to learn about new vulnerabilities.
   - **Plan ahead:** Schedule regular times to install patches and fix problems before they are exploited.

Good patch management is key to a safe and stable IT system.

14.11 What Patch Management Involves

Patch management means finding, getting, installing, and checking updates for software and firmware. A good patching process keeps all systems updated to the latest version approved by the vendor.

There are three main types of patches:

- **Security patches**—Fix known security issues.
- **Bug fixes**—Repair system errors or problems.
- **Feature updates**—Add new functions or improve performance.

Patches are important for fixing vulnerabilities. However, not all patches work perfectly with every system. Sometimes they can cause crashes or errors if they don't match the software or hardware.

According to RFC 2616, the PATCH method can change data in ways that are not clearly shown in the request. It is not considered "safe" and may not always produce the same result unless handled carefully.

Vulnerability management is the process of finding and fixing weaknesses. **Patch management** is part of that—it focuses on applying updates to fix those weaknesses.

14.12 How to Prioritize Patch Management

Patch management should be one of the top tasks for keeping IT systems secure. This includes updating critical machines like database servers, engineering workstations, and HMIs as quickly as possible.

When choosing what to patch first, consider how risky the vulnerability is. Patching high-risk systems first helps reduce exposure to threats.

Keeping key systems updated helps prevent attacks and supports compliance with industry rules like PCI-DSS and HIPAA. Organizations can prevent cyberattackers from exploiting known vulnerabilities by actively updating their important systems. A strong cybersecurity strategy includes this proactive step, which helps protect important data and systems. Patches are updates made to fix security issues in

Fig. 14.2 Prioritizing patch management

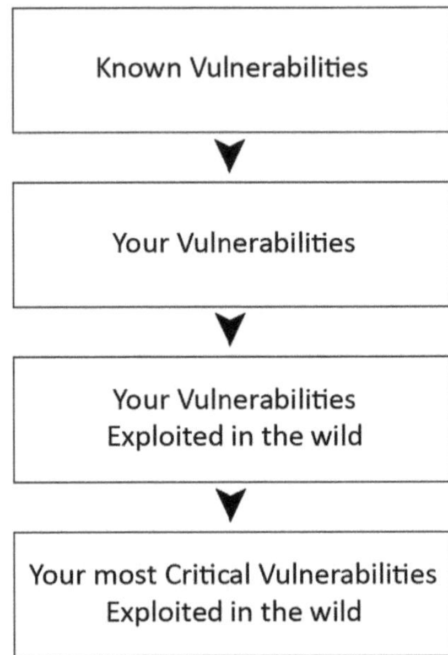

software and operating systems (OS). Software companies release these updates to improve both performance and security. Patch management generally includes three main types of patches:

- **Security patches:** Fix security flaws that attackers could use to harm systems.
- **Bug fixes:** Repair software errors to improve stability.
- **Feature updates:** Add new features or improve usability.

Overall, patch management is an ongoing process that helps secure important systems and reduce risks. It helps organizations quickly respond to new threats and defend against unauthorized access, data leaks, and other cyberattacks. Keeping systems updated is a key part of protecting infrastructure and sensitive data. Figure 14.2, adapted from [154], shows how patching helps protect systems.

Figure 14.2 shows different types of system vulnerabilities.

14.12.1   Real-World Example: HAVEX Malware Attack

1. **HAVEX:** HAVEX is malware that targets SCADA and ICS systems, which are often used in vital sectors like water, power, and manufacturing. It can enter through email attachments, unsafe websites, or fake software updates. Once inside, it can steal data, stop operations, or damage systems.
2. **Patch Infestation:** Sometimes malware is disguised as a software update. If users think it's a real patch, they might install it unknowingly. If attackers change

the way a patch is delivered, they can launch a fake update attack. This is known as "patch infestation."

3. **Secure Patch Hash Verification:** An "out-of-band" method, like phone or separate email, can be used to send a patch's unique code (hash). Users can compare this code with the actual patch to confirm that it hasn't been tampered with.
4. **Bulk Email Risks:** Vendors sometimes send update notifications via mass emails. While this is useful for reaching many users, it also poses risks. Users should verify the source and avoid clicking on suspicious links, especially for important patches.

14.13   Reducing the Attack Surface to Improve ICS Security

Reducing the attack surface is one of the most important steps to protect Industrial Control Systems (ICS).

Many cyberattacks come from the Internet. Limiting ICS exposure to untrusted networks helps strengthen the system's security. Once a hacker gets access, they can steal or change data, gain control, and cause major harm.

By reducing the number of ways an attacker can get in, companies can better protect their systems. This practice is a key part of ICS security. It helps prevent unauthorized access and follows best practices for securing critical systems. Figure 14.3 is adapted from [118].

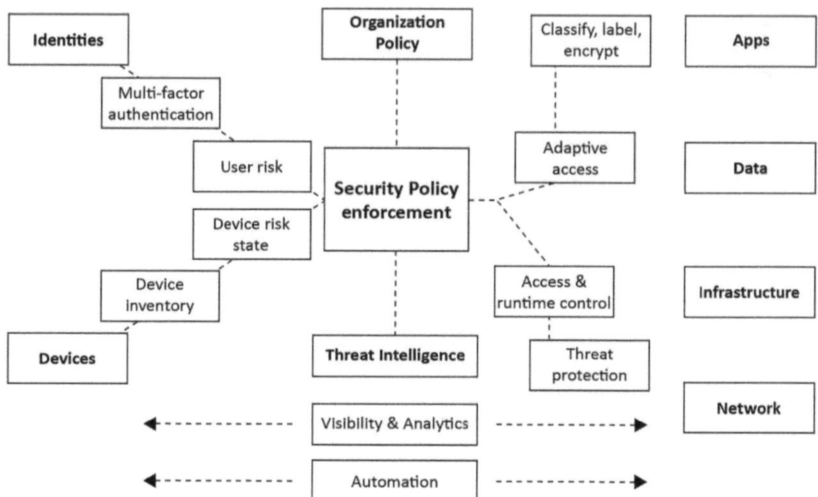

Fig. 14.3 Cyber-security strategy

14.14 Reducing the Attack Surface to Improve ICS Security

Figure 14.3 shows the key parts of an organization's security policy. It includes data classification, encryption, two-factor login, user access control, threat detection, activity monitoring, and automated network tools.

Reducing the attack surface is a key step to protect critical systems. This means limiting entry points that attackers could use. By carefully checking ICS network connections and who can access them, organizations can better defend their systems from cyber threats.

14.15 Key Ways to Reduce the Attack Surface

Closing unused access points and fixing weak spots in the system are major ways to protect sensitive data. Managing user access and updating permissions are essential steps. Here are the main techniques:

1. **Separate from Untrusted Networks:** ICS networks should not be directly connected to the Internet or other untrusted networks. Using firewalls helps block unwanted access.
2. **Disable Unused Ports:** Turn off any network ports that are not in use. This helps reduce entry points for attackers.
3. **Turn Off Unneeded Services:** Stop services that are not required. This lowers the chances of an attack through unused software.
4. **Check Real-Time Connections:** Allow real-time connections to outside networks only when needed for business or control purposes. Limit all unnecessary links.
5. **Use Data Diodes:** A data diode allows data to flow in only one direction. This is useful for sending data out without allowing anything back in, reducing risk.
6. **Limit Two-Way Communication:** If two-way communication is needed, allow only one specific port through a controlled path to lower the risk of attacks.

14.15.1 Example: Public Internet Access to ICS

In 2014, ICS-CERT reported 82,000 ICS systems connected to the public Internet. In many cases, attackers could easily gain access due to weak or no protection. Examples include systems in an Olympic stadium, a dam, a crime lab, and water utilities.

14.16   Using Dynamic Defenses

Dynamic defenses analyze files or programs by running them in a safe test environment to see how they behave. Instead of only checking for known threats, this method can detect new or unknown attacks.

Cybersecurity aims to protect systems and data. This includes firewalls, encryption, secure backups, and more. Firewalls control network traffic. Intrusion Detection Systems (IDS) watch for suspicious activity, and Intrusion Prevention Systems (IPS) stop threats before they spread.

1. **Adaptive and Real-Time:** Dynamic defenses respond to threats as they happen. Tools like IDS, IPS, and SIEM help detect and stop threats. Network segmentation also limits the spread of attacks. Machine Learning (ML) and Artificial Intelligence (AI) help detect unusual behavior in real time.
2. **Based on Behavior:** These systems track activity to find unusual behavior. Signature-based systems match known threats but may miss new ones. Anomaly-based systems catch new threats but may trigger false alarms. Behavior-based methods learn from ongoing activity and offer flexibility, though they also need careful tuning.
3. **Alert and Respond:** When dynamic defenses detect a threat, they send alerts or act right away to stop the attack.

14.17   Benefits of Dynamic Defense in Network Security

The Dynamic Defense Framework is a flexible method to create or choose security solutions that meet specific needs.

1. **Adaptability:** These systems change their defense strategies based on new threats or network changes.
2. **Smart Resource Use:** Resources are shared wisely by filtering what is needed and only using secure actions.
3. **Minor Adjustments:** Most networks only need small changes to stay secure. Safe devices can keep running with little change.
4. **Fast Updates:** Once a strategy is ready, it is quickly applied using filters to speed up the process.
5. **Clear Roles:** Each part of the system—strategy creation, sharing, control, and resource management—has its own job.
6. **Efficient Communication:** Interactive links connect parts of the system so they can respond quickly and adjust to new threats.
7. **Quick Reaction to Threats:** If a threat is detected, the system alerts the right tools to act and protect the network [152].

14.17.1   Restricting Exposure to Threats

Cybersecurity helps reduce the chance of an attack by limiting the number of ways threats can enter a system.

This preventive action follows best practices and helps protect important systems. By reducing the number of entry points, systems become stronger and more secure.

It helps stop data leaks, blocks unauthorized access, and protects systems from being damaged. That is why it is important for organizations to regularly check and manage their system's exposure to threats.

This is a proactive step that focuses on preventing problems rather than just responding to them. By limiting how attackers can get in, organizations are better protected against both known and unknown threats.

Overall, this approach helps keep critical systems and sensitive data safe, even as new threats continue to appear. Figure 14.4 is taken from [74].

Figure 14.4 shows a process related to risk management. It includes identifying risks, understanding how threats can affect systems, analyzing and managing risks, tracking them over time, and reducing their impact.

Protecting Industrial Control Systems is a top priority for securing critical infrastructure. These systems are protected through a layered and complete approach. External attacks are prevented by dividing the network into separate parts. All system components are kept updated through strict patch management.

Strong access control is enforced using multiple steps of identity verification. This limits who can enter the system and what actions they can perform. Special monitoring tools are used to watch for harmful behavior. These tools can detect signs of attack and stop threats automatically.

Along with technical measures, physical security and staff training are also important for better protection. Lastly, having a clear plan to respond to security incidents, doing regular security checks, and protecting the supply chain help reduce risks and prepare the organization for future threats.

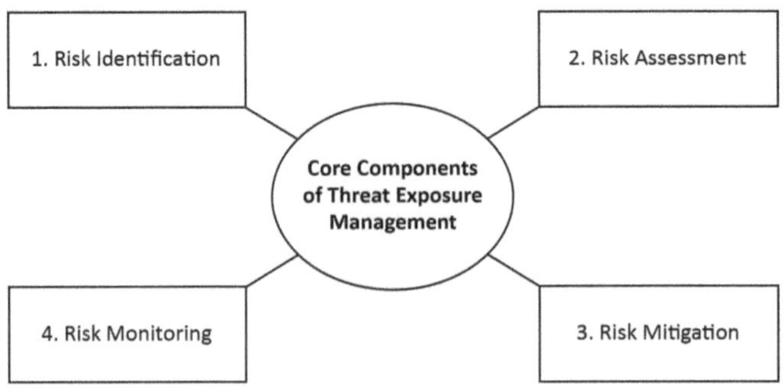

Fig. 14.4 Minimizing exposure to threats

14.18   Network Segmentation

Network segmentation means dividing a computer network into smaller parts. It is also known as network partitioning or network isolation. The main goals are to improve both security and performance.

This is especially useful for organizations that need to follow data protection rules. By keeping important and private information in separate and secure parts of the network, network segmentation helps stop people who should not have access from getting in. Figure 14.5 is taken from [74].

Figure 14.5 shows how a network can be separated using network segmentation. The diagram includes two Demilitarized Zones (DMZs) and labels such as Internet, Firewall, Internal Servers, Application Servers, and Database Servers. It illustrates how a network can be split between internal and external areas.

Example Consider a large bank with many branches. The bank's policy does not allow branch employees to access the financial reporting system. Network segmentation helps enforce this rule by blocking traffic from the branches to the financial system.

Network segmentation provides three main benefits: better security, easier compliance with regulations, and improved network performance.

1. Better Security: By dividing the network into separate zones, it becomes harder for attackers to move around if they gain access. Access controls can also be applied to allow only approved users or devices into certain areas, reducing the risk of unauthorized access or data changes.

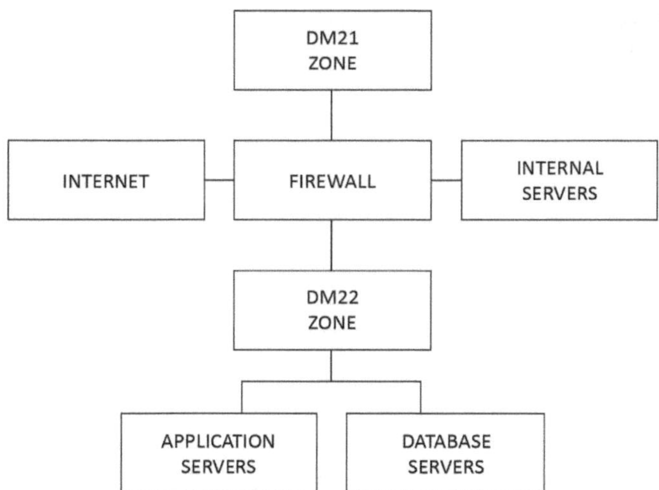

Fig. 14.5   Network segmentation

2. Regulatory Compliance: Many security laws require sensitive data, like customer financial records, to be kept separate. By isolating this data within its own network segment, organizations can follow these laws more easily and support compliance checks.

3. Improved Performance: Segmentation can improve performance by keeping high-bandwidth activities (like video streaming or large file transfers) separate from important business operations. This helps key applications run smoothly and gives users a faster and more reliable network.

14.19 Types of Network Segmentation

1. **Physical Segmentation:** This method uses separate physical devices like switches or routers to create different parts of the network. Each segment is completely separated from the others using hardware.
2. **Logical Segmentation:** This method creates segments using software rather than physical devices. It often uses Virtual Local Area Networks (VLANs) or subnetting to divide the network into smaller parts.

Differences Between Physical and Logical Topology
A network has different segments based on user roles, departments, or locations. These segments can follow either a physical or logical layout. Table 14.1 shows the key differences between physical and logical topology.

Table 14.1 Physical and logical topologies

Physical topology	Logical topology
Depicts the physical layout of the network.	Depicts logistics of the network concerned with transmission of data
The layout can be modified based on needs	There is no interference and manipulation involved here
It can be arranged in star, ring, mesh, and bus topologies	It exists in bus and ring topologies
This has a major impact on cost, scalability, and bandwidth capacity of the network based on the selection and availability of devices	This has a major impact on speed and delivery of data packets. It also handles flow control and ordered delivery of data packets
It is the actual route concerned with transmission	It is a high-level representation of data flow
Physical connection of the network	Data path followed by the network

14.20   Device Hardening

Device hardening means making hardware, network devices, and computers more secure [62]. It includes using different security steps to reduce risks, control access, and protect against cyberattacks. Without hardening, attackers can exploit weak or outdated systems, gain access without permission, steal data, or stop systems from working properly [18].

Main Steps for Device Hardening
- **Change Default Passwords:** Devices often have factory-set passwords that are easy to guess. Replace them with strong and unique passwords.
- **Disable Unused Ports:** Open ports can let attackers in. Close any ports or services that are not needed to reduce risk.
- **Update Firmware and Software:** Keep software and firmware up to date to fix known security problems.
- **Set Access Control Lists (ACLs):** Use ACLs to allow only trusted users or systems to connect to the device.
- **Use Security Policies:** Apply rules to limit what a device can do, such as requiring encryption, authentication, and activity logs.

The main goal of Device hardening is to reduce the chance of attacks. This is done by removing programs, services, and features that are not needed. It also means limiting user access so that people only use what they need. Important settings on the operating system, applications, and firmware are then made more secure. This includes turning off unused ports, making passwords stronger, and using firewalls and intrusion detection systems to monitor the system.

Keeping the system updated is also very important. Patch management helps fix known problems before hackers can take advantage of them. Physical security measures are added to stop people from tampering with devices. Monitoring tools are used to watch system activity and alert if something unusual happens. System hardening is not a one-time task. It must be checked, adjusted, and maintained regularly to keep digital systems safe and stable. Figure 14.6, adapted from [62], illustrates the essential principles of device hardening and its role in securing IT environments.

Figure 14.6 shows device hardening, which protects IT systems. It includes steps like using strong passwords, enabling multi-factor authentication, raising security awareness, and installing antivirus and anti-malware software.

14.20.1   Importance of Device Hardening

By closing unused access points and removing unnecessary items like accounts, apps, ports, and permissions, device hardening helps reduce security risks. The goal

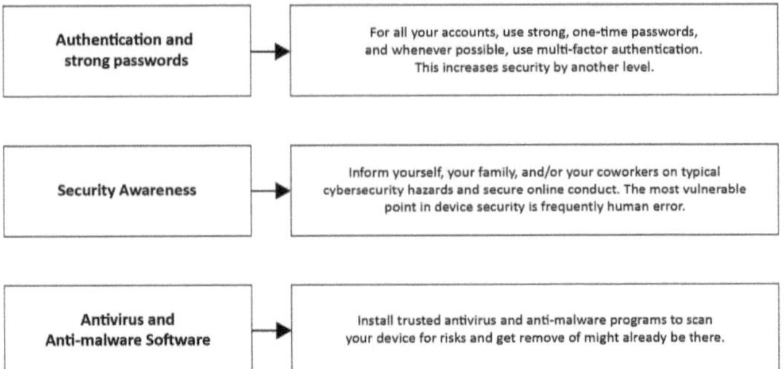

Fig. 14.6 Device hardening protects IT infrastructure

Fig. 14.7 Importance of device hardening

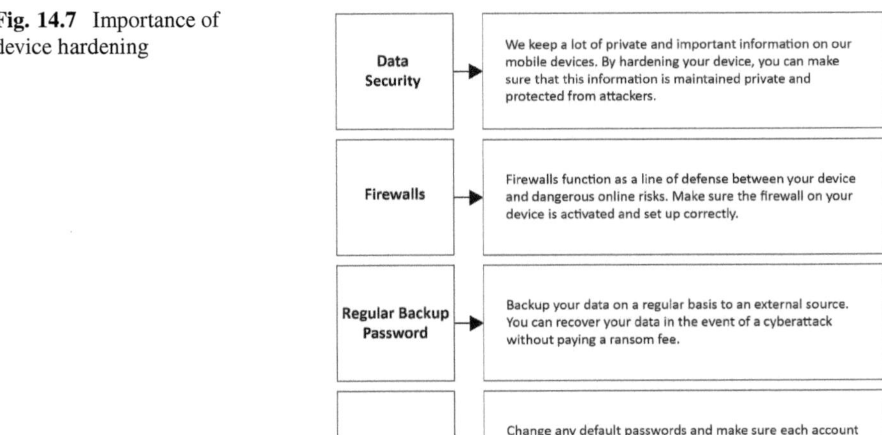

is to make it harder for attackers to get in. Figure 14.7 shows this clearly, as seen in [62].

Figure 14.7 shows important practices for keeping data safe. It highlights the use of firewalls, regular data backups, and strong, unique passwords. The main advice includes turning on and properly setting up firewalls, changing default passwords, and backing up data often to protect against cyberattacks.

1. **Protection from Cyberattacks:** Hackers are always finding new ways to break into systems. Device hardening helps stop threats and lowers the chance of attacks.
2. **Secure Data:** Many devices hold sensitive information. Device hardening helps keep this data safe from hackers [62].

3. **Follow Rules and Standards:** Different industries have different security and legal rules. Device hardening helps meet these standards and avoids fines or damage to reputation.
4. **Longer Device Life:** Secure devices last longer. By avoiding security problems, device hardening [62] helps extend the life of devices and saves money on replacements.

14.21 System Hardening

System hardening is a basic cybersecurity step that helps protect software, operating systems, and IT systems. It means reducing the number of ways attackers can get in, fixing known weaknesses, and using stronger security controls.

Some common system hardening practices include:

- **Turn Off Unused Services:** Stop running programs or background processes that are not needed. This removes extra targets for attackers.
- **Install Security Updates:** Keep software and firmware up to date to fix known problems and protect against attacks.
- **Set Firewall Rules:** Use firewalls to control what traffic can enter or leave the system, and block anything not allowed.
- **Follow Security Policies:** Apply security rules for things like passwords, encryption, and login checks across the organization.
- **Limit Access Rights:** Give users and apps only the access they need. This helps stop misuse or attacks that try to gain higher access.

The main goal of system hardening is to build a secure and stable system that blocks unauthorized access, protects sensitive data, and lowers the risk of cyber threats like malware and data leaks. When done well, it also helps organizations meet security rules, keep data correct, and keep their business running smoothly. A strong cybersecurity plan should always include system hardening as a key step. Figure 14.8, taken from [18], shows the main ideas and steps in system hardening.

Figure 14.8 shows how software systems can be protected from intrusions. It includes prevention methods, system hardening, patch management, malware protection, and attack detection [4].

System hardening is not a one-time task but a continuous process to improve the security of a computer system. This multi-step process starts by checking the system to find and fix weak points in both hardware and software. It begins with listing all parts of the system. The main goal is to remove unneeded programs and services to reduce possible ways an attacker could get in. Then, system settings are changed to be more secure, like tightening the system's defenses. Patch management is also important—it helps fix known security problems on time. Physical security controls help by preventing unauthorized access to the hardware. Constant monitoring is also necessary to spot unusual behavior early and stop attacks quickly. These steps together make the system stronger and more secure against cyber threats.

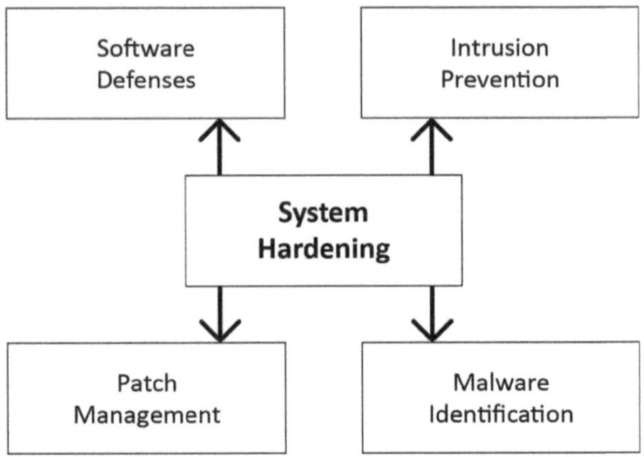

Fig. 14.8 System hardening

14.21.1 *System Hardening and Attack Surface Reduction*

System hardening helps secure a system by reducing its "attack surface," which means all the possible weak spots that attackers could use. Below are common attack surface issues:

1. **Default Passwords:** Many systems use factory-set passwords. If not changed, attackers can easily guess them using automated tools. This makes the system more open to attacks.
2. **Hardcoded Credentials:** Saving usernames or passwords in plain text files or code can create hidden entry points. If not removed or hidden, these can be exploited.
3. **Unpatched Software and Firmware:** When updates or patches are not applied, old problems remain that attackers can take advantage of. Some patches may be hard to install or might not work correctly.
4. **Weak Access Controls:** Poor control of user permissions—especially for admin-level accounts—can be risky. This includes both human and machine accounts, especially in cloud environments.
5. **Bad System Configuration:** Incorrect settings in servers, firewalls, ports, or BIOS can create security gaps. For example, misconfigured cloud systems may leak private data or create security holes.
6. **Unencrypted Data:** If data is sent or stored without encryption, attackers can read or change it. This increases the risk of data theft and tampering.

System hardening reduces these risks and helps create a stronger system that can better defend against cyberattacks.

14.21.1.1 Why System Hardening Is Important

System hardening is a key part of keeping information secure. However, many organizations ignore it or treat it as a low-priority task for the IT team. This section explains why system hardening should be a top priority in any IT security plan.

Reason 1: Technology Is Growing Fast
As digital technology grows, organizations use more systems, such as cloud platforms, IoT (Internet of Things), apps, e-commerce, and smart devices. These systems exchange large amounts of data, making security more important. System hardening helps protect these complex systems.

Reason 2: Remote Work Is Common Now
With more people working from home after COVID-19, employees often share sensitive data outside the office. System hardening helps keep this data safe and supports compliance in remote work environments.

Reason 3: Cyber Threats Are Increasing
Cybercrime is rising as more IT systems are connected, and attackers use smart tools like AI-based malware. Hardening systems helps reduce the risk of attacks by keeping software updated and reducing weaknesses.

Reason 4: New Laws Require Better Protection
Laws like GDPR require organizations to protect personal data. Following these rules helps companies stay legal and protect themselves. As technology changes, more laws are being created to protect data.

Reason 5: Lack of Skilled IT Workers
There are not enough trained IT professionals. As automation grows, companies need to use existing skills wisely. Automating tasks like system hardening can help fill this gap.

14.22 Compliance and Cybersecurity

Following rules and regulations helps reduce security risks. It can prevent data leaks and protect a company's reputation and money.

Laws like SOX (for finance) and HIPAA (for healthcare) were created to fix past issues. Over time, laws have expanded to cover privacy and cybersecurity.

Global laws like the GDPR and CCPA set high standards. Not following them can lead to big penalties. Frameworks like ISO 27001 and NIST give guidance on how to stay secure.

To follow the rules, organizations should:

- Assess risks
- Use good security practices
- Audit systems regularly
- Manage data properly

These steps help keep systems secure and support legal and ethical use of data.

14.23 NERC CIP Compliance

NERC CIP (North American Electric Reliability Corporation—Critical Infrastructure Protection) standards help secure the Bulk Electric System (BES). They focus on cybersecurity for the energy grid.

- NERC CIP protects critical energy systems.
- Network segmentation and limiting system access help improve security.
- It follows a risk-based approach to reduce weaknesses.
- System hardening and application whitelisting reduce attack risks.
- Following these rules helps companies prepare for and respond to cyberattacks.

14.24 HIPAA Compliance

HIPAA (Health Insurance Portability and Accountability Act) requires organizations to protect health data.

- HIPAA safeguards Electronic Protected Health Information (ePHI).
- It covers data like patient records and medical histories.
- Following HIPAA protects patient privacy and builds trust.
- Healthcare systems must follow strict rules to protect sensitive data.
- Security measures like whitelisting and network segmentation reduce risks.

- Strong defenses prevent unauthorized access to patient data.
- These steps support patient privacy and build trust in the system.
- Good security also meets legal requirements and boosts public confidence.

14.25 Why Compliance Matters

HIPAA and other regulations help protect data and build trust.

- All industries must follow their own rules.
- Not following rules can lead to fines and damage to reputation.
- Laws like HIPAA and NERC CIP protect critical data and services.

- Following them keeps systems secure and services running.
- Good compliance also follows best security practices.
- It lowers the chances of cyberattacks and helps keep operations safe.

14.26 Defense-in-Depth for ICS

Industrial Control Systems (ICS) are essential to the operation of critical infrastructure, including power grids, water treatment facilities, manufacturing plants, and transportation networks. As these systems become increasingly interconnected with corporate IT environments and external networks, the potential attack surface also expands. To mitigate the growing risk of cyber threats, a comprehensive and strategic security model known as *Defense-in-Depth* is essential.

Defense-in-Depth is a layered security strategy that applies multiple protective measures across various points in a system, thereby reducing the likelihood that a single point of failure could lead to a successful attack. This approach is inspired by traditional military defense tactics, where multiple layers of defense are deployed to delay, disrupt, or prevent enemy penetration. In the context of ICS security, these layers include physical security, network segmentation, access control, endpoint protection, monitoring, and incident response planning.

The primary objective of Defense-in-Depth is to manage risk by distributing security functions across different components of the ICS environment. If one layer is breached, additional layers continue to provide protection, minimizing the potential impact. This structure is particularly effective against sophisticated, multi-stage cyberattacks that target ICS vulnerabilities through methods such as phishing, stolen credentials, malware injection, or unauthorized remote access.

Key components of a Defense-in-Depth strategy for ICS include:

- **Physical Security:** Controlling physical access to critical ICS hardware and facilities to prevent tampering or sabotage.
- **Network Segmentation:** Isolating control networks from business networks using firewalls and demilitarized zones (DMZs) to limit lateral movement by attackers.
- **Access Control:** Enforcing role-based access, multi-factor authentication (MFA), and least privilege principles to limit user access to only what is necessary.
- **Endpoint Security:** Deploying antivirus software, host intrusion detection systems (HIDS), and application whitelisting to protect ICS devices.
- **Monitoring and Detection:** Using intrusion detection systems (IDS), security information and event management (SIEM) platforms, and log analysis to detect suspicious activity.
- **Patch and Configuration Management:** Ensuring timely updates and secure configurations to reduce vulnerabilities and maintain system integrity.
- **Incident Response and Recovery:** Establishing detailed plans for responding to and recovering from security incidents to ensure business continuity.

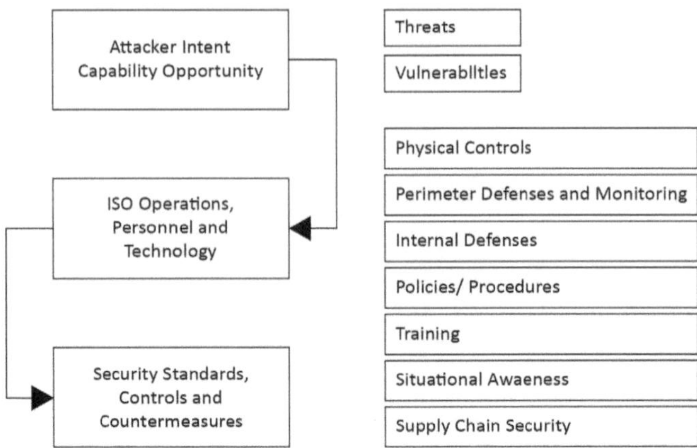

Fig. 14.9 Defense-in-depth strategies for ICS

The Defense-in-Depth model is particularly important in ICS environments due to the high stakes involved. Unlike traditional IT systems, failures in ICS can lead to physical consequences, including equipment damage, environmental hazards, or risks to human safety. Implementing a Defense-in-Depth strategy is not optional but essential. It enables ICS operators to build robust, flexible, and adaptive security frameworks that are capable of responding to the evolving landscape of cyber threats.

Furthermore, attackers are increasingly targeting ICS with advanced persistent threats (APTs) that are difficult to detect and remove. Defense-in-Depth not only strengthens the overall resilience of control systems but also supports compliance with industry standards such as NIST SP 800-82, ISA/IEC 62443, and NERC CIP. Figure 14.9, adapted from [157], illustrates how layered defenses are structured within an ICS environment to safeguard critical infrastructure.

Figure 14.9 shows different items related to attacker goals and security protections. It includes threats, attacker skills, weaknesses in the system, chances for attack, ISO operations, people, technology, physical protections, perimeter defenses, security rules, internal defenses, countermeasures, training, awareness, and supply chain safety.

14.26.1 Proactive Security Model

A proactive security model is important for strong cybersecurity. Unlike reactive methods that only act after an attack, a proactive model includes setting clear rules, identifying threats, checking risks, mapping system design, and listing all key assets. This helps in better risk analysis, knowing which assets are most important, and applying strong security steps. It also includes handling incidents and making

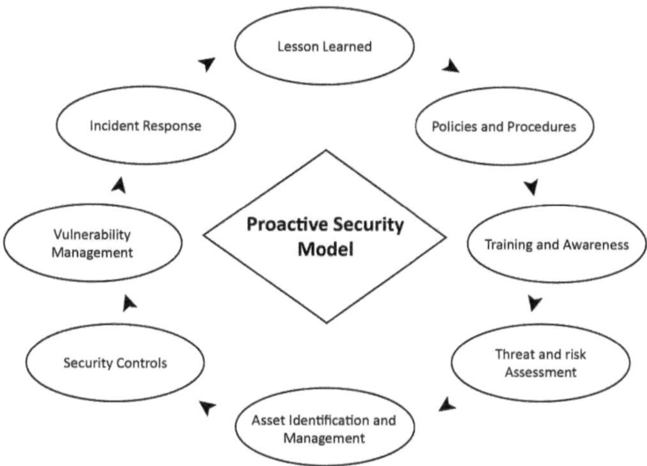

Fig. 14.10 Proactive security as an iterative process

regular updates, so the system keeps improving and stays ready for new threats. Figure 14.10 is adapted from [124].

Figure 14.10 shows a complete security system. It includes items such as lessons learned, security policies, incident response steps, training, awareness, managing vulnerabilities, checking threats and risks, security measures, and asset tracking.

14.26.2 Benefits of a Proactive Security Policy

A proactive security policy offers many advantages:

1. **Smart Vulnerability Management:** Regular testing and scanning help find and fix problems before an attack happens. This prevents risks and makes the system stronger.
2. **Compliance:** Good security follows rules like PCI DSS, HIPAA, SOX, and GDPR. Reports from security checks show that the organization is careful and follows these rules, helping avoid big fines.
3. **Cost and Disruption Avoidance:** Security breaches are expensive and harmful. Proactive steps help stop them before they happen, saving money and keeping the business running smoothly.
4. **Protecting Trust and Reputation:** A strong security plan builds trust with customers and protects the organization's image.

14.26.3 Risk Management and ICS

Improving security in Industrial Control Systems (ICS) starts with knowing the risks the organization faces. It includes spotting cyber threats, adding security to daily work, and adjusting protections to the right level. A well-designed ICS security plan should meet both technical needs and practical realities. Every employee should understand ICS risks and take part in managing them.

14.26.4 Multitier Risk Management Integration

A three-level risk management approach is used to cover all areas of an organization—top-level management, daily business tasks, and the technical systems. The three levels are described below:

Tier 1: Organization
This level focuses on the top leaders and their decisions. It includes:

- **Executive Leadership:** Senior leaders who decide the direction of the organization.
- **Corporate Strategy Policy:** High-level rules that guide company goals and how risks are handled.
- **Actionable Policy and Procedures:** Clear rules and steps to manage risks and follow strategy.
- **Guidance and Constraints:** Industry rules and standards the company must follow.

Tier 2: Mission and Business Processes
This level deals with the main work the company does. It includes:

- **Mission and Business Processes:** The key tasks needed to meet company goals.
- **Monitoring Feedback:** Information gathered from checking how processes work, used to improve them.

Tier 3: Information Technology and Industrial Control Systems
This level covers the technical systems that support operations. It includes:

- **IT and ICS:** The computers, software, and networks that control industrial systems.
- **Monitoring Feedback Results:** Data collected from these systems to improve safety and performance.

Summary of Multitier Risk Management
- **Unified Risk Management Process:** Managing risks across all three levels creates a connected and effective risk system.

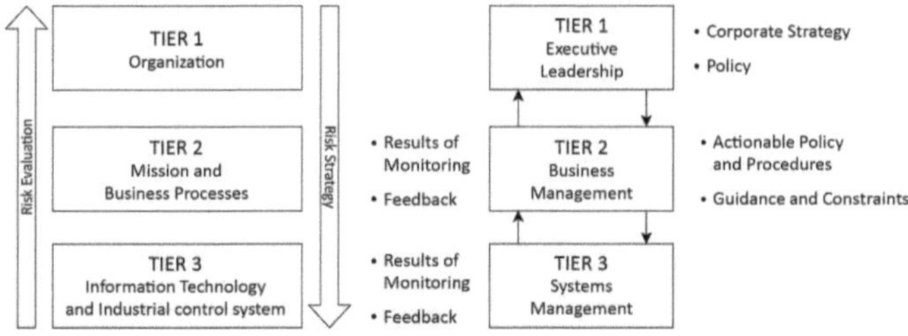

Fig. 14.11 Risk management tiers

- **Aligning with Business Goals:** This approach makes sure security supports the company's mission.
- **Managing ICS Risks:** Because ICS systems are very sensitive, they need layered and detailed risk management.

This helps connect risk handling to business goals. Figure 14.11 is taken from [32].

Figure 14.11 presents the approach used to manage risks in Industrial Control Systems (ICS). Throughout the entire life cycle of an ICS, key activities such as maintaining an asset inventory, classifying system components, identifying threats, and accepting certain levels of risk all contribute to effective, ongoing risk management.

14.26.5   Risk Management Approach

Risk management in ICS plays a critical role in ensuring the reliability and security of essential infrastructure. ICS environments face unique challenges compared to traditional IT systems. Attacks on Operational Technology (OT) can lead to far more severe outcomes. In particular, targeted cyberattacks may result in environmental damage, public health risks, or the failure of critical services.

An example involves Safety Instrumented Systems, which are designed to detect unsafe conditions and shut systems down safely. If these systems are compromised, such as in the case of a gas pipeline, the result could be large-scale power outages or hazardous situations.

A well-planned risk management strategy reduces the exposure of key assets to threats. The process begins with identifying all commercial and industrial assets. Once the assets are known, security teams assess existing threats, system vulnerabilities, and the potential impact of various risks. Based on this assessment, security controls are carefully applied to protect operations without interrupting their essential functions.

Risk management is not a one-time task. Real-time monitoring and continuous improvement are necessary to respond to new threats as they arise. Adjusting security controls based on changing conditions helps organizations stay prepared

Fig. 14.12 Risk management approach

and ensures the ongoing protection of critical infrastructure. Figure 14.12 is reproduced from [12].

Figure 14.12 outlines a process for managing inventory assets within an organization. It includes steps such as monitoring, categorizing, adjusting asset criticality, implementing security controls, identifying risks, and tailoring potential controls.

ICS security teams should adopt an aggressive defensive stance within the Sliding Scale of Cybersecurity to counter these attacks. Skilled analysts use technology to monitor, respond to, and learn from risks within the control network. To enhance overall security, active defense works alongside passive defense and relies on a well-documented asset inventory.

Moreover, to improve resilience, focus must be given to enhancing recovery procedures in addition to prevention. This involves collaborating with engineering teams to improve recovery procedures and ensure systems are designed with redundancy and fail-safes. A strong risk management plan must include both efficient incident recovery and proactive ICS cybersecurity measures.

Assigning controls to identified risks is another crucial aspect of effective ICS risk management, especially when these risks are related to internal business activities within the organization. Continuous development, frequent updates to risk management protocols, and constant monitoring of ICS networks ensure that security measures remain effective against evolving threats.

14.27 Risk Mitigation and Monitoring

The safety and performance of ICS systems depend on reducing risks and monitoring systems. It is important to focus on the most serious vulnerabilities and fix them through updates or configuration changes. At the same time, systems must constantly check for issues that could impact the safety, availability, or privacy of ICS operations.

Centralized logging systems should collect all events in the ICS network. Penetration tests should be done regularly to see if attackers could break in and if existing security measures work well. Logs should be reviewed to make sure all important events are captured and alerts are triggered on time.

Good practices include strong perimeter defenses, separating ICS from corporate IT systems, and following basic cybersecurity hygiene to reduce weaknesses. In ICS security, staying ahead of threats is essential in a fast-changing digital environment.

Risk mitigation means planning and creating strategies to reduce possible harm to a project or business. Teams working on projects, like building a new product, use risk mitigation to identify and manage possible issues before they cause real problems. Actions taken to reduce or solve these problems are part of the risk mitigation process.

Risk management is not a one-time task. It is a continuous process that needs regular checking and improvement to manage both known and unknown risks properly.

14.28   Most Typical Reactions to Risk

1. **Risk Avoidance:** This means avoiding activities that might create risks. For example, a company may choose not to launch a risky new product.
2. **Risk Reduction:** This method does not remove risk completely but lowers its impact. For instance, regular health check-ups under an insurance plan are an example of reducing medical risks.
3. **Risk Sharing:** Risk is spread among different people or organizations. For example, investors in a company share the risks of doing business.
4. **Risk Transferring:** Risk is moved to a third party, such as buying insurance to cover damage to property.
5. **Risk Acceptance and Retention:** Some level of risk always remains even after all actions. Accepting and managing this remaining risk is part of the overall strategy.

14.29   Risk Analyst

1. **Identify the Risks:** Understand what risks exist for the organization.
2. **Record Network Architecture:**

   - Record the technologies used by the organization.
   - List the software applications involved in business operations.

3. **Human Resources and Operations:**

   - Identify the people involved in the system.
   - List the resources needed to run the business.
   - Estimate how likely it is that a threat will be carried out.

4. **Intrusion Forecasting:** Predict how someone might exploit current weaknesses.
5. **Prioritize Risk Scenarios:** Focus on the most important risks first, based on how likely and how harmful they are.

14.30   Enforce Security Controls

Security controls are tools and processes used to protect data, systems, and infrastructure. These controls help stop or reduce security threats. High-risk ICS systems should be protected first, especially systems using outdated software. Training staff in cybersecurity reduces the chance of human errors. Systems must also include full lifecycle protections. While layered defenses can reduce risk, no system is fully secure. Some ICS systems may need special rules or exceptions due to their unique nature.

1. **Physical Security Controls:** These protect physical locations and equipment. Examples include door locks, access cards, surveillance, and temperature control. Buildings might need blast protection. Cabling should be well-planned to avoid tampering or damage.
2. **Digital Security Controls:** These include usernames and passwords, two-factor authentication, antivirus software, and firewalls.
3. **Cybersecurity Controls:** These are aimed at preventing cyber threats, like DDoS protection and intrusion detection systems.
4. **Cloud Security Controls:** If using cloud services, companies must follow best practices and policies to keep workloads and data safe.

14.31   ICS Network Architectures

The combination of traditional IT systems with ICS has made system management easier. However, this also brings new risks. Not following ICS standards, outdated technology, and insecure network links increase the chance of failure or attacks.

Figure 14.13 is reproduced from [81].

Figure 14.13 shows different parts of a network, such as web servers, email servers, enterprise zones, controllers, workstations, servers, SCADA systems, and field devices.

In the past, industrial control systems (ICSes) were stand-alone and used in trusted and limited environments, so they had weak security. But when modern IT systems were added, network connections brought new risks. Shared systems gave attackers more ways to reach important parts of the network. To deal with this, new ICS security methods like "zones and conduits" were introduced. These use secure, encrypted communication between trusted areas.

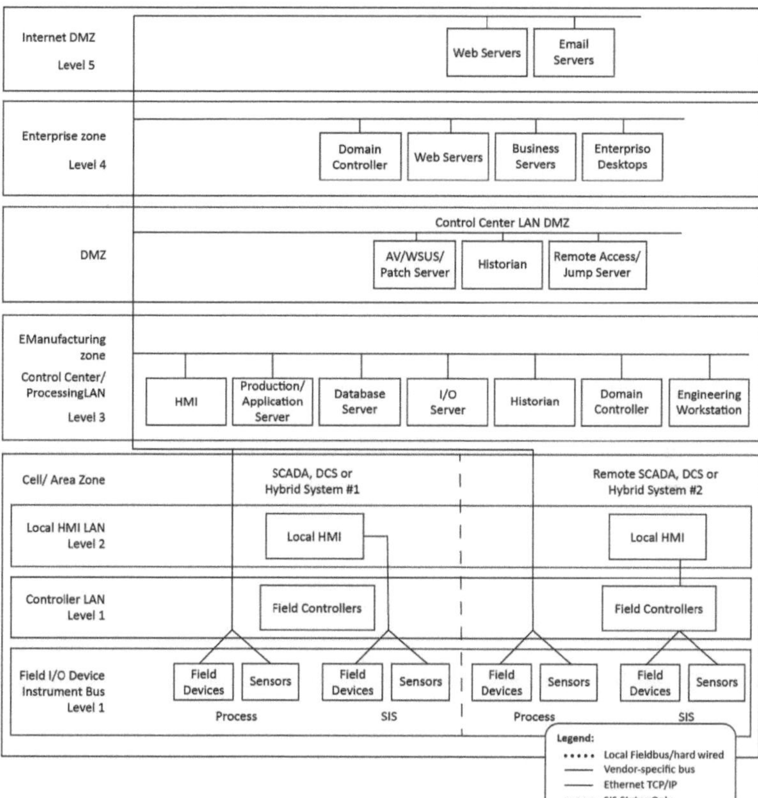

Fig. 14.13 ICS network architectures

14.32 Common Architectural Zones

ICS security uses different models to divide systems into zones. Two well-known models are the IEC 62443 Standard Zones and the Purdue Model. The Purdue Model is a conceptual framework that separates physical processes, sensors, control systems, operations, and business tasks. It introduces the idea of an "air gap" to separate IT systems from OT (Operational Technology) systems. This helps limit access without interrupting business activities. It defines six levels in ICS architecture:

- **Level 4/5 (Enterprise Zone):** Standard IT network functions like scheduling and production planning.
- **Level 3.5 (Demilitarized Zone):** Uses firewalls and proxies to stop threats from moving between OT and IT.

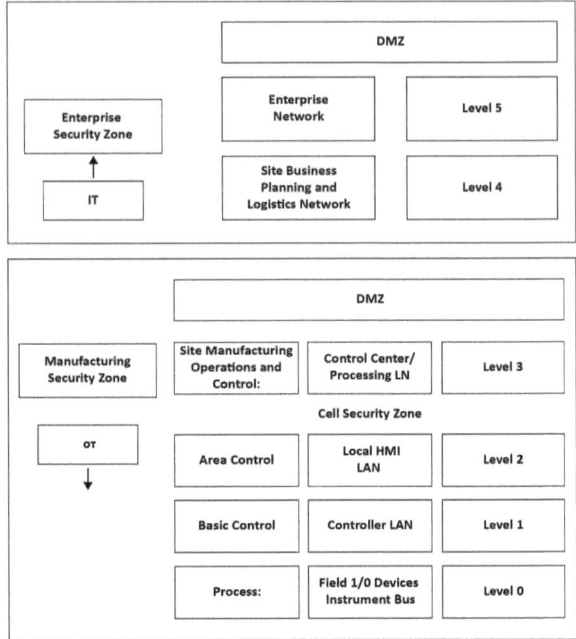

Fig. 14.14 Common architectural zones

- **Level 3 (Manufacturing Operations):** Includes OT devices that control production on the factory floor.

Problems at these levels can cause financial loss and harm to infrastructure. That's why it's important to protect these systems. The IEC 62443 model defines zones based on how critical they are:

- **Zone 4:** Less critical; includes administrative and business systems.
- **Zone 0:** Most critical; includes direct control of physical processes.

These zones are protected using layers of defense. To design this properly, we need to understand the technology and how the systems depend on each other. Dividing the system into zones makes it easier to add multiple layers of security. One important part of designing security zones is network segmentation. This means dividing the network into smaller parts to separate the control system from other systems. Segmentation helps improve security by making it harder for attackers to move around the system. It also makes it easier to monitor and control each part. Figure 14.14, based on [107], shows these architectural zones.

Figure 14.14 shows a network architecture and security zones in an enterprise. It includes different areas such as the Enterprise Network, DMZ (Demilitarized Zone), Manufacturing Site, Security Operations, and Control Center.

14.33 Access and Authentication Controls

ICS networks need proper access control to support smooth operations. Since ICS must work all the time and allow fast access, normal authentication tools used in business networks are often not suitable. Many ICS vendors do not offer built-in authentication, which makes it difficult to manage users across different sites. Sometimes, systems have to share user credentials.

Access control can be centralized or distributed. In a distributed system, each device checks user access on its own. This works well for small systems but does not scale for larger ones. Big ICS networks often use a central server like Active Directory or LDAP for authentication. Other tools like Kerberos, RADIUS, and TACACS help with secure communication. However, if the main server is attacked, the whole ICS could be at risk. Using backup servers adds reliability but increases cost.

To protect ICS and SCADA systems, strong access and authentication controls are needed. Authentication checks who the user is. It can use passwords, key cards, digital certificates, or biometrics like fingerprints or eye scans. Strong methods help block unauthorized access. Authorization controls what each user can do based on their role. Security can be improved with context-aware authentication, which checks location, time, and behavior. Using multi-factor authentication (MFA) adds extra protection. Good access control combines strict rules, risk analysis, and smart policies.

14.34 Security Monitoring

Effective security monitoring is a critical component of protecting Industrial Control Systems (ICS) from cyber threats. Given the increasing complexity and connectivity of ICS environments, continuous observation of network traffic, system activity, and user behavior is essential to detect and respond to anomalies before they escalate into serious incidents. Unlike traditional IT networks, ICS environments often prioritize system availability and operational continuity over security monitoring. This can result in limited visibility into system behavior and delayed detection of malicious activity. Attackers who gain access to an ICS may remain undetected for extended periods, allowing them to map network topology, escalate privileges, and disrupt operations. To address this gap, ICS operators must implement dedicated monitoring strategies tailored to the unique requirements and constraints of industrial systems.

Security monitoring involves the use of both automated tools and human oversight to track the health and integrity of the control environment. Technologies such as Security Information and Event Management (SIEM) systems and syslog servers are commonly deployed to aggregate logs, correlate events, and generate alerts for suspicious activities. These tools can identify unusual patterns—such as unexpected login attempts, configuration changes, or unauthorized data flows—that may signal the presence of a threat actor. In addition to traditional monitoring tools, organizations may deploy honeypots and canary tokens—decoy systems or files designed to lure attackers and trigger alerts upon access. These can be particularly useful in detecting insider threats or lateral movement within the network.

Monitoring should also include:

- **Network Traffic Analysis:** Continuously examining communication between devices to detect protocol anomalies, abnormal traffic volumes, or unauthorized access attempts.
- **Host-Based Monitoring:** Tracking activity on critical endpoints such as Human-Machine Interfaces (HMIs), Programmable Logic Controllers (PLCs), and engineering workstations.
- **Log Management:** Collecting, storing, and analyzing logs from devices, servers, and applications to support forensic analysis and compliance reporting.
- **Dashboards and Visualization:** Providing real-time visibility into system status and alerts to help operators and security teams respond quickly to incidents.

Real-time monitoring enhances an organization's ability to maintain operational integrity and defend against attacks. It also supports incident response by enabling faster identification of compromised systems and providing contextual data for remediation efforts. Furthermore, regular auditing and log analysis can help refine security policies and improve overall situational awareness. Security monitoring is not just a technical control but a strategic necessity for ICS environments. It enables early threat detection, supports compliance with cybersecurity standards, and helps maintain the reliability and safety of critical infrastructure.

14.35   Issues in ICS Cybersecurity

ICS are used to manage critical services like energy, water, transport, and manufacturing. As digital systems are added, cybersecurity becomes more important. This section focuses on common problems and how static defenses can help prevent attacks.

1. **Legacy Systems:** Many ICS devices were built before cybersecurity became important. These older systems have weak security and are easy targets.
2. **Interconnectivity:** As ICS connects more with business networks and the Internet, the risk of attacks like ransomware and malware increases [149].
3. **Human Factors:** Mistakes, negligence, or insider threats can expose systems. Weak passwords, bad setup, or lack of training can lead to security failures.
4. **Supply Chain Risks:** ICS depends on many vendors for hardware and software. Fake or tampered parts from the supply chain can introduce security holes.

To solve these problems, organizations must focus on better cybersecurity. Static defenses, like system hardening and patching, can help reduce risks and protect critical systems.

14.36 Static Defense Techniques

Here are some common static defense methods for ICS:

- **Application Whitelisting (AWL):** Only allows approved software to run. Useful for fixed systems like HMIs and database servers.
- **Patch Management:** Updating systems with the latest security patches. Attackers often target outdated software.
- **Configuration Management:** Keeping systems set up properly to avoid weaknesses. Important for both hardware and software.
- **Limiting External Connections:** Reducing laptop use or using tested devices for vendors can improve security.

Working with vendors to set standards and safely apply updates is also important.

14.37 Importance of Sensor Fault Mitigation

Sensor faults in ICS can affect performance, safety, and reliability. They can be caused by aging, environment, or physical and electrical issues. Common faults include drift, noise, and stuck readings.

Faults can lead to bad data, unsafe conditions, and system shutdowns. To detect faults, ICS uses backup sensors, self-checks, regular calibration, and tools like machine learning.

Good fault management helps systems stay safe and efficient. It is important to use strong detection tools and clear repair plans. Documents like the National Instruments white paper give more information on this topic.

14.38 Sensor Fault Detection: Exploratory Analysis

In this section, we perform an exploratory data analysis (EDA) of the sensor fault detection data. We import and visualize the data using various plots to understand the distribution and relationship between the sensor readings. We begin by importing the necessary libraries and defining functions for plotting the data.

Importing Libraries and Defining Functions

The following Python code imports the required libraries and defines functions for plotting: referred from: https://github.com/Sunzidasiddique1/ICS.

Importing Required Libraries

```python
from mpl_toolkits.mplot3d import Axes3D
from sklearn.preprocessing import StandardScaler
import matplotlib.pyplot as plt  # plotting
import numpy as np               # linear algebra
import os                        # accessing directory structure
import pandas as pd              # data processing, CSV file I/O
    (e.g. pd.read_csv)
```

Listing Files in the Dataset Directory

```python
for dirname, _, filenames in os.walk('/kaggle/input'):
    for filename in filenames:
        print(os.path.join(dirname, filename))
```

Expected Output

```
/kaggle/input/sensor-fault-detection.csv
```

Plotting Functions

In this step, we define several functions to visualize the data, which is crucial for gaining insights into the underlying patterns and relationships. These functions include the creation of a distribution plot, which helps in understanding the spread and frequency of data points within a specific variable. Additionally, we define a function to plot a correlation matrix, allowing us to visualize the relationships between multiple variables and identify any strong correlations or potential multi-collinearity. Finally, scatter and density plots are implemented to provide a clearer view of the bivariate relationships between variables, making it easier to detect any trends, outliers, or clustering patterns. Together, these plotting functions offer a comprehensive set of tools to explore and analyze the dataset visually.

14.38.1 *Distribution Graphs (Histogram/Bar Graph)*

The following function generates distribution plots for each column in the dataset:

```python
def plotPerColumnDistribution(df, nGraphShown, nGraphPerRow):
    nunique = df.nunique()

    # Pick columns that have between 1 and 50 unique values for
    better visualization
```

```
5    df = df[[col for col in df if 1 < nunique[col] < 50]]
6
7    nRow, nCol = df.shape
8    columnNames = list(df)
9    nGraphRow = (nCol + nGraphPerRow - 1) // nGraphPerRow
10
11   plt.figure(figsize=(6 * nGraphPerRow, 8 * nGraphRow), dpi
     =80, facecolor='w', edgecolor='k')
12
13   for i in range(min(nCol, nGraphShown)):
14       plt.subplot(nGraphRow, nGraphPerRow, i + 1)
15       columnDf = df.iloc[:, i]
16
17       if not np.issubdtype(columnDf.dtype, np.number):
18           valueCounts = columnDf.value_counts()
19           valueCounts.plot.bar()
20       else:
21           columnDf.hist()
22
23       plt.ylabel('Counts')
24       plt.xticks(rotation=90)
25       plt.title(f'{columnNames[i]} (Column {i})')
26
27   plt.tight_layout(pad=1.0, w_pad=1.0, h_pad=1.0)
28   plt.show()
```

14.38.2   *Correlation Matrix*

The function below generates a correlation matrix for the dataset:

The following function generates a correlation matrix plot for a given dataframe. This code is referenced from: https://github.com/Sunzidasiddique1/ICS.

```
1    import matplotlib.pyplot as plt
2
3    def plotCorrelationMatrix(df, graphWidth):
4        filename = getattr(df, 'dataframeName', 'Dataset')
5
6        # Drop columns with NaN values
7        df = df.dropna(axis=1)
8
9        # Keep only columns with more than one unique value
10       df = df[[col for col in df if df[col].nunique() > 1]]
11
12       if df.shape[1] < 2:
13           print(f'No correlation plots shown: The number of non-
     NaN or constant columns ({df.shape[1]}) is less than 2')
14           return
15
16       corr = df.corr()
```

```
plt.figure(figsize=(graphWidth, graphWidth), dpi=80,
facecolor='w', edgecolor='k')
corrMat = plt.matshow(corr, fignum=1)

plt.xticks(range(len(corr.columns)), corr.columns, rotation
=90)
plt.yticks(range(len(corr.columns)), corr.columns)
plt.gca().xaxis.tick_bottom()
plt.colorbar(corrMat)
plt.title(f'Correlation Matrix for {filename}', fontsize
=15)
plt.show()
```

14.38.3 Scatter and Density Plots

The following function generates scatter and density plots for numerical columns:

```python
import numpy as np
import pandas as pd
import matplotlib.pyplot as plt

def plotScatterMatrix(df, plotSize, textSize):
    # Keep only numerical columns
    df = df.select_dtypes(include=[np.number])

    # Remove rows and columns with NaN values
    df = df.dropna(axis=1)

    # Keep only columns with more than one unique value
    df = df[[col for col in df if df[col].nunique() > 1]]

    # Reduce the number of columns for matrix inversion in
    kernel density plots
    columnNames = list(df)
    if len(columnNames) > 10:
        columnNames = columnNames[:10]
        df = df[columnNames]

    # Create scatter matrix
    ax = pd.plotting.scatter_matrix(
        df,
        alpha=0.75,
        figsize=[plotSize, plotSize],
        diagonal='kde'
    )

    corrs = df.corr().values
    for i, j in zip(*np.triu_indices_from(corrs, k=1)):
        ax[i, j].annotate(
            f'Corr. coef = {corrs[i, j]:.3f}',
```

```
33          (0.8, 0.2),
34          xycoords='axes fraction',
35          ha='center',
36          va='center',
37          size=textSize
38      )
39
40  plt.suptitle('Scatter and Density Plot')
41  plt.show()
```

14.38.4   Loading and Previewing the Data

We load the dataset and preview the first few rows:

```
import pandas as pd

nRowsRead = 1000  # specify 'None' to read the whole file
df1 = pd.read_csv('inputsensor-fault-detection.csv', delimiter=
    ',', nrows=nRowsRead)
df1.dataframeName = 'sensor-fault-detection.csv'
nRow, nCol = df1.shape
print(f'There are {nRow} rows and {nCol} columns')
```

Output:

Example 14.1 There are 1000 rows and 3 columns

Next, let's take a quick look at the first five rows:

```
df1.head(5)
```

Output:

```
Timestamp;SensorId;Value
0   2017-03-01T23:20:00+03:00;1;18.4798069
1   2017-03-02T04:00:00+03:00;1;19.53911209
2   2017-03-23T06:25:00+03:00;1;19.25019836
3   2017-03-23T19:35:00+03:00;1;18.96128464
4   2017-04-04T15:10:00+03:00;1;25.32162285
```

14.39   Distribution Graphs

We plot the distribution graphs (histogram/bar graph) of the sampled columns using the following code:

```
plotPerColumnDistribution(df1, 10, 5)
```

The output will be a figure showing the distribution of the "Value" column (Fig. 14.15).

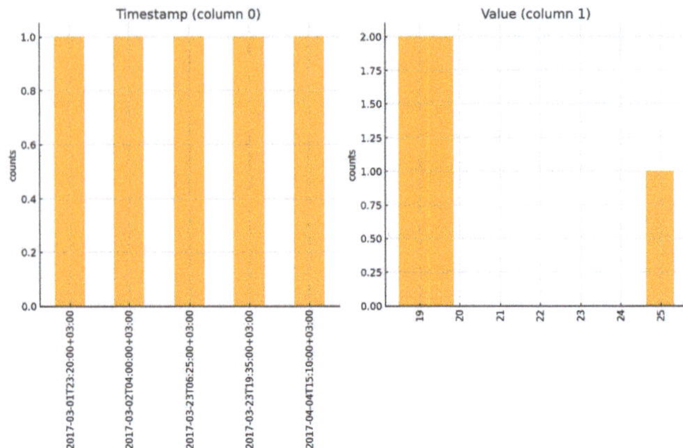

Fig. 14.15 Distribution plot of the value column

14.40   Correlation Matrix

We now plot the correlation matrix of the dataset:

```
plotCorrelationMatrix(df1, 8)
```

However, since the dataset has only one numeric column ("Value"), no correlation plot will be shown. The output will be:

```
No correlation plots shown:
The number of non-NaN or constant columns (1) is less than 2
```

14.41   Scatter and Density Plots

Finally, we plot the scatter and density plots:

```
plotScatterMatrix(df1, 6, 15)
```

Since there is only one numerical column, no scatter matrix plot will be generated.

Explanation of Code
- **Imports and Setup:** The necessary libraries are loaded (e.g., matplotlib, pandas, numpy).
- **Functions for Plotting:** Functions for distribution plots, correlation matrices, and scatter plots are defined in the verbatim environment.

- **Data Loading and Preview:** Code for loading the dataset and displaying basic information (number of rows, columns, and first 5 rows) is provided.
- **Plotting:** Commands to plot distributions, the correlation matrix, and the scatter matrix are included.
- **Output:** The expected outputs from running the code (such as data previews, distribution plots, and messages for correlation and scatter plots) are mentioned.

14.42 Challenges in Using Static Defense Strategy

Industrial Control Systems (ICS) help run important sectors like energy, water, transport, and manufacturing. As more digital technology is added to ICS, keeping them secure from cyber threats has become a serious issue. This section explains the key challenges in using static defense methods, which aim to protect ICS before any attack happens.

1. **Impact on Operations:** Strong security controls can make daily work harder and may reduce the system's speed or performance.
2. **Lack of Resources:** Many organizations do not have enough money, skilled people, or tools to set up strong static defenses for their ICS.
3. **Hard to Fit into Existing Systems:** Adding security tools to current ICS without causing problems or breaking old systems needs careful planning and teamwork.

Overall, solving these problems—operational impact, lack of resources, and system integration—is key to successfully using static defense in ICS. If these issues are handled well, organizations can better protect their systems and keep important services safe and reliable.

14.43 Summary

The complexity of ICSs and their linkages to external and internal networks increase the number of potential security issues and the risks they pose. With many attack routes targeting different resources within control systems, asynchronous attacks can occur over an extended period and target various weaknesses in the ICS environment. Organizations cannot address every security issue with a single countermeasure. A combination of countermeasures is required to effectively protect ICSs from cyberattacks. By integrating multiple security mitigation techniques, the overall risk can be reduced. It should be noted that not all vulnerabilities in the ICS environment can be addressed by Defense-in-Depth techniques. These techniques are primarily used to either slow down an attacker long enough for IT and OT workers to detect and resolve ongoing threats or to make the attacker's job so difficult that they may focus on easier targets.

Chapter 15
Intrusion Detection Systems (IDS) in ICS: Supervisory Frameworks, Signature Versus Anomaly-Based Detection, and Architectural Design

Abstract In this chapter, we provide a detailed analysis of Intrusion Detection Systems (IDS) in Industrial Control Systems (ICS). The discussion covers types of misuse and anomaly detection, addressing physical, system, and remote intrusions, as well as abuse detection mechanisms. The chapter presents Network-based, Host-based, and Hybrid IDS, highlighting the importance of each IDS type. It also compares IDS with Intrusion Prevention Systems (IPS), delineating their advantages and features. Furthermore, both signature-based and anomaly-based detection techniques are examined, along with their respective strengths and weaknesses. The architecture of IDS and the related methods for risk supervision in ICS environments are also presented. With an understanding of these aspects, readers will be better equipped to defend critical infrastructure using effective Intrusion Detection Systems.

Keywords Intrusion Detection Systems (IDS) · Physical Intrusions · System Intrusions · Remote Intrusions · Abuse Detection · Anomaly Detection · Network-based IDS · Host-based IDS · Hybrid IDS · Intrusion Prevention Systems (IPS) · Signature-based Detection · Anomaly-based Detection · IDS Architecture · Supervisory Risk Management

15.1 Introduction

An Intrusion Detection System (IDS) helps protect computer networks by watching for unusual or harmful activity [64]. It works like a security guard, checking for anything strange and warning users when something suspicious happens.

A Security Information and Event Management (SIEM) system collects and stores information about these activities. It sends alerts to the main system so that the right people are informed. This is like using a model to tell the difference between normal and unusual connections.

IDS is very useful for keeping networks safe. It keeps checking the data in real time and sends alerts if it finds anything that breaks the rules or seems dangerous.

M. A. Rahman et al., *Securing Industrial Control Systems*,
https://doi.org/10.1007/978-3-032-03018-4_15

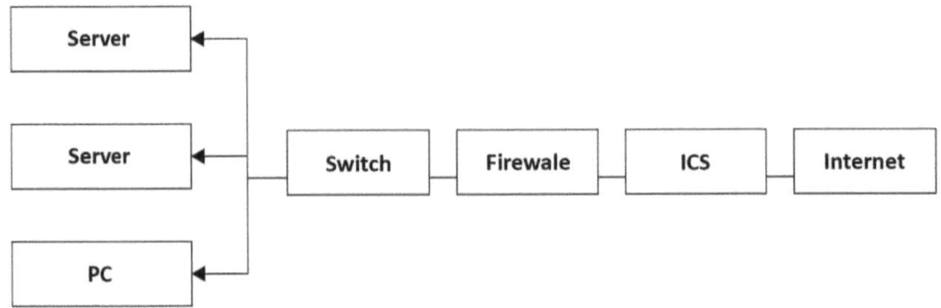

Fig. 15.1 Intrusion detection system

This early warning helps stop attacks before they cause damage. Figure 15.1 is taken from [64].

IDS also helps find hacking attempts, malware, and other threats that can harm important monitoring systems. Because IDS sends alerts right away, security teams can quickly study the problem, take action, and improve the system's protection. This helps important systems in different industries stay safe and work properly.

Figure 15.1 illustrates the Intrusion Detection System. The setup process includes a server, switches, a firewall, an ICS, an Internet connection, and PCs.

15.2   Working Process of IDS

1. It checks computer network traffic to find any suspicious activity.
2. It looks at data patterns to notice anything unusual.
3. It compares network activity with known attack rules or patterns.
4. If it finds a match, it sends a warning to the system administrator.
5. The administrator then investigates the issue and takes steps to stop or fix the problem.

15.3   Overview of Intrusion

An Intrusion Detection System (IDS) helps detect unusual or harmful activity in a system. When it finds something wrong, it sends an alert [49]. A Security Operations Center (SOC) analyst or security team can then check the issue and take action to remove the threat.

There are four common types of IDS [126]:

1. **Network Intrusion Detection System (NIDS):** Monitors network traffic 24/7 in local or cloud setups to detect harmful activity like data theft or policy violations.

NIDS usually works in passive mode, meaning it watches but doesn't directly stop traffic. It helps identify threats and supports strong network security.

2. **Host-based Intrusion Detection System (HIDS):** Runs on individual computers or servers. It checks for strange behavior caused by users or programs. HIDS watches for signs of attacks or misuse.

3. **Perimeter Intrusion Detection System (PIDS):** Uses sensors placed on fences or walls to detect physical entry attempts. It alerts security staff if someone tries to break into a protected area.

4. **Virtual Machine-based Intrusion Detection System (VMIDS):** Works inside a Virtual Machine (VM) to find intrusions. It acts like other IDS types but is designed for virtual setups and continues to improve over time.

15.4   Intrusion Pathways

To improve security, it is important to know how attackers try to break into systems. These paths are called intrusion vectors. Common intrusion methods include physical access, insider attacks, malware, network breaches, social engineering, wireless attacks, and web app hacking.

To reduce risks, organizations can monitor networks, run audits, train employees, use strong passwords, and keep software updated. Cybersecurity needs constant attention and strong defense plans.

Here are the three main ways intruders access systems:

1. Physical intrusion
2. System intrusion
3. Remote intrusion

Physical Intrusion

Physical intrusion means an attacker enters a company's building or server room. These systems use sensors, alarms, and control panels. Sensors on doors or windows detect movement or sound. When triggered, they send data to a control panel that raises an alarm.

1. **Definition:** Gaining access by physically handling the computer or devices.
2. **Close Access Needed:** The attacker must be near the system.
3. **Bypassing Security:** Attackers may enter secure rooms without permission.
4. **Tampering Devices:** They can change device settings or add harmful software.
5. **Tricking People:** Using fake identities or following someone inside (tailgating) to gain access.

System Intrusion

This happens when an attacker breaks into a system without needing to be physically present [20]. It usually happens by exploiting software flaws or weak security settings.

1. **Software Bugs:** Attackers take advantage of known or unknown flaws in software.
2. **Weak Authentication:** Using simple passwords or default logins makes it easier for attackers.
3. **Malicious Software:** Viruses or Trojans can allow attackers to control the system.

Remote Intrusion

Remote intrusion is when someone outside the physical location accesses the system using the Internet or other networks [120].

1. **Network Weaknesses:** Attackers break in through flaws in routers, firewalls, or switches.
2. **Brute Force Attacks:** They try many passwords until one works.
3. **Phishing:** Fake emails or websites trick users into giving away passwords or private information.

15.5   Types of Intrusion

An intrusion refers to any unauthorized or malicious activity aimed at compromising the confidentiality, integrity, or availability of a computer network or system [64]. Intrusions can involve data breaches, system manipulation, unauthorized access, or the disruption of services. In Industrial Control Systems (ICS), such activities can have severe consequences, including operational disruption, equipment damage, and threats to safety.

To safeguard against these risks, organizations implement Intrusion Detection Systems (IDS) that monitor network traffic and system activity for signs of malicious behavior. IDS technologies play a vital role in identifying both external attacks and insider threats by analyzing patterns, behaviors, and data flows that deviate from normal operations. There are several types of intrusions, each with unique characteristics and potential impacts. Figure 15.2 is adapted from [145].

Figure 15.2 shows different types of intrusion methods.

15.5.1   Misuse Detection

Misuse detection is a method used to find threats in a computer system by checking for behavior that matches known attacks. It works by comparing current actions

Fig. 15.2 Types of intrusion

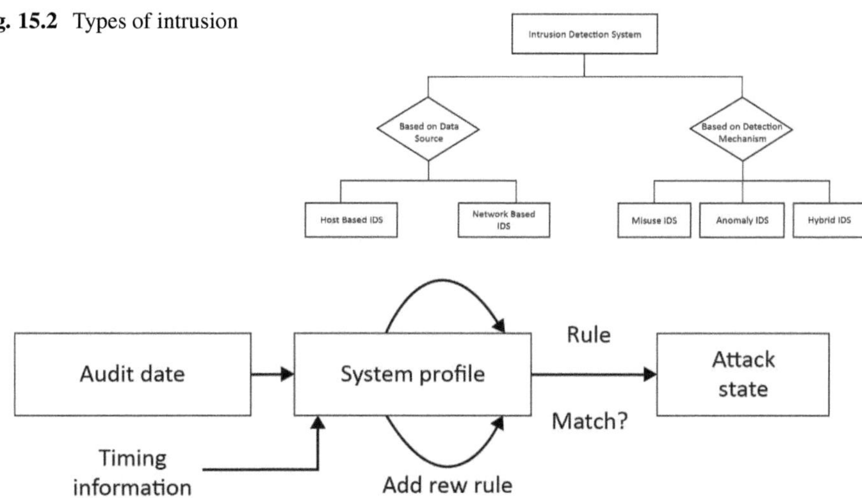

Fig. 15.3 Misuse detection for ICS

with stored patterns of past attacks. This is different from anomaly detection, which looks for anything unusual compared to normal behavior.

Examples of misuse detection methods include rule-based systems and attack signature matching.

Misuse detection works well when the system knows what threats to look for. It can quickly detect and stop known attacks. However, it has some drawbacks. It cannot easily detect new or unknown threats, and it does not adapt well to new types of attacks.

Intrusion Detection Systems (IDS) use misuse detection by collecting data and checking it against a large database of known attack signatures. This helps the system find and stop attacks that have happened before.

Even though misuse detection is useful for catching familiar threats, it may not be able to detect new or unknown attacks.

Figure 15.3 is taken from [104].

Figure 15.3 shows how misuse detection works. It includes audit data, system profiles, detecting attacks, time information, and adding new rules.

1. **Pattern Matching:**

   (a) The system checks network traffic or system actions against a list of known attack patterns.
   (b) If it finds a match, it sends an alert.

2. **Intrusion Signatures:**

   (a) These signatures help detect different types of attacks.
   (b) They often use "if-then" rules or behavior models to describe how attacks look.

3. **Live Monitoring:**

(a) The system watches for signs of attacks in real time using device calls or audit logs.

15.5.1.1 Pseudocode for Misuse Detection in Industrial Control Systems

Misuse detection in Industrial Control Systems is used to find known attacks by looking for specific patterns or signatures.

```python
# Define known attack signatures
attackSignatures = ["Signature1", "Signature2", "Signature3"]

# Function to detect misuse
def DetectMisuse(systemLogs):
    detectedMisuse = []
    for logEntry in systemLogs:
        for signature in attackSignatures:
            if signature in logEntry:
                detectedMisuse.append(logEntry)
                break  # Avoid duplicate detections if multiple
    signatures match
    return detected Misuse
# Main process
misuseIncidents = DetectMisuse(systemLogs)
if misuseIncidents:
    print("Misuse detected:", misuseIncidents)
else:
    print("No misuse detected.")
```

Example Input:

```python
systemLogs = [
    "User login from IP 192.168.0.1- Signature1 detected",
    "User access to control panel- normal operation",
    "Unauthorized command execution- Signature2 detected",
    "System reboot- normal operation"
]
```

Example Output:

```
Misuse detected: ['User login from IP 192.168.0.1- Signature1
    detected',
'Unauthorized command execution- Signature2 detected']
```

The pseudocode is referenced from: https://github.com/Sunzidasiddique1/ICS for misuse detection in Industrial Control Systems (ICS) helps find known attack patterns in system logs.

- **Initialization:** The system starts with a list of known attack signatures like "Signature1", "Signature2", and "Signature3".
- **DetectMisuse Function:** This function reads system logs and checks each log entry to see if it matches any known attack signature. If a match is found, the system adds that entry to the list of detected misuse.
- **Main Process:** The system collects logs and uses the DetectMisuse function to find any misuse. If any misuse is found, it sends an alert with the details. If not, it reports no misuse found.

This pseudocode is a simple model for detecting known attacks using pattern matching.

15.6   Anomaly Detection in Industrial Control Systems (ICS)

Anomaly detection is important for keeping Industrial Control Systems (ICS) secure and reliable. It helps find changes from normal system behavior, which may signal security threats or system failures.

However, it is difficult to define what "normal" looks like in ICS, because these systems use many different devices and communication protocols. As networks grow more connected and advanced, new weaknesses appear, which are harder to spot and understand.

To solve this, researchers use advanced techniques like deep learning. One popular method is the Long Short-Term Memory (LSTM) network, which learns time-based patterns in data. These models can detect small changes that traditional methods may miss.

A key part of anomaly detection is identifying stable patterns, called invariants. These serve as a baseline to detect unusual or dangerous behavior. Once normal patterns are well defined, it becomes easier to detect problems early.

Anomaly detection is useful in many areas:

- Improving system performance
- Detecting fraud and cyberattacks
- Predicting equipment failure (predictive maintenance)

By catching problems in system behavior or sensor data early, it helps avoid downtime and keeps processes running smoothly.

It also works as an early warning system to stop unauthorized activity in critical systems.

Still, anomaly detection has challenges. It can give false alarms, detecting normal actions as threats. Also, finding the real cause behind an anomaly can be hard. Solving the root cause is important so that the problem is fixed properly, not just the result of it.

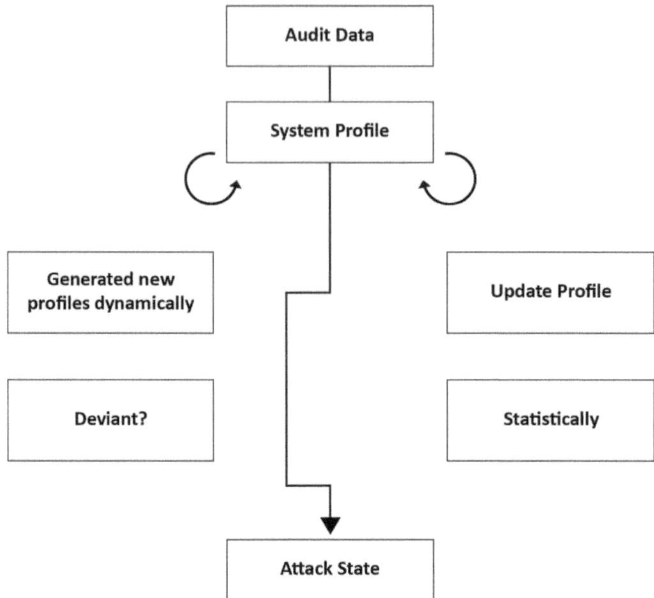

Fig. 15.4 Anomaly detection for ICS

Figure 15.4 and reference [66] support these ideas. With deep learning and better ways to reduce false alerts and find root causes, anomaly detection is becoming a key tool in protecting ICS and ensuring their safe operation.

Figure 15.4 shows how anomaly detection works.

An anomaly-based Intrusion Detection System checks system activity and classifies it as normal or abnormal. This helps detect intrusions or misuse in computers and networks. Unlike signature-based systems, anomaly detection does not rely on known attack patterns. Instead, it uses rules to decide if behavior is normal or not. Network-based anomaly detection adds an extra layer of protection by finding strange network activity or unexpected physical behavior after traffic passes through firewalls or other security tools.

There are three main types of anomaly detection methods: supervised, semi-supervised, and unsupervised.

In Industrial Control Systems (ICS), protocol-based anomaly detection finds unusual behavior by checking if it follows standard communication rules. For smart grids, researchers have used statistics to study SCADA system alarms. In ICS, analyzing system data helps create a "normal behavior" profile, which is then used to spot unusual actions.

Some researchers use the Apriori algorithm to find patterns in past data to detect anomalies. However, behavior-based methods can miss attacks that copy normal activity or take advantage of hidden system weaknesses.

1. **Behavior Analysis:**

   (a) Anomaly detection methods define what normal behavior looks like. Any difference from this is seen as an anomaly.

2. **Static and Dynamic Detection:**

   (a) Static methods raise alerts when parts of the system act differently than expected.
   (b) Dynamic methods check logs or network traffic in real time to find anomalies.

15.7   Network Intrusion Detection Systems

Network-Based Intrusion Detection Systems (NIDS) are important for protecting enterprise networks [64]. NIDS watch network traffic all the time to find and stop possible attacks. They are built to detect harmful activity and unusual traffic patterns that may show signs of a threat.

NIDS work in real time or close to real- time using sensors that watch data packets as they move through the network. They check for harmful actions and protect important data and the network's safety.

A strong security system should include NIDS to reduce the risk of attacks and data leaks. By spotting threats early, NIDS can help prevent major damage.

When a NIDS finds something suspicious, it creates an alert to warn system administrators. It checks the data flow across the network and compares current traffic to past patterns. This helps detect attacks. NIDS also store detailed records that can be used to investigate and understand security incidents.

NIDS are usually placed with routers and other network tools to give full coverage and protection. With NIDS, organizations can improve their network security, spot threats quickly, and respond in time to prevent harm. This approach helps keep sensitive data safe and ensures strong overall security.

Figure 15.5 shows a network security system. It includes tools like NIDS, routers, flow data, forensic tools, and databases.

Some important features of Network Intrusion Detection Systems (NIDS) are:

1. **Packet Analysis:**

   (a) NIDS checks network packets to find devices, servers, or signs of intrusions.
   (b) It looks at network data to find patterns that suggest an attack.
   (c) By inspecting the contents of packets as they move through the network, NIDS can detect suspicious or harmful actions.
   (d) NIDS looks for problems like unauthorized access, port scans, or unusual traffic by checking packet headers and contents. The goal is to stop threats before they harm the system.

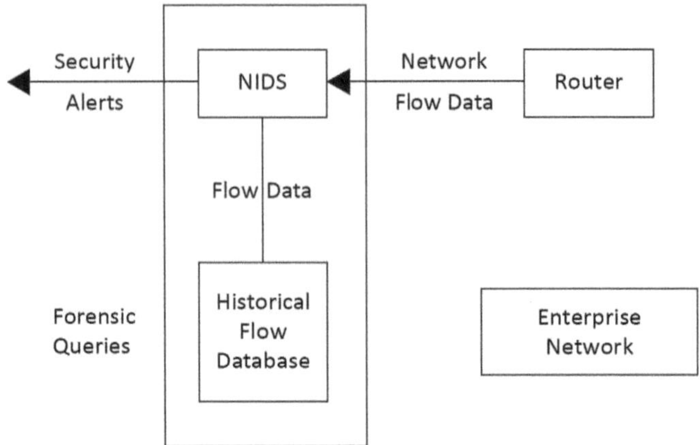

Fig. 15.5 Network intrusion detection systems for ICSs

2. **Active Component:**

   (a) NIDS not only monitors network traffic but also acts when it finds a threat.
   (b) Unlike systems that just watch and record, NIDS can react to danger in real time.
   (c) When NIDS finds abnormal activity, it can alert system admins, block traffic, or start defensive actions.

15.7.1   Hybrid IDS

Hybrid Intrusion Detection Systems (Hybrid IDS) combine the features of Network-Based IDS (NIDS) and Host-based IDS (HIDS) to provide better protection.

NIDS watches network traffic to find attacks, while HIDS checks activity on specific computers or servers to find harmful behavior.

By using data from both NIDS and HIDS, Hybrid IDS gives a more complete view of what is happening in the system. This helps detect complex attacks that one system alone might miss.

Hybrid IDS uses two main detection methods:

- **Signature-based detection:** This method finds known attacks by comparing activity to a database of known attack patterns.
- **Anomaly-based detection:** This method finds new or unknown threats by looking for behavior that doesn't match what's normal.

Signature-based detection is good for finding common attacks, while anomaly-based detection helps discover new threats, though it may raise more false alarms.

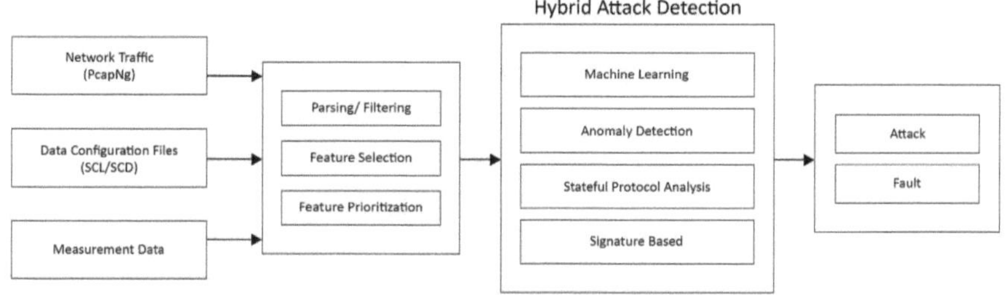

Fig. 15.6 Hybrid IDS for ICS

By using both methods, Hybrid IDS can detect both familiar and unfamiliar threats. This makes it stronger and more flexible.

The Hybrid IDS approach uses:

- **AIDS (Anomaly-Based IDS):** To detect unknown attacks by noticing unusual activity.
- **SIDS (Signature-Based IDS):** To catch known attacks by matching against saved patterns.

Using the strengths of both NIDS and HIDS, Hybrid IDS offers a stronger and more complete way to detect and respond to attacks. It improves the overall security of the network. Figure 15.6 and reference [92] support this discussion about Hybrid IDS.

Figure 15.6 shows a Hybrid Attack Detection System. It includes modules like Network Traffic, Machine Learning, Anomaly Detection, Feature Selection, Stateful Protocol Analysis, and Data Configuration Files.

1. **Total Detection:** A Hybrid Intrusion Detection System (Hybrid IDS) combines anomaly-based and signature-based detection methods. Signature-based detection finds known threats by comparing network data with predefined attack patterns, while anomaly-based detection looks for unusual behavior in network traffic. By using both methods together, Hybrid IDS improves threat detection, finding new threats, and provides better security.
2. **Enhanced Security:** Hybrid IDS improves security by reducing false negatives—when an IDS misses an actual attack. Combining signature-based and anomaly-based methods increases the accuracy of detection, helping to catch more threats and offer stronger protection.
3. **Flexibility:** The hybrid approach is flexible enough to adapt to new and changing threats. When cybercriminals create new attack methods, Hybrid IDS can adjust its detection techniques to stay effective. This ability to adapt makes the system useful even as threats evolve over time.

Hybrid Intrusion Detection Systems (Hybrid IDS) combine two or more detection methods to offer a strong and adaptable network security solution. They are a key defense against cyberattacks.

Fig. 15.7 Host-based IDS

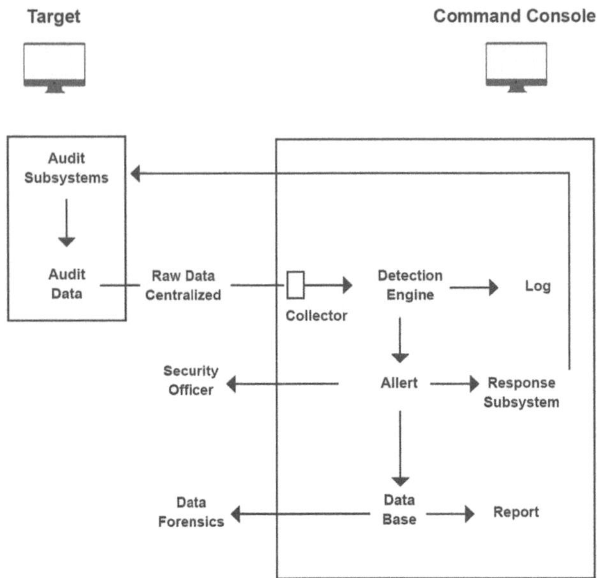

15.7.1.1   Host-Based IDS

A Host-based Intrusion Detection System (HIDS) monitors the activity of individual systems to detect intrusions. It checks system integrity, application behavior, file changes, network traffic from that host, and system logs. HIDS is software used to watch over specific hosts, like computers or servers, and keeps track of various activities, such as file changes, user activity, and network connections. When it finds suspicious activity, HIDS alerts the system administrators.

The advantages of HIDS are that it adds protection to Network-Based IDS, provides detailed security information, improves internal host security, and can be customized. Its drawbacks include needing maintenance, the potential for false alarms, and higher resource use. However, HIDS is an important tool for identifying potential security issues on specific hosts. Figure 15.7 is from [71]. The most common type of Host-based IDS is antivirus software, which uses signature-based detection methods.

Figure 15.7 shows a Host-based IDS, which includes components like the Command Console, Audit Data, Raw Data Centralized Collector, Detection Engine, Log Collector, Security Officer, Alerts, Response Subsystem, Data Forensics, Database, and Report.

1. **Monitoring and Analysis:**

    (a) HIDS operators observe the host's behavior and compare it to expected norms.

    (b) Methods used to check the host's state include:

- File hashing
- Timestamp analysis
- System log inspection
- Network interface monitoring

2. **Alerting and Response:**

   (a) HIDS alerts users and central management servers about unauthorized activity or changes.
   (b) Responses may include blocking, logging, or other actions based on set policies.

Key Features of HIDS:

1. **Log Analysis:**

   (a) HIDS processes logs from the operating system and applications.
   (b) It detects patterns of malicious activity or security events.

2. **File Integrity Checking:**

   (a) HIDS compares critical system files and settings to a baseline to ensure they haven't been altered.
   (b) Alerts are sent if unauthorized changes are detected.

3. **System Call Monitoring:**

   (a) HIDS tracks system calls, which are requests for services made by programs.
   (b) This helps detect anomalies and unauthorized actions at the system level.

4. **Registry Monitoring (on Windows Systems):**

   (a) HIDS tracks changes to the Windows registry.
   (b) The registry stores system configurations, user profiles, and application settings.
   (c) Registry changes may indicate:

   - Normal activity, such as software installations or system changes.
   - Malicious activity, like modifications made by rootkits or malware.
   - Suspicious activity that needs further investigation.

   (d) Regularly reviewing registry logs and setting alerts for important changes can help identify and fix security issues.

5. **Malware Detection:**

   (a) HIDS constantly monitors individual hosts (like workstations or servers) for security threats.
   (b) Malware detection is a key part of HIDS. It works as follows:

   - HIDS scans the host system for signs of malware.
   - If malware is found, HIDS takes action to remove it.

    – Possible responses include quarantining infected files, blocking network access, or notifying administrators.

(c) HIDS improves the host's security by detecting and containing malware.

15.7.1.2   Stack-Based Intrusion Detection System (IDS)

A stack-based Intrusion Detection System (IDS) represents a modern and efficient approach to network security. Unlike traditional IDS solutions that analyze data after it has been processed by the operating system or stored in logs, a stack-based IDS integrates directly with the TCP/IP protocol stack. This allows it to inspect data packets in real time as they move through the layers of the Open Systems Interconnection (OSI) model, from the network layer to the application layer.

By monitoring traffic at such a foundational level, stack-based IDS can detect and block malicious packets before they reach the operating system or higher-level applications [64]. This early detection capability significantly reduces the potential for damage, especially in environments where real-time threat mitigation is crucial, such as in Industrial Control Systems (ICS), critical infrastructure, and cloud-based services.

Unlike Host-based IDS, a stack-based IDS does not usually rely on the host operating system to store audit trails or perform post-event analysis. However, it can be used in conjunction with operating system-level logging and audit features for enhanced visibility and security. While the IDS itself may not generate persistent logs, integration with external logging mechanisms like Security Information and Event Management (SIEM) systems ensures that relevant events are captured and analyzed for forensic or compliance purposes. Stack-based IDS is especially effective at detecting threats like:

- Protocol violations and anomalies at the network level
- Malicious payloads embedded in TCP/UDP packets
- Real-time network-based attacks such as SYN floods or port scans
- Packets that bypass application-level security controls

This type of IDS is designed for high-speed environments and is capable of operating with minimal impact on system performance. Its placement within the protocol stack makes it a critical component of layered defense strategies, particularly in networks where high-speed packet inspection and immediate response are necessary. Figure 15.8, adapted from [93], illustrates how a stack-based IDS operates within the TCP/IP architecture, highlighting its position relative to the network stack and its role in preemptive threat detection.

Figure 15.8 shows a stack-based IDS system, which includes components such as a filter, logic, agent protocol machine, and a central manager. These parts work together to monitor and manage network activities.

Fig. 15.8 Stack-based IDS

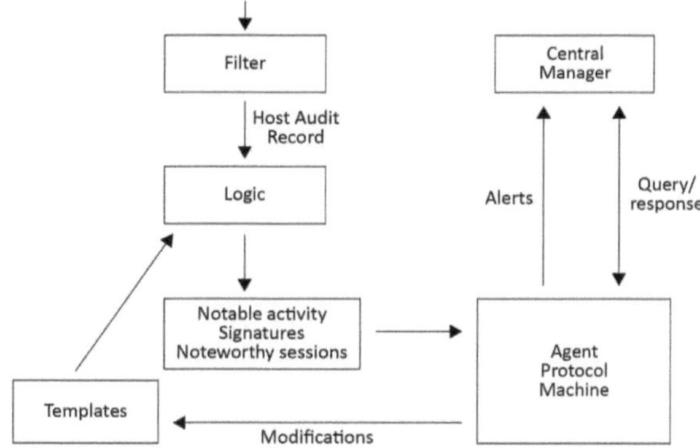

15.7.1.3   Stack-Based IDS Characteristics

Stack-based IDS monitors network traffic across different layers of the protocol to detect attacks. They can inspect traffic in detail and identify both known and unknown threats by analyzing data from the physical layer to the application layer. By focusing on specific protocols like TCP/IP or HTTP, stack-based IDS improves detection accuracy. These systems correlate traffic data from various layers, providing a complete view of network activity, which helps in detecting complex, multi-layered attacks.

1. Basic Operation

   (a) **Stacks:** Stacks are data structures that follow the Last In, First Out (LIFO) principle.
   (b) **Storage:** Stack-based IDS store and manage network activity data.
   (c) **Push and Pop Operations:** Network traffic events are added to the stack, with the most recent events being pulled out and analyzed.

2. Event Handling

   (a) **Event Processing:** The IDS monitors network events like packet transfers, login requests, and system calls.
   (b) **Event Stack Operations:** After each event is added to the stack, the IDS checks the stack for unusual patterns or abnormalities.

3. Pattern Recognition

   (a) **Behavioral Analysis:** Stack-based IDS uses behavioral analysis to establish a baseline of normal network traffic.
   (b) **Pattern Matching:** Normal behavior patterns are stored in the stack. Any deviations from this normal behavior trigger alarms or prompt further investigation.

4. Scalability and Flexibility

   (a) **Adaptability:** Stack-based IDS can adjust to new threats by changing detection rules and patterns.
   (b) **Scalability:** The system can handle increasing network traffic without affecting its performance.

15.7.2 Pseudocode for Anomaly Detection in Industrial Control Systems

Anomaly Detection focuses on identifying unusual activity based on normal behavior within the ICS system.

```
BEGIN AnomalyDetection
  // Define normal behavior thresholds
  INITIALIZE normalBehavior = {
      "temperatureRange": [50, 100],
      "pressureRange": [20, 80],
      "flowRateRange": [10, 50]
  }

  // Function to detect anomalies
  FUNCTION DetectAnomalies(sensorData):
      detectedAnomalies = []
      FOR EACH dataPoint IN sensorData:
          IF dataPoint.temperature < normalBehavior.
    temperatureRange[0]
              OR dataPoint.temperature > normalBehavior.
    temperatureRange[1] THEN
              detectedAnomalies.append(dataPoint)
          ELSE IF dataPoint.pressure < normalBehavior.
    pressureRange[0]
              OR dataPoint.pressure > normalBehavior.
    pressureRange[1] THEN
              detectedAnomalies.append(dataPoint)
          ELSE IF dataPoint.flowRate < normalBehavior.
    flowRateRange[0]
              OR dataPoint.flowRate > normalBehavior.
    flowRateRange[1] THEN
              detectedAnomalies.append(dataPoint)
          END IF
      END FOR
      RETURN detectedAnomalies
  END FUNCTION

  // Main process
  sensorData = GET sensor data
  anomalies = DetectAnomalies(sensorData)
  IF anomalies IS NOT EMPTY THEN
```

```
      ALERT "Anomalies detected:", anomalies
  ELSE
      PRINT "No anomalies detected."
  END IF
END AnomalyDetection
```

Example Input:

```
Example Input:
sensorData = [
    {"temperature": 55, "pressure": 25, "flowRate": 15},
    {"temperature": 45, "pressure": 85, "flowRate": 20},
    {"temperature": 60, "pressure": 30, "flowRate": 55},
    {"temperature": 70, "pressure": 50, "flowRate": 40}
]
```

Example Output:

```
Anomalies detected: [
    {"temperature": 45, "pressure": 85, "flowRate": 20},
    {"temperature": 60, "pressure": 30, "flowRate": 55}
]
```

The pseudocode is referenced from: https://github.com/Sunzidasiddique1/ICS for Anomaly Detection in Industrial Control Systems identifies deviations from normal behavior based on sensor data. The system starts by setting threshold values for important parameters like temperature, pressure, and flow rate. The Detect Anomalies function compares each sensor reading with these thresholds. Any reading outside the acceptable range is flagged as an anomaly. The main process reads sensor data, runs the anomaly detection function, and raises an alert if any anomalies are found. If no anomalies are detected, it simply prints a message.

- **Initialization:** Sets normal ranges for temperature, pressure, and flow rate.
- **Anomaly Detection:** Identifies data points outside the normal range.
- **Alerting:** Alerts are raised when anomalies are detected; otherwise, the system reports no anomalies.

15.8   Intrusion Detection and Prevention Systems

Intrusion Detection Systems (IDS) help monitor and detect unusual network traffic patterns in Industrial Control System (ICS) environments. IDS passively watches network traffic and uses ICS vendor traffic signatures, making them very useful in these environments.

Intrusion Prevention Systems (IPS) work alongside IDS but also block malicious traffic, mainly in severe cases of traffic anomalies in ICS. The Defense-in-Depth approach combines IDS and IPS to detect and react to problems like misconfigurations and intrusion attempts.

IDS doesn't work alone; it needs a mix of tools and approaches to provide complete event information. Most systems also include file integrity-checking software. Host-based IDS, such as Open Source Security tools, perform security checks. However, they can use a lot of system resources, which is a concern in ICS environments, where speed and performance are critical. IDS systems often combine anomaly-based and signature-based detection methods to improve security [25].

15.9 Intrusion Actions Form

An Intrusion Detection System is designed to capture and detect malicious activities, including policy violations. An Intrusion Prevention System not only detects but also prevents malicious actions.

1. **Unauthorized Access:** Hackers and compromised credentials can give unauthorized access to systems, networks, or accounts.
2. **Malware Infections:** Intruders can install malware, such as viruses, worms, Trojans, ransomware, and spyware.
3. **Denial of Service:** Denial of Service or Distributed Denial of Service attacks flood a system or network with traffic, making it unavailable to legitimate users.
4. **Data Breaches:** Intruders steal sensitive data for unauthorized use, compromising privacy and security.
5. **Phishing Attacks:** Intruders use phishing techniques to steal usernames, passwords, and credit card information.
6. **Social Engineering:** Intruders trick individuals or employees into revealing sensitive information or granting access.
7. **Exploiting Weaknesses:** Attackers take advantage of security flaws, like unpatched software, to breach systems.
8. **Internal Threats:** Authorized users misuse their access to sensitive systems or data.

1. **Total Detection:**

    (a) The hybrid method combines signature-based and anomaly-based intrusion detection for better coverage.
    (b) It can detect both known and unknown threats, improving overall security.

2. **Improved Security:**

    (a) Hybrid Intrusion Detection System boosts security by detecting more threats and reducing false alarms.

3. **Adaptability:**

   (a) This method adapts to new threats, ensuring continuous protection.

15.10 Overview of Intrusion Detection Systems

Intrusion Detection Systems for Industrial Control Systems are security tools used to monitor and analyze activities within Industrial Control System environments. They help detect security threats and breaches that could endanger data integrity, affect operations, or jeopardize public safety.

1. **Comprehensive Industrial Control System Security:**

   (a) **Protection of Critical Infrastructure:** Intrusion Detection Systems provide multiple layers of protection, such as network and host monitoring, to safeguard critical infrastructure in industries like energy, water, and transportation.

   (b) **Data Integrity:** Intrusion Detection Systems track data flow and system activity to detect unauthorized changes, ensuring data remains accurate and reliable.

   (c) **Operational Continuity:** Intrusion Detection Systems identify security risks and prevent disruptions to maintain smooth operations.

   (d) **Public Safety:** Industrial Control System security helps prevent accidents, environmental damage, and worker injuries, especially in the energy and utilities sectors.

2. **Detection of Unusual and Malicious Activity:**

   (a) **Signature-Based Detection:** Intrusion Detection Systems detect known threats by comparing activities against a database of attack signatures.

   (b) **Anomaly-Based Detection:** Intrusion Detection Systems detect new threats by identifying behavior that deviates from the established norm.

   (c) **Behavior-Based Detection:** Intrusion Detection Systems track system and user activities to identify abnormal behavior, such as unauthorized access or privilege escalation.

3. **Fast Warning and Response:**

   (a) **Real-Time Alerts:** Intrusion Detection Systems send instant alerts about potential threats, their location, and possible impacts.

   (b) **Triage and Prioritization:** Security experts use alerts to prioritize and investigate critical incidents quickly.

   (c) **Automated Incident Response:** Intrusion Detection Systems automate responses, such as collecting evidence, isolating infected devices, and blocking malicious traffic.

(d) **Post-Incident Analysis:** Intrusion Detection Systems logs and alerts support forensic analysis and compliance reporting after an incident.

15.11   Principle of Intrusion Detection Systems

An Intrusion Detection System monitors network traffic to detect malicious activities and known attack patterns. The intrusion prevention engine scans the data stream to identify these patterns.

1. The Intrusion Detection System must operate continuously for long periods without supervision.
2. The Intrusion Detection System must always be running and secure.
3. The Intrusion Detection System should be able to identify abnormal behavior.
4. The Intrusion Detection System should have minimal impact on system performance.
5. The Intrusion Detection System must be configurable.

15.12   Function of Intrusion Detection Systems

An Intrusion Detection System reacts to attacks as soon as they are detected. It does not prevent attacks but provides information to help security personnel respond. It is a reactive tool, alerting users to potential threats and supporting incident response.

The Intrusion Detection System helps in cybersecurity by providing real-time monitoring of network and system behavior. It acts as an informer, helping security teams respond to threats quickly and efficiently, while also enabling post-incident analysis.

15.13   Intrusion Detection System Detection Methods

Intrusion detection systems can be hardware appliances or software installed on endpoints. They use two main approaches for threat detection:

1. Signature-based detection
2. Statistical anomaly-based detection

Fig. 15.9 Signature
detection for IDS

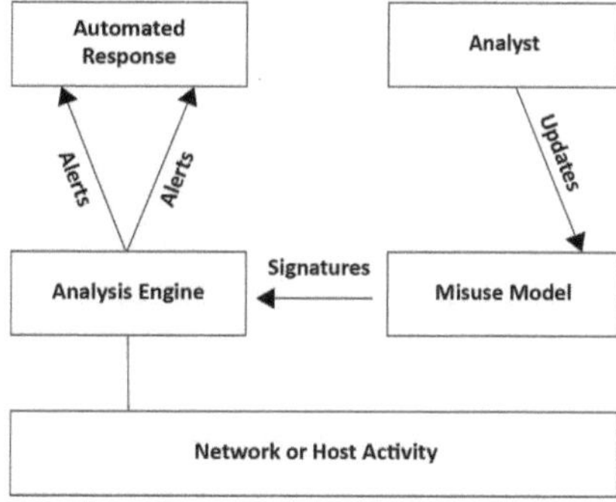

15.13.1   *Signature Detection for Intrusion Detection Systems*

Signature detection identifies attack patterns or anomalies by comparing network or system data to known threat signatures. It is commonly used in Intrusion Detection Systems and other security software like antivirus programs. When a match is found, an alert is sent to allow action to be taken. Signature detection quickly identifies and responds to known attacks by comparing current activity to past information. However, it has limitations, especially against new or advanced threats that do not have preexisting signatures. While signature detection is still used in cybersecurity, it is typically combined with other advanced methods to provide more comprehensive protection against various types of cyberattacks. Data from the network or system is checked against a signature database to identify known threats. Figure 15.9 shows an example from [73].

Figure 15.9 shows how signature detection in Intrusion Detection Systems works. It includes components like an Automated Response Analyst System, Analysis Engine, Network or Host Activity, Analyst, and Misuse Model.

15.13.2   *Statistical Anomaly-Based Intrusion Detection Systems*

Statistical anomaly-based Intrusion Detection Systems detect potential security threats by monitoring unusual patterns in user activity. These systems collect data on normal user behavior over time and establish a baseline of what is considered typical. They then use statistical methods to identify deviations from this baseline and flag potential threats.

Regular monitoring of user activity is required to detect suspicious behavior that could signal a security issue. By comparing current activity to the baseline, the system can spot anomalies, such as unauthorized access. The benefits of statistical anomaly detection include early identification of threats, allowing for prompt action, and flexibility, as the system can adapt to changes in user behavior. This approach

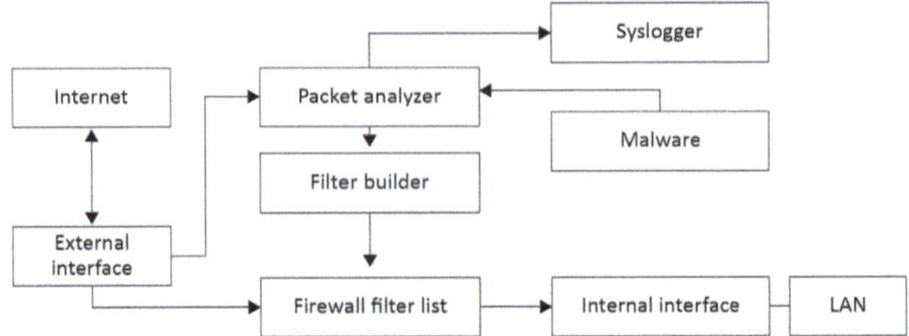

Fig. 15.10 Statistical anomaly-based IDS

improves network security by using statistical methods to identify possible threats, as shown in visual aids and references (Fig. 15.10, [22]).

Figure 15.10 illustrates the components of statistical anomaly detection, including the Syslogger, Internet, Packet Analyzer, External Interface, Filter Builder, Malware Database, Local Area Network, Internal Interface, Firewall, and Filter.

Statistical anomaly detection identifies security threats by analyzing typical and acceptable system or network behavior. It is broadly categorized into:

1. Threshold Detection
2. Profile-Based Anomaly Detection

15.14 Simple Intrusion Detection System in Python

```python
import socket
import struct

# List of suspicious patterns
sus_patterns = ['danger.com', 'unauthorized_access']

def sniff_packet():
    sock = socket.socket(socket.AF_PACKET, socket.SOCK_RAW,
    socket.ntohs(0x0003))
    while True:
        raw_data, addr = sock.recvfrom(65535)
        eth_header = raw_data[:14]
        eth_data = struct.unpack('!6s6sH', eth_header)
        payload = raw_data[14:]

        for pattern in sus_patterns:
            if pattern.encode() in payload:
                print(f'Suspicious activity detected: {pattern}
    in {addr}')

if __name__ == '__main__':
```

```
21    print('Starting packet sniffer, intrusion detection!')
22    sniff_packet()
```

This Python script is referenced from: https://github.com/Sunzidasiddique1/ICS implements a simple Intrusion Detection System that captures network packets and checks for suspicious activity. Using raw sockets and the **struct** module, the script listens for incoming packets, extracts the Ethernet header and payload, and searches for predefined suspicious patterns like **'danger.com'** and **'unauthorized_access'**. When a match is found, the system prints a message indicating potential intrusions. This script serves as a basic method to monitor network traffic for specific anomalies.

15.15 Threshold Detection

Detecting attacks early in real time can prevent further attacks and stop unauthorized access to systems. By selecting the right features and setting the right thresholds, Intrusion Detection Systems can identify network anomalies. Proper threshold configuration is essential for detecting security threats and reporting them. Administrators set thresholds to alert the system when abnormal behavior or attacks are detected. Finding the right threshold is important to balance threat detection and avoid false alarms.

False positives—misreporting harmless activities—can overwhelm administrators. Too many false positives waste resources and distract from real security concerns. On the other hand, a threshold set too high may miss real attacks. Therefore, thresholds must be set carefully to minimize false positives while ensuring threats are detected.

Administrators must adjust thresholds based on network behavior and the organization's needs to achieve the best detection accuracy. While minimizing false positives, the system must remain effective at detecting real threats.

15.15.1 Profile-Based Anomaly Detection

Intrusion detection systems use profile-based anomaly detection to identify unusual behavior by creating a model of normal activity. This model is continuously monitored for deviations that might indicate a security issue.

In profile-based anomaly detection, system or network behavior is monitored over time to create a baseline. This includes parameters like login frequency, data transfer volume, and application use. Statistical methods or machine learning algorithms are used to spot deviations from the baseline. When an anomaly is detected, security administrators are alerted. The baseline is updated regularly to adapt to changes in the environment. However, the challenge lies in minimizing false positives and creating an accurate baseline. Profile-based anomaly detection can detect new threats without relying on pre-known signatures.

Profile-Based Intrusion Detection Parameters

Profile-based intrusion detection uses several parameters to build a baseline of normal user behavior and detect anomalies. Some important parameters in this approach are:

1. **Counter:** A counter tracks specific events over time.

    (a) **Logins:** Tracking how often users log in.
    (b) **Password Failures:** Recording failed login attempts.
    (c) **Command Execution Count:** Counting the number of commands executed during a user session.

    Counters help measure activity frequency, which aids in detecting unusual behavior.

2. **Interval Timer:** The interval timer tracks the time between two consecutive events to analyze activity frequency and timing.

    • Measuring time between repeated login attempts on an account.
    • Monitoring time intervals between commands executed in a user session.

    Interval-based measures add a temporal aspect to intrusion detection. Deviations from expected time intervals, such as repeated login attempts or unusual command execution patterns, can signal suspicious activity.

15.16 Signature-Based Versus Anomaly-Based Intrusion Detection Systems

The main differences between anomaly-based and signature-based Intrusion Detection Systems are shown in Table 15.1 [108].

15.17 Intrusion Detection System Architecture

An Intrusion Detection System monitors network traffic for suspicious activity and sends immediate alerts. It checks systems and networks for malicious software and policy violations. If it detects any illegal activity, it logs the event in a central system (like a Security Information and Event Management system) or notifies the system administrator. An Intrusion Detection System helps protect a network or system from unauthorized access, including from internal threats, by identifying malicious activity. The key function of an Intrusion Detection System is to differentiate

Table 15.1 Comparison between signature-based and anomaly-based IDS

Feature	Signature-based IDS	Anomaly-based IDS
Detection method	Identifies known threats using predefined signatures or patterns	Identifies unknown threats by comparing behavior to a baseline
Approach	Matches network activity against a database of known threat signatures	Utilizes machine learning to establish a baseline of normal behavior
Indicator of Compromise (IOC)	Relies on pre-programmed IOCs for known threats	Does not rely on predefined IOCs; detects deviations from the baseline
Sensitivity to unknown threats	Limited; effective for known threats only	High; can detect unknown or zero-day threats
False positives	Generally lower due to specificity	Can be higher due to atypical behavior triggering alerts
Resource intensity	Lower resource usage for known threats	Higher resource usage, potential for more false positives
Speed	High processing speed for known threats	Slower processing speed, especially with large datasets
Zero-day exploits	Limited ability to detect zero-day exploits	Can effectively detect zero-day exploits
Complementary use	Works well in identifying known attacks quickly	Effective in catching unknown or novel threats
Commonly used in tandem	Often used alongside anomaly-based IDS	Often used alongside signature-based IDS

between "bad connections" (such as attacks) and "good connections" (normal behavior). Intrusion Detection Systems are similar to auditing systems that automate this process. There are three main components in an Intrusion Detection System:

1. **Agent:** Collects data from various sources, processes it, and removes unnecessary information. Network-based agents monitor network traffic for threats, while host-based agents check system and application logs.
2. **Director:** Analyzes log records using different techniques to determine whether an attack is happening. It operates on a separate system for security reasons. Adaptive directors can update rules and profiles based on the results of multiple analysis methods.
3. **Notifier:** Reacts to the information received from the director and responds to attacks by alerting responsible parties. Alerts can be sent through emails, alarms, or log entries. For example, SATAN (1993) was a system that monitored logs, detected attacks, notified administrators, and triggered recovery processes.

The agent functions as a logger, gathering data from the computer system. The director, acting as an analyzer, processes this data to determine if an attack is occurring or has already taken place. The notifier then decides whether and how

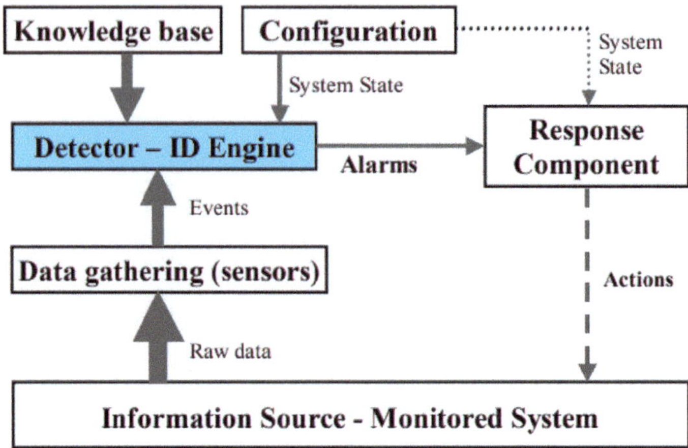

Fig. 15.11 Intrusion detection systems architecture

to alert the relevant entity. Additionally, the notifier may instruct agents to modify logging settings. Figure 15.11 is adapted from [95].

Figure 15.11 shows the structure of an Intrusion Detection System, with its main components working together to detect and respond to security threats. The sensor collects data from the system being monitored. This data is then analyzed by the detector to identify possible intrusions. The database stores attack patterns and expert-filtered data, which helps improve detection accuracy. The configuration of the device displays the current status of the system, allowing administrators to monitor system activity. Lastly, the response component triggers automatic or manual actions to stop any threats detected, applying a proactive approach to security.

15.18 Comparison of Intrusion Detection System and Intrusion Prevention System

The main difference between an Intrusion Prevention System and an Intrusion Detection System is that an IPS is used for controlling traffic, while an IDS is used for monitoring traffic [75]. Both IDS and IPS check network traffic against known attack patterns or normal behavior.

An IDS is a system that watches network traffic, identifies patterns that match known attack patterns, and sends alerts when it detects unusual activity. However, the traffic continues to flow without interruption. On the other hand, an IPS not only examines network traffic but also takes action. If it detects suspicious traffic, the IPS can stop the flow of traffic until an investigation is done and a decision is made about whether to allow traffic to continue. A comparison table is shown in Table 15.2.

Table 15.2 Comparison between IDS versus IPS

Feature	Intrusion Detection System (IDS)	Intrusion Prevention System (IPS)
Primary function	Detect and alert suspicious activities	Detect and prevent/block malicious activities
Operational mode	Passive observer, does not disrupt network flow	Performs in-line to prevent threats
Automated response	Relies on human involvement to block or prevent dangers	Detects risks and blocks harmful traffic automatically.
Alert generation	Alerts for suspected behavior or predetermined attack signatures	Sends real-time notifications and automates danger prevention
Deployment focus	Monitoring network traffic for security threats	Active defense, blocking known threats based on rules
Use cases	Important for forensic analysis and post-incident investigations	Especially useful for real-time threat mitigation and defense
Level of intervention	Monitors and notifies, leaving security to intervene	Automatically blocks or contains threats without human interaction.
Response time	Human analysis and intervention determine response time	Actively blocks hazards as they emerge
Immediate action	Lacks real-time threat prevention	Real-time, automated threat prevention

15.19   Advantages of Intrusion Detection Systems

Intrusion Detection Systems help analyze the types and number of attacks, allowing organizations to improve their security measures. IDSs also help identify weaknesses in the configuration of network devices, which strengthens overall system security. Figure 15.12 is taken from [93].

The main goal of an IDS is to ensure secure and reliable information transfer. It is an important part of the security system, providing protection when other security systems fail. An IDS plays a crucial role in detecting security threats and improving the cybersecurity strength of an organization.

Figure 15.12 illustrates some benefits of Intrusion Detection Systems. An Intrusion Detection System (IDS) enhances network security by identifying suspicious behavior and alerting administrators before significant damage occurs. It also improves network performance by detecting and addressing issues efficiently. Additionally, IDS helps ensure regulatory compliance by monitoring network activity and generating reports. Furthermore, it provides valuable insights by analyzing traffic, identifying vulnerabilities, and strengthening overall security. Table 15.3 describes IDS vs. Firewall [27].

Fig. 15.12 Benefits of IDS

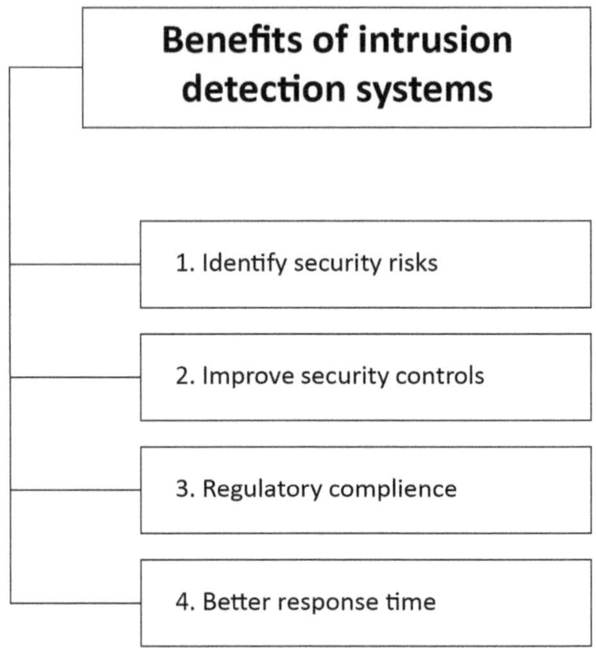

Table 15.3 Description of parameters for Firewall and IDS

Parameter	Firewall	IDS
Philosophy	Filters incoming and outgoing network traffic based on predetermined rules	Monitors traffic for malicious activity and sends alerts on detection.
Principle of working	Filters traffic based on IP address and port numbers	Detects real-time traffic and looks for traffic patterns or signatures of attack and then generates alerts
Configuration mode	Layer 3 mode or transparent mode	Inline or as an end host (via span) for monitoring and detection
Placement	Inline at the Perimeter of Network	Non-inline through port span (or via tap)
Traffic patterns	Not analyzed	Analyzed
Placement wrt each other	Should be first line of defense	Should be placed after the firewall
Action on unauthorized traffic Detection	Block the traffic	Alerts/alarms on detection of anomaly
Related terminologies	• Stateful packet filtering • permits and blocks traffic by port/protocol rules	• Anomaly-based detection • Signature detection • Zero day attacks • Monitoring • Alarm

15.20 Intrusion Prevention System

An intrusion prevention system represents a significant advancement in securing control systems. Building upon an Intrusion Detection System, it employs proactive and automated measures to prevent and neutralize potential threats and security vulnerabilities.

The additional security provided by an intrusion prevention system enhances monitoring capabilities. It promptly triggers predefined responses upon detecting a threat, preventing attacks before they can cause harm. This automated response adds an extra layer of protection, strengthening the overall security posture against emerging cyber threats. Figure 15.13 is adapted from [64].

Figure 15.13 illustrates an intrusion prevention system connected to a router in the network setup, including a firewall linked to the Internet.

Organizations can significantly reduce response time to attacks by using an intrusion prevention system, minimizing potential damage and enhancing security. This proactive approach is essential for ensuring the availability, security, and privacy of critical monitoring systems that support various business operations and activities.

Fig. 15.13 Intrusion detection system

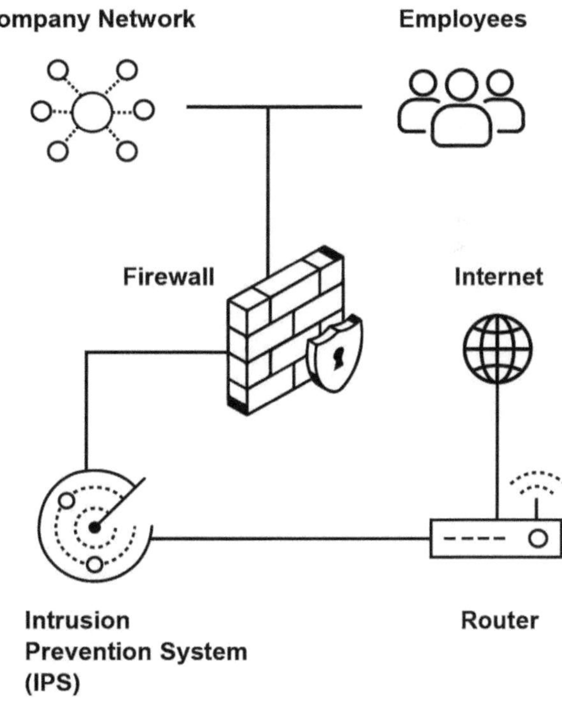

15.21 Supervisory Framework

The supervisory framework is a well-organized and risk-focused approach designed to help monitor institutions. It aims to assess institutions proactively, making it adaptable to changes in the industry and the economy. This allows supervisors to detect and manage new risks as they emerge.

The framework provides a clear process for understanding and assessing significant risks linked to an organization's activities, regardless of its risk management efforts. This includes identifying, evaluating, measuring, tracking, controlling, and reducing risks. It also includes rules set by regulators (e.g., guidelines on capital adequacy and large exposures for international banks).

Supervisors use this framework to check if an organization's risk management efforts are enough, and if its earnings, capital, and liquidity are sufficient to handle its risks and resist negative shocks. Since 19 March 2015, regulators have used a risk-based approach to supervise institutions, and this approach has been tested and improved since then.

This management structure must be supported by legal requirements for regulated institutions, which clarify these obligations further.

15.21.1 Data Integrity and Confidentiality Risks

Data integrity and confidentiality are key parts of information security, ensuring that sensitive information is accurate, reliable, and private. Data integrity risk is when unauthorized changes, deletions, or additions are made to data, which can harm its reliability. This can lead to wrong decisions, interruptions in business, and loss of trust. When data integrity is compromised, organizations may have to make decisions based on incorrect or altered facts, leading to financial losses, damaged reputations, and legal issues. To prevent these risks, organizations can use security measures such as access controls, encryption, and active monitoring. Regular audits help find and fix any data inconsistencies.

Confidentiality risk, on the other hand, deals with unauthorized access to or sharing of confidential information. This can result from major privacy violations and financial losses. Confidential information, such as personal or financial data, should be protected from unauthorized access. Breaching confidentiality can cause privacy violations, significant financial loss, reputation damage, and legal penalties. To maintain confidentiality, organizations use encryption, strict access controls, and clear security policies to ensure that only authorized people can access sensitive data. These measures should be regularly reviewed and updated to keep up with new threats.

Data integrity and confidentiality risks are often closely linked. For example, unauthorized access to a system can lead to changes or deletions of data, which directly affects its integrity. To protect against both risks, a comprehensive secu-

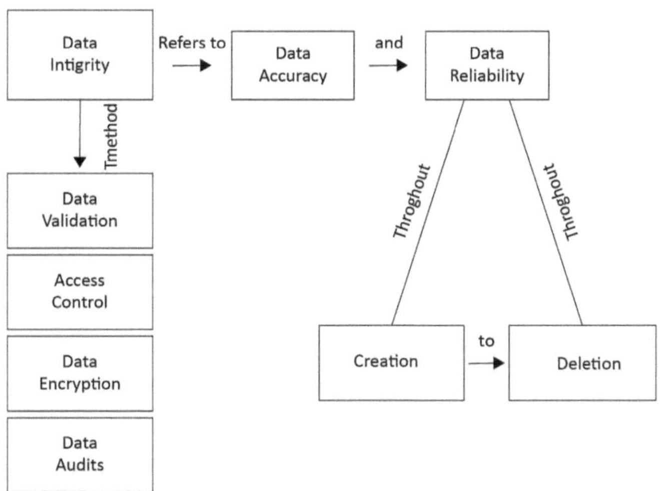

Fig. 15.14 Data integrity

rity approach is needed. This should include technology-based defenses, clear procedures, and ongoing staff training to make everyone aware of their role in securing data. A successful strategy combines preventive, detective, and corrective actions to safeguard against both internal and external threats, ensuring that sensitive information is accurate, secure, and only accessible to authorized users.

Figure 15.14 illustrates the concept of data integrity, including elements such as data visualization, integrity, accuracy, reliability, methods, validation, access control, encryption, deletion, creation, and data audits.

Ensuring data security is crucial for maintaining the accuracy and reliability of information, which is essential for informed decision-making. This involves detecting and preventing unauthorized modifications using tools such as encryption, access controls, and regular data audits. To protect sensitive information, strict access controls, encryption, and other security measures are necessary to ensure that only authorized systems or individuals can access the data. As a result, the risk of unauthorized access to critical information is minimized.

15.21.2 Entity-Level Risk Assessment

Entity-level risk assessment encompasses all threats to an organization's objectives, operations, and overall security. This assessment provides a comprehensive view of organizational risks, enabling management to make informed decisions and allocate

Table 15.4 AML/CFT oversight framework components

Framework component	Details
Sectoral assessment	Identify sector-specific AML/CFT control risks and weaknesses. Group entities by subsector for risk assessment
Entity-level assessment	Assess individual entity ML/TF risk based on business type, size, client profile, and high-risk jurisdictions
Risk mitigation	Tailor supervisory engagement based on entity-level risk assessment. Rate the quality of mitigation measures considering sectoral and entity-level inherent risks
Residual risk rating	Develop a risk matrix for residual risk rating considering inherent ML/TF risks and the quality of AML/CFT mitigation
Aggregation of assessments	Identify common ML/TF risks by aggregating entity-level risk assessments at the sectoral level.
Impact	Inform supervisory actions and prioritize efforts. Serve as the basis for sectoral and national regime improvements

resources effectively. Key factors in entity-level risk assessment are summarized in Table 15.4, which outlines the AML/CFT Oversight Framework Components.

15.22   Management of Supervisor's Risk

Supervisors have an important responsibility to make sure their work is effective and honest. To do this, they must be aware of and manage risks that can affect their decisions and performance. One key risk is cognitive bias, which is a mental shortcut or habit that can influence decisions without the person realizing it. If not controlled, these biases can lead to unfair decisions and affect the neutrality of supervisory actions.

To manage this risk, supervisors can use structured decision-making methods that help reduce bias. They should also promote diverse opinions in teams and encourage open discussion and critical thinking. Additionally, supervisors must be careful about outside influences, like financial or political pressure, or internal conflicts of interest, that may affect their judgment.

Such pressures can stop supervisors from acting in the best interest of the organization or the public. To prevent this, it is important to set strong ethical rules, provide continuous training, and use independent monitoring for extra accountability. By following these steps, supervisors can improve public trust, ensure fairness, and strengthen the supervision process.

Figure 15.15 illustrates these ideas and is adapted from [153].

Fig. 15.15 Managing
supervisor's risk

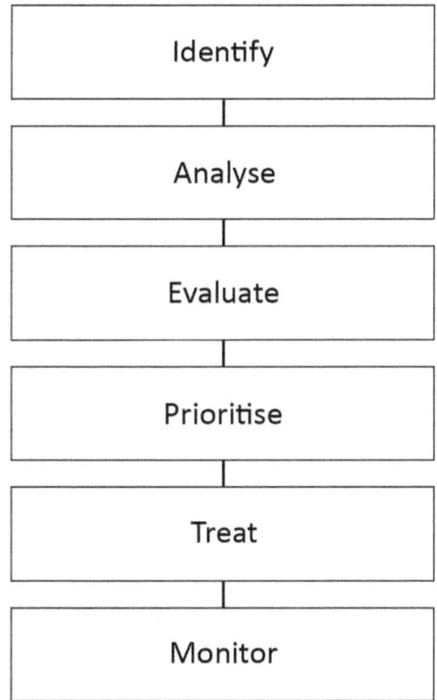

15.23 Cross-Cutting Issues in Supervisory Systems

Cross-cutting issues are common problems that affect different parts of supervisory systems across industries. Solving these problems helps improve overall supervision and regulation.

1. **Consistency and Standardization:** Rules should be applied in the same way across different industries. This avoids confusion and ensures fair treatment for all organizations.
2. **Technological Adaptability:** Regulations should keep up with new technologies. Using modern tools helps monitor and manage risks better.
3. **Interagency Coordination:** Agencies need to work together and share information to avoid gaps or overlaps in supervision.
4. **Human Capital and Expertise:** Skilled professionals with knowledge in finance, technology, and law are needed to manage supervision effectively.
5. **Data Privacy and Confidentiality:** Sensitive data must be protected. Following privacy rules ensures data security.
6. **Cybersecurity Resilience:** Strong cybersecurity protects supervisory systems from digital threats and keeps processes running safely.
7. **Agility and Flexibility:** Supervisory systems should quickly respond to new risks and changes in the industry.

8. **Transparency and Accountability:** Open processes and clear responsibilities help build trust in regulatory systems.

15.24 Consistency and Standardization

Consistency and standardization in supervisory systems ensure that all entities are treated equally and evaluated according to a unified framework. This approach removes ambiguity, builds trust among stakeholders, and helps organizations understand what is expected of them. Without clear and consistent standards, there may be discrepancies in enforcement, leading to confusion, unfair treatment, and gaps in regulatory compliance. Standardization supports better benchmarking, facilitates audits, and enhances cooperation across agencies and sectors. It also simplifies the implementation of automated tools for risk assessment and compliance verification, making the system more scalable and efficient.

15.25 Technological Adaptability

As new technologies emerge and evolve, supervisory systems must keep pace to remain effective. Technologies such as artificial intelligence (AI), blockchain, and big data have transformed financial services, introducing new opportunities and risks. Traditional supervisory models that rely on static reports and scheduled reviews are not equipped to detect fast-moving digital threats. Modern regulatory frameworks must incorporate real-time monitoring, automated analysis, and predictive tools to identify risks as they develop. Adaptability allows supervisors to respond quickly to innovations while maintaining oversight, fostering a balance between innovation and security. Regulatory sandboxes, pilot programs, and flexible policy frameworks are tools that support this adaptability.

Interagency Collaboration
Collaboration between regulatory and supervisory agencies is crucial for addressing cross-sectoral risks and ensuring cohesive enforcement. Different agencies bring diverse expertise, legal mandates, and operational strengths. When they coordinate efforts, they can close oversight gaps, reduce duplication, and develop a clearer, more unified set of expectations for regulated entities. Interagency communication enables shared threat intelligence, joint audits, and harmonized responses to crises. It is especially important in a globalized economy where financial activity crosses borders. A collaborative approach enhances institutional resilience and promotes international regulatory alignment.

15.26 Human Capital and Expertise

People are the foundation of any supervisory system. Skilled and knowledgeable personnel are essential for identifying, analyzing, and managing risks. Supervisory staff must understand complex financial products, regulatory requirements, cybersecurity principles, and emerging technologies. Continuous professional development programs, certifications, and international exchanges help build and maintain expertise. Institutions should invest in recruitment, training, and retention strategies to ensure that staff can adapt to evolving challenges. Having the right people in the right roles also enables proactive supervision, quicker response times, and more informed decision-making.

15.27 Data Privacy and Confidentiality

Supervisory agencies handle sensitive financial and personal information, and protecting this data is essential to maintain trust and ensure compliance with data protection laws. A strong data privacy policy includes limiting access to sensitive data, using encryption, applying secure authentication methods, and conducting regular audits. Data breaches not only harm reputation but can also result in significant legal penalties and disrupt regulatory operations. Supervisory systems must comply with local and international data protection frameworks, such as the GDPR or HIPAA, and maintain transparency about how data is collected, stored, and used. A trustworthy data governance framework is a cornerstone of effective supervision.

15.28 Cybersecurity Resilience

Cybersecurity is a critical aspect of supervisory systems, which are increasingly reliant on digital platforms for data collection, analysis, and reporting [30]. A cyberattack on a supervisory authority can compromise sensitive information, disrupt regulatory processes, and shake public confidence. Cyber resilience involves more than just prevention—it requires preparedness, detection capabilities, rapid incident response, and post-incident recovery [16]. This includes deploying firewalls, Intrusion Detection Systems, secure communication channels, and regular vulnerability assessments. Integrating cybersecurity into every layer of the supervisory architecture ensures uninterrupted operations and strengthens institutional credibility.

15.29   Agility and Flexibility

The ability to adapt quickly to new threats, technologies, or regulatory challenges is vital in today's dynamic environment. Agility means being able to modify oversight strategies, update compliance checklists, or reallocate resources in response to real-time developments. Flexibility in policy design allows supervisors to manage unexpected crises—such as market crashes or cyber incidents—without compromising long-term goals. Agile systems are supported by modular tools, dynamic reporting platforms, and decision-making processes that can accommodate changing circumstances. This ensures that supervision remains effective, relevant, and forward-looking.

Transparency and Accountability

Transparency and accountability are essential in retaining public trust and holding supervisory bodies responsible to the public. Transparency calls for the open exposition of objectives, methodologies, and decisions. Agencies are equally expected to account for their decisions and actions, take responsibility for their decisions, and their outcomes. These principles reduce corruption, influence behavioral intent and philosophy, and ensure quality through continuous improvement. Reports, public consultation, and stakeholder involvement are ways through which the supervisory authorities can achieve transparency. Independent reviews, audit trails, and performance measures enable accountability such that regulators remain responsible for their work, enhancing the integrity of the financial system.

15.30   Challenges of Intrusion Detection Systems

Intrusion Detection Systems (IDS) are a core part of cyber infrastructure, but they have issues with working with them. However, one of the persistent issues with these so-called SIEMs is a high false positive rate or alert fatigue, which subsequently leads to a large number of events that are being triggered for benign activity. The result is often a flood of false positives that distract security teams, drain resources, and produce alert fatigue, which may cause real threats to go unchecked. Configuration, tuning, and integrating context-aware detection models in the system are important to increase accuracy.

Additionally, IDS tools often produce alerts that lack clear context or severity ratings. Analysts must correlate IDS data with logs, traffic patterns, and other indicators to understand the threat's true impact. This process is time-consuming and requires specialized expertise. Many organizations also struggle with response readiness—detecting an intrusion is only part of the solution. Responding effectively demands incident response protocols, trained teams, and proper forensic tools.

Wireless IDS adds another layer to preventative or detective security, and an improperly secured network will bypass the detection layers completely. Security teams should incorporate IDS with SIEM systems, utilize machine learning to minimize the noise, and adopt ongoing training like the Certified Ethical Hacker (CEH) certification as a solution for these obstacles. Such efforts increase the ability to detect threats, respond to incidents, and ultimately, the resilience of an organization.

15.31 Summary

Intrusion detection systems significantly enhance the cyber posture of industrial control systems. Intrusion detection systems (IDS) monitor network traffic, system logs, and behavioral patterns in real time, enabling them to identify security problems and inform businesses of such in a matter of seconds. It is a proactive approach in response to threats and cyberattacks, mitigating risks and minimizing damage to the functions of critical infrastructure.

Intrusion detection systems also yield important context intelligence of the kind and scale of cyber threats that are being aimed at industrial control system environments. It provides a way to evaluate the detected anomalies and threat signatures, allowing organizations to understand their cybersecurity exposure and where they could improve. This intelligence-driven approach assists stakeholders to better craft their security strategies, deploy resources more judiciously, and apply focused countermeasures against emerging threats.

While the placement of sensors and tuning detection criteria certainly play a significant role in success, there are many other factors that come into play when installing an Intrusion Detection System in a control system, like how it is integrated with incident response protocols. For this reason, Intrusion Detection Systems must be carefully adapted and implemented for the particularities of different infrastructures and their operational characteristics in industrial control system environments not to disturb critical encryption operations.

In summary, cybersecurity plans for industrial control systems must incorporate Intrusion Detection Systems as vital components. By utilizing Intrusion Detection System technology, organizations can strengthen their defenses, enhance situational awareness, and protect critical infrastructure assets from cyberattacks. Maintaining the resilience, integrity, and reliability of industrial control systems in an increasingly digital world requires adopting a proactive and intelligence-driven approach to the implementation and maintenance of Intrusion Detection Systems.

Chapter 16
Attack Recovery Mechanisms for ICS: Common Cyberattacks, Cryptographic Key Management, and Host-Based Mitigation Strategies

Abstract This chapter discusses ICS attack recovery strategies, which are essentially targeted through information, IT, operational technology (OT), and offensive cybersecurity. It describes standard recovery measures for cyberattacks, focusing on cyber resiliency, as well as the timeline of a cyberattack in the ICS vertical. The chapter explores lessons learned from recoveries of some attacks, including CRASHOVERRIDE, and also discusses ICS vulnerabilities and threats that also apply to embedded systems. It also includes risk management and prevention methods, network security, and best practices for ICS cyberdefense. Additionally, the chapter examines how Public Key Infrastructure (PKI) can be leveraged for improved security. This is the context that will allow readers to build resilient recovery techniques to protect their ICS environment from cyberattacks.

Keywords Information Security · IT Security · Operational Technology (OT) · Offensive Cybersecurity · Cyber Resilience · Cyberattack Timeline · CRASHOVERRIDE · ICS Vulnerabilities · Embedded Systems · Risk Management · Prevention Strategies · Network Security · Public Key Infrastructure (PKI)

16.1 Introduction

Recovery methods in industrial control systems (ICS) are the steps and tools needed to get things back to normal following a cyberattack. These systems manage key areas like factories, water treatment plants, and power plants. As cyberattacks get better, it's important to have solid recovery protocols in place to limit damage and get back to normal operations.

A lot of ICS environments are using new technologies like robotics and the Internet of Things (IoT), but they often don't have good security measures in place [91]. Cybercriminals like to target these people. Having ways to recover is more vital than ever in the age of Industry 4.0.

The following are the main reasons to use recovery methods in ICS:

1. **Physical Safety Assurance:**

   (a) **Why It Matters:** Keeps workers and nearby people safe.
   (b) **Details:** Without proper security, ICS failures can lead to accidents and injuries. Recovery helps avoid such risks by restoring safe operations quickly.

2. **Protection of Critical Infrastructure:**

   (a) **Why It Matters:** Supports key services like electricity, water, and transportation.
   (b) **Details:** A cyberattack on these services can cause major problems for hospitals, homes, and public services. Recovery plans help bring systems back online quickly.

3. **Incident Command System (ICS) Integration:**

   (a) **Why It Matters:** Ensures organized response to incidents.
   (b) **Details:** The incident command system defines roles and communication paths during a crisis. It helps teams act fast and reduce the impact of cyberattacks.

4. **Cybersecurity Control Implementation:**

   (a) **Why It Matters:** Helps prevent, detect, and reduce cyber threats.
   (b) **Details:** Security controls include physical security (like cameras and guards) and technical tools (like firewalls and encryption). Using both improves system protection.

5. **Wireshark Use:**

   (a) **Why It Matters:** Analyzes network traffic to detect threats.
   (b) **Details:** Wireshark captures and studies network data. It helps experts find unusual behavior, fix issues, and prevent future attacks.

16.2 Components of Cybersecurity

ICS security keeps both the software that operates the devices (such as sensors and controllers) and the devices themselves safe. It also includes the individuals who run the systems. If these technologies don't have strong security, they could break down, cost money, and even hurt people.

Cybersecurity involves a variety of technologies and methods to keep networks, systems, and data safe. Since threats are continually changing, defense needs to have many levels of protection.

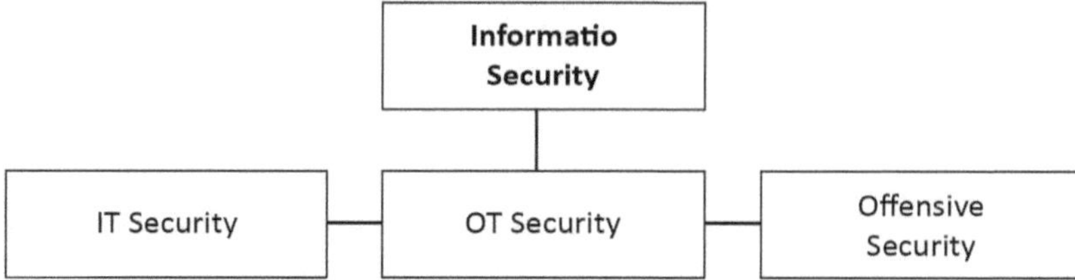

Fig. 16.1 Components of cybersecurity

Three important areas of cybersecurity are:

- **Governance:** Sets rules and ensures compliance with laws and standards. It establishes a robust foundation for security initiatives.
- **Technology:** Uses mechanisms like firewalls, encryption, and secure communication to protect systems from attacks.
- **Operations:** Focuses on risk management, incident response, and ongoing system monitoring.

These components collaborate to establish robust cybersecurity. Organizations must consistently revise their governance, technology, and operations to address emerging risks. Figure 16.1, adapted from [63], shows how these three parts connect in a strong cybersecurity strategy.

Figure 16.1 displays information related to security, including IT security, OT security, and offensive security.

16.3 Information Security

Information security deals with protecting information so that it stays private, remains accurate, and is available whenever it is needed. This means putting measures in place so that only authorized people can get access, data can be restored if something goes wrong, and it stays secure both when stored and when sent over a network. In industrial control systems (ICS), the aim is not just to defend against cyber threats but also to make sure the process continues without unnecessary interruptions. Typical measures include keeping an updated list of all assets, fixing known weaknesses, securing networks and devices, applying regular updates, and carefully managing user permissions.

1. **Confidentiality:** Classifying data helps apply the right security measures. For example, data can be marked as public, internal, or highly confidential.

   Access controls limit who can see or change sensitive information. These include setting user roles, strong passwords, fingerprint scans, and multi-factor authentication. Technologies like firewalls also help control access.

Encryption changes readable data into unreadable code that only authorized users can decode. It protects data stored in devices, moving through networks, or being used by software.

2. **Integrity:** Digital signatures confirm that files and messages haven't been changed. They use special keys to prove that the data is original and trustworthy.

   Hash functions create a short code (hash value) from data. If the data changes, the hash will change too. This helps detect any tampering.

   Data validation also plays a role. It checks that the data is correct, in the right format, and reasonable for its purpose. This helps keep systems running without errors caused by bad input.

3. **Availability:** Backups are extra copies of data stored in safe places. They help recover information after hardware failure, cyberattacks, or accidents.

   Redundancy means having extra systems ready to take over if the main ones fail. This includes spare power units, extra storage drives, or backup network connections.

   Disaster recovery plans explain how to return to normal operations after something goes wrong. These plans cover restoring data, fixing systems, and keeping services running.

16.4 IT Security

IT security is the practice of preventing misuse, disruption, and loss of digital information from servers, networks, and endpoint devices. It is designed to prevent unauthorized access, eliminate the Web's vulnerability to cyberattacks, and ensure that if such an attack or serious incident were to occur in one of its functions, it will not disable operations or access data.

At its base, the field is driven by three key principles: confidentiality, which means only authorized parties should be able to access sensitive data; integrity, which maintains the accuracy and trustworthiness of information; and availability, which ensures systems remain operational when needed.

Industrial security priorities are different because they are more than just about protecting the information. More importantly, however, ensuring the safety of workers (and in many cases, live machine operation) and preventing operational losses due to downtime. This is where modern security technologies serve a dual purpose: to help predict and prevent future risks from manifesting, and they also provide real-time detection and responses when incidents do occur.

1. **Firewalls and Intrusion Detection/Prevention:** A firewall is the network security barrier that decides under what conditions data comes in and out of a network. They act as a safety wall between secure internal systems and the outside Internet. Firewalls come in two basic forms: hardware and software.

Intrusion Detection Systems (IDS) monitor and look for anything out of the ordinary or potential threats [64]. Intrusion Prevention Systems (IPS) block such activity. Product by-product rule or behavior is detected as a threat, and these systems are identified by Name Phishing.

2. **Endpoint Security:** Each device can have its own firewall. These control which apps can access the Internet and help block threats from coming in or going out.

   Antivirus software finds and removes harmful programs like viruses or ransomware. It checks files, apps, and downloads to keep devices safe.

3. **Security Tools:** Antivirus tools scan data to find known threats and stop dangerous programs.

   Security Information and Event Management (SIEM) systems collect logs and alerts from many devices. These systems help teams watch for threats, fix problems, and follow laws and rules. SIEM tools are helpful for tracking attacks and responding quickly.

16.5 Operational Technology Security

OT stands for Operational Technology, which is the hardware and software used to monitor and control devices throughout many industrial functions: Think of a manufacturing plant, power grid, or transportation systems. Essentially, the main OT constituents are programmable logic controllers (PLCs) and supervisory control and data acquisition (SCADA) systems that are responsible for encompassing real-time processes. Since OT systems directly affect objects such as machines and production processes, the security of this IT component is important. These might include things like frequent monitoring, dividing your network into segments to limit any breach, and machine learning–based anomaly detection. The increasing deployment of smart sensors and connected devices (the Industrial Internet of Things (IIoT)) is then enabling OT further.

1. **Industrial Control Systems:** PLCs are specialized industrial computers designed to control machinery. They receive input from sensors, execute programmed logic, and send commands to actuators or machines to perform specific tasks.

   SCADA systems, on the other hand, provide operators with a centralized interface to monitor and control industrial processes. They display real-time operational data and allow technicians to diagnose and address issues remotely, reducing downtime and improving response efficiency.

2. **Network Segmentation and Isolation:** OT networks are primarily dedicated to controlling and automating industrial processes, whereas IT networks support business functions such as communication, data storage, and enterprise applications.

   Segmentation involves dividing a network into smaller, manageable zones. This helps control data flow, limits access between different parts of the system, and enhances security. Routers, managed switches, and access control rules are common tools for implementing segmentation.

Isolation goes a step further by keeping certain systems physically or logically separate. This approach ensures that if one network segment is compromised, the others remain protected. Techniques such as virtual LANs (VLANs) and restrictive firewall configurations are widely used for this purpose.

3. **Detecting Anomalous Behavior in OT:** In OT environments, machines, sensors, and PLCs work together to perform precise, real-world actions. Detecting anomalies means identifying operational patterns that deviate from expected norms—for instance, a sudden temperature spike, unexpected pressure changes, or irregular motor speeds.

   Anomaly detection methods combine real-time data analysis with domain-specific rules and expert knowledge, allowing operators to identify potential faults or security incidents early, thus preventing equipment damage or operational shutdowns.

16.6 Offensive Security

Offensive security refers to the practice of proactively identifying and addressing security issues before they become a target for malicious attackers. Penetration testing is the practice of simulating an attack on a network or web application in order to test security. This enables organizations to identify their vulnerabilities and strengthen protection. Pen Testing (e.g., red team operations) and manual checking for known security problems. The primary focus will be on identifying and resolving issues in networks, computer systems, and software solutions so that real attacks on the solution cannot pass.

1. **Penetration Testing:** Penetration tests, or ethical hacking, are performed by trained security experts. They pretend to be attackers and try to break into computer systems, networks, or applications. The aim is to determine what are the weak areas and how much harm a real attacker might do. This helps businesses assess their security and make changes to be better able to protect themselves.

2. **Red Team Operations:** Red team operations are a detailed trial when a team of skilled experts mirrors real attackers to understand how skilled the company is at protecting itself. Unlike regular penetration testing red team also practices what the firm's response to an attack looks like. This team uses sophisticated and smart procedures to replicate actual threats. This provides businesses with a clear picture of their security and assists them in improving detection, response, and recovery capabilities.

3. **Vulnerability Assessment:** Vulnerability assessment is the procedure of checking systems, networks, and applications for security issues. It entails scanning computers and networks for vulnerable installation flaws or popular defects that

assailants exploit. This approach allows organizations to identify and resolve problems before they are targeted. Companies are supplemented by regular assessments in order to keep up with regulations, stay close, and shut down potential attacks prematurely.

16.7 Recovery Mechanisms for Common Cyberattacks

Recovery mechanisms are also a vital part of any cybersecurity strategy framework and must be included in any consideration regarding Industrial Control System (ICS) environments or other mission-critical infrastructure. Although prevention and detection are important, a company's ability to quickly respond following a cyber incident is crucial to allow continued operation, minimize the financial losses and reputational damage, and trust from all affected parties.

Common cyberattacks—such as ransomware, Distributed Denial of Service (DDoS), data breaches, and insider threats—require clearly defined and tested recovery processes. These mechanisms should be proactive, flexible, and updated regularly to reflect evolving threat landscapes.

Key recovery strategies include:

- **Backup and Restoration:** Secure and frequent data backups are fundamental to recovery. Backups must be stored in protected, preferably offline or immutable locations to prevent compromise during an attack. Restoration procedures should be routinely tested to ensure data integrity and rapid recovery.
- **Incident Response Planning:** A well-documented Incident Response Plan (IRP) provides structured guidance for identifying, containing, eradicating, and recovering from cyber incidents. It includes predefined roles, communication workflows, and escalation procedures to manage the crisis effectively.
- **System Reimaging and Patching:** Infected systems should be wiped and reimaged with verified clean software environments. Applying the latest patches during the recovery process addresses security gaps that may have been exploited by attackers.
- **Business Continuity and Disaster Recovery (BC/DR):** BC/DR strategies ensure that essential operations can either continue or be rapidly restored. This involves the use of redundant systems, alternative sites, and defined recovery time objectives (RTOs) and recovery point objectives (RPOs).
- **Forensic Analysis and Root Cause Identification:** Conducting a forensic investigation after containment allows organizations to understand the nature and scope of the attack. Root cause analysis identifies vulnerabilities and helps refine security measures to prevent future incidents.
- **Post-Incident Review and Process Improvement:** Following recovery, organizations should carry out a post-mortem review to evaluate the response. Lessons learned must be translated into actionable improvements in policy, training, and technology.

- **Cyber Insurance:** Cyber insurance policies can help organizations recover from financial losses related to cyberattacks, covering costs such as legal services, notification requirements, system restoration, and reputational repair.

Successful recovery depends not only on technology but also on coordination across departments, informed leadership, and preparedness. Establishing a resilient recovery plan enables organizations to restore critical functions quickly, minimize downtime, and protect essential infrastructure against evolving cyber threats.

16.8   Common Cybersecurity Threats and Attacks

1. **Zero-Day Attack:** A zero-day attack happens when hackers exploit a security problem before the developer can fix it [70, 89]. These attacks are dangerous because traditional antivirus programs may not recognize them. They can cause serious damage, like stealing data, losing money, or stopping services. To reduce risk, organizations should regularly update all software and only install needed programs. Modern antivirus software that uses behavior checks and machine learning can also help. Tools that watch over the network and computers can detect strange behavior and stop attacks early.
2. **Domain Name System Tunneling:** This attack hides harmful data in domain name lookups. Since most systems trust domain name traffic, the attacker can sneak in or steal data without being noticed. To stop this, watch domain traffic for strange patterns, use DNS security tools, and block unnecessary requests [96]. Intrusion detection systems can also help spot unusual behavior.
3. **Business Email Compromise:** In this attack, hackers pretend to be someone trustworthy, like a company boss or vendor, and trick employees into sending money or data. To prevent this, use two-factor login methods, train staff to spot fake emails, and verify payment requests through phone calls or other methods. Watch for odd email behavior to stop problems early.
4. **Cryptojacking:** Cryptojacking uses someone's computer without permission to mine cryptocurrency. It slows systems down and uses a lot of electricity. It often spreads through infected websites or files. To stop it, block ads, check for strange network traffic, and teach users to avoid risky websites and links.
5. **Drive-by Attack:** This happens when visiting a harmful website causes a virus to install on device without knowing. These attacks use weak points in browsers or plugins. To prevent them, update software, block pop-ups, and limit downloads. If an attack is found, remove the harmful code and separate infected systems quickly.
6. **Cross-Site Scripting Attack:** Hackers insert bad code into websites that users visit. The code runs in the user's browser and can steal passwords or do other damage. To protect against this, clean user input, set strict content rules, keep apps updated, and use safe coding practices.

7. **Password Attack:** Hackers guess or steal passwords to get into accounts. They may use tools to try many password combinations or record what users type. To stop this, require strong passwords, use two-factor login, and watch for unusual login attempts. If a password is stolen, change it right away.
8. **Eavesdropping Attack:** This attack happens when someone listens in on private communication, either physically or online. It can expose private or financial information. Use encrypted communication tools like secure messaging apps or VPNs. Also, avoid using public Wi-Fi for sensitive tasks and look out for strange network activity.
9. **Insider Threats:** These threats come from people inside the organization, like employees, who either by mistake or on purpose cause harm. To reduce this risk, limit access to important data, train staff on safe practices, watch for odd behavior, and have a clear plan to respond to incidents.
10. **Internet of Things (IoT) Attacks:** Connected devices like smart cameras or sensors can be easy targets if not secured properly. Use strong passwords, update software regularly, and keep IoT devices separate from important systems [91]. Watching network traffic can also help find and stop attacks early.

16.9 Cybersecurity Key Components and Relationships

Cybersecurity is the practice of protecting systems, networks, and programs from digital attacks with a mix of technical ingredients. It is with these components working in concert that a strong cybersecurity strategy takes form as an integrated security solution.

On the other hand, network security prevents data from being carried away through firewalls, intrusion detection systems (IDS), and onsite intrusion prevention systems, among others; endpoint security ensures that devices are protected with antivirus software as well as global health-check and remediation.

Data security is, including encryption, data loss prevention mechanisms and access controls while identity and access management involves verifying who you are (authentication) and checking what you can do (authorization).

Others raise user awareness with security training and orchestrate incident response preparations for security incidents based on an incident response plan, forensic toolset, or a security operations center.

Organizational policies are informed by security, risk, and compliance frameworks, while vulnerability management alleviates system weaknesses.

For use cases that require applying secure design principles and controls, Security Architecture can be used as part of the solution to defend from cyberattacks. Figure 16.2 is adapted from [63].

Figure 16.2 presents cybersecurity key components and relationships, including training, policies, software, encryption, data recovery, monitoring, vulnerability assessment, and remediation.

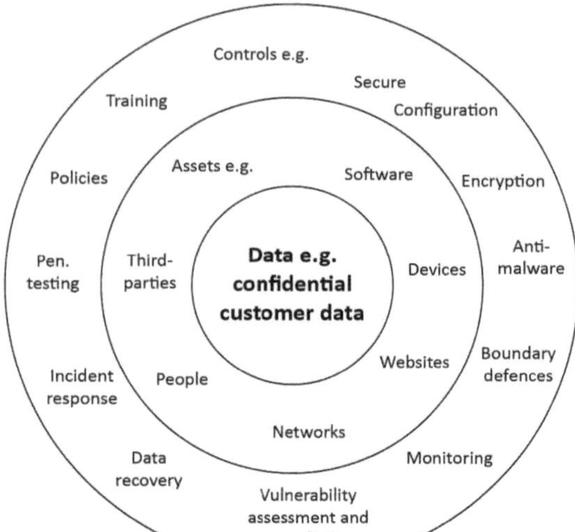

Fig. 16.2 Cybersecurity key components and relationships

16.10   Data Security and Cyber Resilience

The Data Security Alliance as a Specific Protection Against Emerging Cyber Threats. This partnership unites organizations, governments, and experts who are dedicated to cybersecurity in a world where the digital environment is becoming significantly more complex. It is a platform designed to make sure that your data remains safe and available and private regardless of what new version of ever-changing cyber threats is being released.

The Data Security Alliance is a community of security professionals dedicated to sharing ideas and providing insight, expertise, knowledge, technology, and business resources to provide the best solutions possible for any data security challenge. By working together, it provides the opportunity to create additional technologies and standards for maintaining privacy of data in terms of confidentiality, integrity, and availability. In addition, the alliance promotes cross-sector cooperation on data protection solutions that satisfy legal requirements and regulatory standards.

In addition to its technical efforts, the Data Security Alliance also represents an important information sharing and skill building platform. It is where individuals from various sectors come together for a discussion on how to deal with data leaks, cyberattacks, and threats to security. This open sharing helps organizations keep pace with new threats, making them stronger against looming attacks.

The alliance serves a variety of sectors, including healthcare, finance, government, and telecommunications, in which the security of data is paramount to protect personal information and maintain public confidence. Through this initiative, the Data Security Alliance is advancing both efforts to promote collaboration and continuous improvement and broader international cooperation to develop robust and mature policies and technologies for data protection.

Cybersecurity continues with the Data Security Alliance, facing pathogens and machine learning powered threats to make sure data security measures are potent.

16.11   Industrial Control Systems (ICS) and Cybersecurity

16.12   Cyber Resilience and Data Security

Cyber resilience is the ability to survive, recover, and adapt after a cyber event or attack. It involves managing risks, using security measures, responding effectively to incidents, and ensuring business continuity. The benefits include lower costs, more trust, and less downtime. Industrial Control Systems are central to the functioning of critical infrastructure such as energy production and distribution, water treatment, transportation systems, manufacturing plants, and other vital services. These systems are responsible for the oversight and management of physical processes that promote public health and safety while also sustaining the economy. Considering the crucial role played by ICS in society, more threat actors are turning their attention to the systems in an attempt to cause extensive disruptions, gain unauthorized access, or steal sensitive information. Unlike the standard IT systems, the components of ICS are ancient and were created without security considerations in mind. While ICS shares the need for cybersecurity, their IT cousins prioritize trustworthiness and real-time performance, making the application of the regular security directives challenging. ICS also features long-lifecycle parts that lack the current update capacities and containers and are regularly based on traditional security archetypes. Cyberattacking ICS can have devastating repercussions. A cybersecurity incident can cause operational shutdowns, physical equipment harm, environmental damage, financial losses, or, in extreme cases, human life destruction. The Stuxnet worm, Triton malware, which is designed to sabotage safety systems, and various ransomware assaults against the pipeline owners are examples of prominent peril instances. Given the growing risk, organizations from the Data Security Alliance and different national and sector-specific cybersecurity agencies have focused on implementing effective safeguarding strategies designed for ICS. Such techniques involve executing multilayered security arrangements, establishing sector-specific cybersecurity structures, and improving the incident response procedures. Information sharing, cross-sector harmonization, and fostering IT and OT teams' collaboration are additional efforts. By comprehensively incorporating cybersecurity into the ICS architecture, stakeholders can boost nationwide critical infrastructure safety and reducing the undue exposure of essential services to ever-changing cyberattacks are critical. Ultimately, the utilization of cybersecurity in ICS layout, implementation, and maintenance is no longer optional. It is a proactive security measure designed to safeguard national and public security in an interconnected modern societal context. For example, a cyber-resilient company can use communication plans and data backups to handle the effects of a ransomware attack.

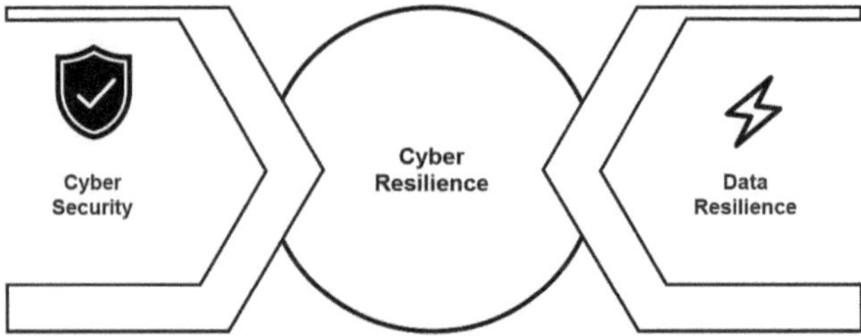

Fig. 16.3 Data security alliance

Data security, on the other hand, is about protecting sensitive information from unauthorized access, disclosure, or misuse. This includes access control, encryption, authentication, and regular security checks. Data breaches can lead to financial losses and harm to a company's reputation. For example, online retailers use encryption and access control to protect their customers' payment data.

Both data security and cyber resilience are important. In today's digital world, organizations must prioritize protecting data, build strong defenses, and be ready for security challenges.

Figure 16.3 illustrates cyber resilience and data security which are adapted from [63]. The Data Security Alliance serves as a unifying platform for stakeholders, industry leaders, and experts dedicated to addressing the growing challenges of data security.

16.13   Cyber Resilience: Withstand and Recover

Cyber resilience is the ability of an organization to survive, adapt, and recover from cyber threats. In today's digital world, where everything is connected and cyberattacks are becoming more common and complex, cyber resilience is crucial. It ensures that an organization can continue its work even after a cyberattack. It is not only about protecting data and systems but also about being ready to recover quickly when something goes wrong. This proactive approach strengthens the organization's security and readiness against short-term and long-term cyber threats.

At the heart of cyber resilience is the ability to prevent attacks before they cause damage. This requires strong security measures, like real-time monitoring, threat intelligence, and advanced cybersecurity tools that can spot and stop threats before they affect key systems. By watching network traffic, user behavior, and system performance, organizations can identify weaknesses and act fast to reduce risks. Quick action helps prevent major damage and keeps business activities running smoothly.

Despite the best defenses, no organization is fully immune to cyberattacks. A key part of cyber resilience is being able to recover quickly if an attack does happen. Recovery means restoring affected systems, apps, and data to their original state, so business operations can continue with minimal disruption. This stage also involves using business continuity plans and preparing staff to handle the aftermath of an attack.

Cyber resilience is more than just surviving attacks—it's about having a well-organized recovery process. This minimizes downtime, reduces data loss, and gets the organization back to full function as quickly as possible. Achieving this requires careful planning, including regular backups, system backups, and a detailed response plan.

Cyber resilience and cyber recovery are related but different concepts. Cyber resilience focuses on reducing weaknesses, improving defenses, and keeping operations running during and after an attack. On the other hand, cyber recovery focuses on getting systems and data back to normal as quickly and efficiently as possible. By combining both, organizations can improve their ability to handle and recover from cyber threats.

Ultimately, by building strong and adaptable security systems, organizations make it much harder for attackers to break through. This lowers the chance of serious breaches and ensures that if an attack happens, the organization can recover quickly with little disruption. When implemented properly, cyber resilience moves an organization from a reactive to a proactive stance, ensuring long-term security and stability despite changing cyber threats.

Figure 16.4 illustrates cyber resilience and recovery using white and black boxes, highlighting the phases before, during, and after an attack which is adapted from [63].

Cyber resilience means the ability of an organization to stop, handle, and recover from cyberattacks. It involves taking early action to block threats, keeping important services running during attacks, recovering quickly, and adjusting to new types of threats. Cyber resilience helps improve safety, reduce damage, and build trust in the organization. Getting certified in cyber resilience can also give companies an advantage in the digital world.

Fig. 16.4 Cyber resilience and recovery

16.14   Components of Cyber-Physical Attacks

Cyber-physical systems are systems where physical parts and computer-based systems work closely together. These systems use many types of sensors and devices to send and receive information. Examples include touchscreens, cameras, Global Positioning System chips, microphones, light sensors, and motion sensors. These tools collect real-time data and allow the system to respond to what is happening around it. Because of this, cyber-physical systems are fast and can change based on their environment.

To help these systems share information, they use communication methods like Wireless Fidelity, Fourth Generation mobile network, Enhanced Data rates for GSM Evolution, and Bluetooth. These allow cyber-physical systems to connect to the Internet and other devices, forming large networks known as the Internet of Things. These connections make systems work better and faster but also open up new ways for cyberattacks to happen.

A cyber-physical attack is a harmful action against the digital part of a cyber-physical system that causes real-world problems. These attacks can stop machines from working, break equipment, put people in danger, or harm the environment. Figure 16.5, adapted from [85], shows how these attacks work and where they happen in the system.

Figure 16.5 shows how a cyber-physical attack can happen. It includes actions like changing how a system works, stopping the system from reacting, tricking the operators and systems, and modifying the controllers and processes.

Recovering from a cyber-physical attack starts with quickly finding and stopping the breach. It is important to save evidence and call the team responsible for handling such events. The next steps are to use security tools, fix the systems and data, and follow a proper recovery plan. Regular system checks, staff training, good communication, and working with authorities help improve recovery. After recovery, reviewing the event helps to improve plans and stay ready for future threats.

Fig. 16.5 Constituents of cyber-physical attack

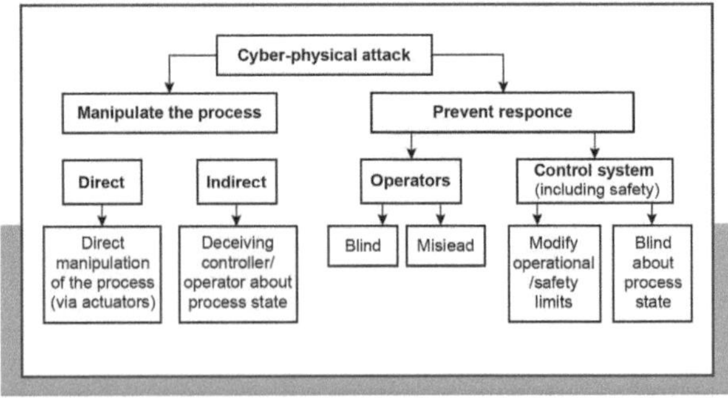

16.15   Host-Based Man-in-the-Middle Attack

In a host-based Man-in-the-Middle attack, the attacker changes processes and hides their actions on the affected computer or device. One common target is the programmable logic controller, which controls important machines. These controllers often lack strong security, making them easier to attack. Some attacks also use code that already exists in the system, which makes them harder to detect.

An example is the Pin Control Attack, where the attacker changes the input and output of a programmable logic controller. Public wireless networks, especially those without proper security, are easy targets for this kind of attack. Attackers can create fake access points with familiar names to trick users into connecting.

In these attacks, hackers secretly listen to or change the communication between two people or systems. In email hijacking, they take over real email accounts, watch messages, and send fake ones. In wireless spying, fake wireless networks capture the communication between users and real services. These attacks can lead to spying, fake messages, and stealing identity, so it is important to use strong security steps.

16.16   Recovering from a Host-Based Attack

After a host-based attack, the first step is to disconnect the affected system. Then, a team should be called to investigate how the attack happened. Systems and data must be restored from clean backups, and all weak points must be fixed. The process should also include monitoring systems and training staff to prevent future attacks.

Talking clearly with authorities and partners is also important. Cybersecurity experts can help strengthen defenses, and reviewing what happened will help improve future response plans. A full recovery includes technical tools, trained people, and always being alert.

Unsecured wireless networks are easy targets for attacks, as hackers can read data shared between devices. It is better to use a cellular connection or a secure wireless network that asks for a password. A virtual private network adds another layer of protection when using public wireless connections.

To protect against host-based attacks, take these important steps: check all security certificates, change any stolen passwords, and reset affected devices and network settings. Scan the network to look for changes, and watch for unusual activity. Use tools like virtual private networks, secure Internet connections, and regular software updates. Always being careful is the best way to protect data and privacy.

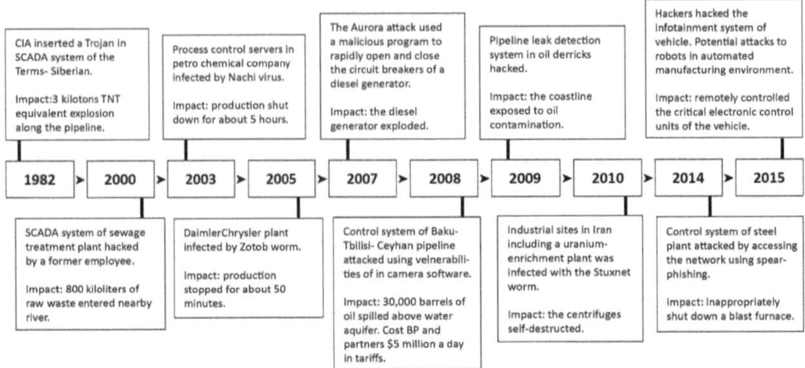

Fig. 16.6 Cyberattacks on ICS and physical impacts

16.17 Timeline of Cyberattacks on Industrial Control Systems and Physical Impacts

Industrial control systems help run important services like power, water, and transport, but they are being attacked more often. These attacks can stop operations, cause money loss, and even put lives in danger. Many of these events go unreported or are hard to track because some organizations do not want to share them, and some attackers use advanced methods to hide.

Some well-known cases include computer viruses that shut down chemical plants, malicious programs that stopped safety systems, and fake emails that broke into industrial networks. These attacks show real-world damage. Figure 16.6, based on [85], shows some of these major attacks. To protect against such threats, organizations should divide their networks, keep watching for problems, follow strong security rules, prepare response plans, and train their teams regularly.

Figure 16.6 illustrates various cyberattacks and hacking incidents, highlighting their impact on different systems and industries from 1982 to 2015.

16.18 Security Assessment of Industrial Control Systems

The security assessment of industrial control systems means checking the organization's systems to find any weak points or security problems. It also includes making sure there are good rules and steps in place to protect against online attacks. Reviewing control systems and operational technology helps improve safety and makes the systems stronger against cyber threats.

This process finds system weaknesses, checks how secure the network is, and looks for problems related to specific equipment or software from different vendors. Besides meeting legal requirements, these security checks help avoid

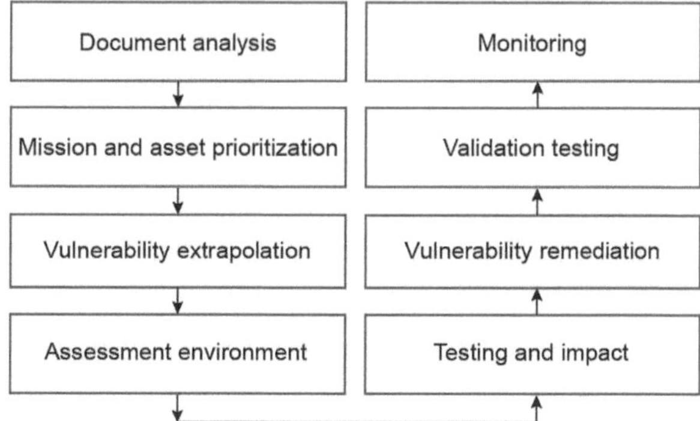

Fig. 16.7 Cyberattacks on ICS and physical impacts

risks and allow companies to continue working without interruption. To protect important infrastructure, it is necessary to carry out cybersecurity checks by trained professionals or trusted organizations. Industrial control systems must work safely. Both the physical devices and the software need to be protected, and the people who run the systems must also follow secure practices. Figure 16.7, based on [117], shows more details.

Figure 16.7 illustrates cyberattacks on industrial control systems and their physical impacts.

An organization's operational technology and industrial control systems must be checked carefully to make sure they are safe from cyberattacks. These checks help make the systems stronger and reduce the chance of disruptions. If critical infrastructure is attacked, it can cause serious problems like system downtime or legal trouble. Security assessments look at safety rules, possible risks, and whether the system follows industry standards. Packetlabs offers expert cybersecurity services to help with these types of assessments.

16.19 Attacks and Incidents on Industrial Control Systems

Cyberattacks on industrial control systems are becoming more common and dangerous. These systems control important services in energy, factories, transport, and utilities. Knowing how these attacks happen helps us stop them or react quickly. The Industrial Control System Cyber Kill Chain is a step-by-step method used to understand and stop these attacks. It is based on a similar model first used in general cybersecurity.

Some well-known examples of such attacks include:

1. HAVEX malware attack on industrial systems (2011–2015)
2. Cyberattack on the German government (2014)
3. Cyberattack on Ukraine's power grid (2015–2016)
4. TRISIS malware targeting Saudi oil and gas plants (2017)
5. Attacks on energy systems in the United States, United Kingdom, and Germany (2017–2018)

Attackers often follow steps. First, they try to get into the system and then move through the network. These early steps need different tools and methods than the final steps that cause harm. It is better to stop attacks in the early stages before damage happens.

16.20 Attack on Industrial Control Systems

Industrial systems can be attacked in many ways—through computer viruses, remote access, inside help, or targeted hacking. These attacks can happen in nuclear power plants, transportation systems, electricity networks, factories, buildings, and even space programs. Hackers can take control, change how systems work, or steal sensitive data using common hacking tricks.

There are three main weak points:

- Operators and their lack of security training
- Weaknesses in programmable controllers
- Problems in sensors, especially smart meters in smart electric grids

Many people wrongly believe these systems are safe because they are hidden or use uncommon tools. But this is not true. Simply adding random security tools or following standard rules is not enough. Old software bugs and weak security rules still exist in many devices. Smart meters can be hacked to steal electricity or to shut down services.

To protect important systems, we need strong safety plans, fix system problems, and stop believing myths about security.

16.21 Recovery from Industrial Cyberattacks

After a cyberattack in an industrial setting, it is very important to recover quickly. Recovery helps avoid loss of money, safety risks, and work delays. Some smart steps include:

- Getting cyber insurance
- Keeping encrypted backups
- Running regular training drills, such as tabletop exercises

Recovery needs constant review and advice from experts like the National Institute of Standards and Technology. New ideas are needed to improve how we recover.

Salvador Technologies, a company from Israel, has made it easier to recover from attacks. Their tool uses air-gapped storage (which is not connected to the Internet) and smart checks for damaged data. It can restore everything in less than 30 seconds. Instead of only trying to stop attacks, Salvador focuses on fast recovery and short response times.

Their system uses physical safety steps, like cutting power to the backup drive. It works for many kinds of devices and is setting new standards for how we protect industrial systems.

16.21.1   CRASHOVERRIDE Attack

CRASHOVERRIDE, also called Industroyer, was the first malware created to attack electricity distribution systems. Unlike older attacks that stole data, this malware was built to take control of the system. It could send commands to power systems and turn off electricity by opening circuit breakers. The malware was dangerous because it was easy to use and could be changed to attack other power systems too.

The attackers broke into the network by stealing passwords or using software bugs. They studied the system to find weak points and then used special software to attack. CRASHOVERRIDE was flexible, so attackers could update it and use it in different environments.

Old defense tools that look for known viruses or unusual system activity did not work well because this malware acted like a normal part of the system.

To stop such attacks, defenders needed to:

- Watch for stolen passwords
- Use strong login systems (like two-factor authentication)
- Monitor the network for unusual behavior

The CRASHOVERRIDE attack showed that cyber threats are always changing. It reminded us of the need for strong safety plans like dividing the network into secure parts, limiting access, and having a plan to respond to attacks. Figure 16.8,

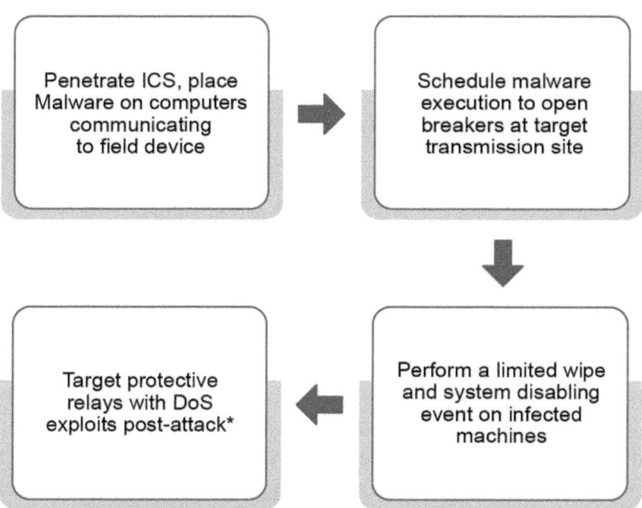

Fig. 16.8 CRASHOVERRIDE attack and event

based on [53], shows the impact of this attack and why strong protection is needed for our electric systems.

Figure 16.8 illustrates the cyberattack process on an industrial control system. The process involves placing malware on targeted computers, executing the malware to establish communication, launching denial-of-service (DoS) attacks on protective relays, and ultimately disabling critical systems.

16.22 Embedded System Applications in Industrial Control Systems

Industrial control systems rely on embedded systems, which are specialized computing devices integrated into both hardware and software. These systems play a crucial role in precise control, monitoring, and automation. They are widely used in industries such as aerospace, manufacturing, energy, and transportation. Ensuring their security requires implementing secure coding practices, effective patch management, and robust access controls to protect critical infrastructure while improving efficiency and connectivity.

Embedded systems are commonly utilized in commercial, aerospace, medical, automotive, home appliances, industrial, consumer, and telecommunications applications. In telecommunications, embedded systems are used in various technologies, from network phone switches to mobile devices.

1. Supervisory Control and Data Acquisition (SCADA)
2. Distributed Control Systems (DCS)
3. Programmable Logic Controllers (PLC)
4. Human-Machine Interface (HMI)
5. Remote Terminal Units (RTU)

16.23 Overview of Vulnerabilities

A vulnerability is a flaw in an information technology system that an attacker can exploit to conduct a successful attack. Attackers often chain together multiple vulnerabilities to reach their objectives, making it vital to address these weaknesses comprehensively. These vulnerabilities may arise from system design flaws, insecure functionalities, or human errors, requiring organizations to implement a multilayered security strategy to mitigate associated risks. Failure to address these vulnerabilities on time can leave critical systems, especially industrial control systems (ICSes), exposed to serious cyber threats.

Industrial control systems are particularly susceptible to multilayered attacks due to various security gaps across components. Many commercial software solutions lack robust security controls, increasing the ease of exploitation. The growing integration of industrial systems with Internet-connected and web-enabled distributed networks has significantly expanded the attack surface, introducing new cyber threat vectors. Within the competitive energy market, ICSes are expected to deliver high performance and reliability, but underlying security weaknesses can compromise operational integrity and create vulnerabilities in market dynamics.

One of the fundamental issues stems from the lack of security considerations during system development. Many ICS architectures were designed without integrating essential protections such as firewalls and intrusion detection systems, rendering them highly susceptible to attacks. Additionally, the use of open protocols like ICCP, CIM, DNP3, Modbus, and Profibus without encryption poses significant risks, as it enables potential eavesdropping and tampering with communication flows. Intelligent devices within these systems frequently operate without encryption, amplifying the security concerns.

Furthermore, some ICSes rely on insecure real-time operating systems that have known vulnerabilities, thereby increasing the chances of unauthorized access and manipulation. Remote access points such as dial-in modems, FTP servers, and unsecured remote desktop applications further expose systems to exploitation, particularly when default vendor passwords are left unchanged in embedded firmware. Control system components, including Supervisory Control and Data Acquisition (SCADA) and Distributed Control Systems (DCS), are at risk of exploitation through insecure interfaces like XWindows and ActiveX, while both wired and wireless communications carrying control signals are vulnerable to spoofing and denial-of-service (DoS) attacks.

The diversity of operating systems—ranging from Windows and Unix to QNX, RTX, and VxWorks—introduces varied and often poorly understood security risks. When these industrial systems are connected to broader corporate networks, the likelihood of cyber intrusions increases significantly. Lastly, poor communication across industrial teams and departments can hinder the successful deployment of cybersecurity policies, leaving critical vulnerabilities unaddressed.

Addressing these issues requires a layered approach, including strong network defenses, timely patch management, security audits, access controls, and continuous security awareness training. Effective risk mitigation also depends on continuous monitoring and proactive detection, ensuring vulnerabilities are identified and remediated before adversaries can exploit them.

16.24   Threats: Who Is Targeting Industrial Control Systems?

Industrial control systems are increasingly being targeted by various cyber threat groups. Many of these attacks are carried out by highly skilled and well-funded organizations, often supported by national governments. These groups have the technical ability and resources to find weaknesses in complex systems and launch large-scale cyberattacks.

Cyber attackers, including both inexperienced individuals and highly trained professionals, also pose serious risks. Those with a deep understanding of industrial operations and energy systems are especially dangerous, as they can design attacks that specifically target these systems and cause significant damage.

While there are no officially confirmed examples of cyberattacks by terrorist groups, intelligence reports suggest that such groups are actively building their ability to carry out cyberattacks. This makes them a possible threat in the future. Threats from people inside an organization also remain a serious concern. Current or former employees who are unhappy may misuse their access to systems to cause disruptions. These insiders often have special knowledge about system operations, which can make their attacks more harmful.

In addition, cyber units backed by national governments are increasingly using attacks on industrial control systems to serve political, economic, or military goals. The seriousness of these threats depends on the goals, skills, and resources of the attackers. Given these risks, it is important to build strong and flexible cyber defenses. Protecting important industrial systems requires security strategies that can keep up with changing and growing threats.

16.25 Industrial Control System Evaluation Criteria

To enhance the security and resilience of energy infrastructure, we need a comprehensive evaluation of the entire system to identify weak points and risks. One crucial step is examining the network architecture to ensure secure and reliable communication, especially for networks like WANs and data centers, which have unique security needs. Tools like Intent-Based Networking (IBN) can help by automating policy applications across the network.

It's important to stay updated with the evolving threat landscape. Penetration testing can simulate real-world attacks to identify and fix vulnerabilities before exploitation. Physical security also plays a major role, so regular assessments of assets and surveillance are vital. Strengthening the security of operational systems and control processes is equally important. Performing an impact analysis helps anticipate potential breaches and prepares us for worst-case scenarios. Understanding system interdependencies can reduce the risk of cascading failures.

Both physical and cyber risks should be assessed for likelihood and impact to prioritize which threats to address first. A risk-based approach leads to better resource allocation and decision-making. Additionally, scenario planning and simulation exercises help identify gaps in response plans. Cybersecurity awareness training is critical, as human error is a leading cause of security breaches. Finally, policies and procedures must be regularly updated to adapt to emerging threats and technologies, ensuring long-term protection for industrial control systems.

Figure 16.9 illustrates various network architectures and security and is sourced from [147]. It includes elements such as threat environment, penetration testing, physical security, operations security, risk characterization, and policies and procedures.

Fig. 16.9 ICS network
architecture possible attack

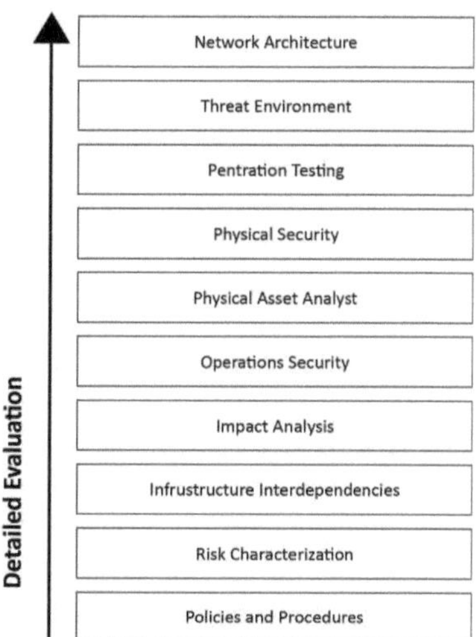

16.26 Prevention Strategies and Recommendations

To safeguard industrial control systems against cyber threats, it's crucial to implement a multilayered defense approach. Organizations must first develop business strategies that identify and address risks, vulnerabilities, and technological challenges specific to their systems. They should also establish platforms for sharing critical information across industries to enhance collective security. Robust encryption methods for communications are essential, along with regular system testing through test beds to spot and resolve vulnerabilities. For legacy systems, it's vital to apply security solutions that are compatible and devise long-term plans to maintain their protection while ensuring secure connections follow best practices. Furthermore, having a well-defined incident response and recovery plan is necessary to minimize damage in the event of a breach. Automated security alerts, clearly defined security standards, and strong perimeter defenses add extra layers of protection. Evaluating the system architecture to ensure encrypted communication via SSL VPNs, SSH, and Public Key Infrastructure (PKI) is critical. Access control can be reinforced with one-way connections using diode firewalls, along with secure authentication methods like dial-back modems and RSA SecurID tokens. Organizations should also disable unnecessary services, change default passwords, and incorporate real-time security tools designed for specialized operating systems. Regular cybersecurity audits help identify potential gaps, and embedded systems should undergo formal verification for quality assurance. Finally, having a comprehensive business continuity and disaster recovery plan ensures the continuity of operations, even during a cyber incident.

16.27   Risk Management Checklist

This checklist offers a detailed approach to securing critical energy infrastructure, starting with identifying essential functions and assets. It emphasizes understanding the consequences of losing these assets, followed by evaluating the current protective measures in place. From there, the focus shifts to assessing the potential threats and vulnerabilities to the system, ranking them based on their severity and likelihood of impact.

A thorough risk assessment helps determine which assets require immediate attention and protection. Once these priorities are clear, the next step is selecting the most effective mitigation strategies and evaluating their costs to ensure they are both practical and affordable. The process then moves toward implementing a security enhancement program that focuses on the identified priorities.

Collaboration with plant operators is highlighted as crucial to ensure that everyone is on the same page when it comes to protecting the infrastructure. This cooperative effort strengthens the overall security posture. Finally, continuous monitoring and regular reassessment of security measures are critical to ensure the infrastructure remains resilient, adaptable, and secure against evolving threats.

Overall, this checklist provides a holistic and proactive approach to managing risk, balancing protection with cost-effectiveness, and ensuring long-term security.

16.28   Network Security

Network security refers to the protection of a computer network's infrastructure. This includes hardware and software devices, such as printers, scanners, and software applications, which transfer data over the network. Security measures encompass a comprehensive set of actions designed to prevent and reduce security breaches.

These security measures enhance network resilience and protect against intrusions and unauthorized access. Any malicious activity that threatens the stability of the network or compromises user data privacy is considered a security attack. To prevent and recover from security breaches, both proactive and reactive security strategies are employed.

To protect the network from specific threats, a range of procedures are integrated across various protocol layers, where network communication and data sharing occur. Since network security solutions are diverse, they address vulnerabilities that emerge during information transmission. As technology advances, network security continues to evolve, developing sophisticated strategies and tools to counter new attacks and vulnerabilities in computer networks.

16.29   Types of Security Mechanisms

A security mechanism is a device or function specifically designed to provide one or more security services, typically evaluated based on its service strength and design confidence. Sources refer to procedures or systems used to generate specific outcomes. Different types of security mechanisms are outlined below: Figure 16.10 is sourced from [65].

Figure 16.10 illustrates various aspects related to security mechanisms.

1. **Encipherment:**

   (a) Involves using mathematical techniques to conceal data.
   (b) Protects privacy by transforming data into an unintelligible form.
   (c) The selected algorithm determines the level of encryption.

2. **Access Control:**

   (a) Prevents unauthorized access to data during transmission.
   (b) Applied through methods such as personal identification numbers, firewalls, and passwords.

3. **Notarization:**

   (a) Uses a trusted third party to mediate communication.
   (b) Reduces the possibility of disputes by logging requests from senders to recipients.

4. **Data Integrity:**

   (a) Adds a value generated by the data itself to ensure its integrity.
   (b) Similar to sending a known data packet that is verified both before and after transmission.

5. **Authentication Exchange:**

   (a) Focuses on verifying identity during communication.
   (b) Two-way handshakes are used at the transport layer to confirm data transfer.

6. **Bit Stuffing:**

   (a) Adds extra bits to transmitted data to allow verification at the recipient's end.
   (b) Achieved through the use of even or odd parity.

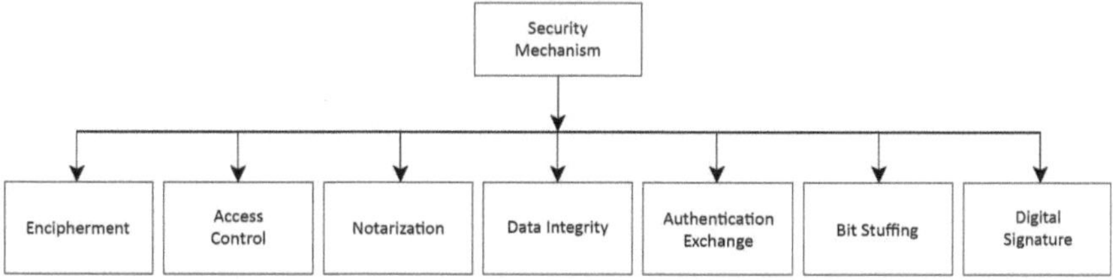

Fig. 16.10 Types of security mechanisms

7. **Digital Signature:**

   (a) Involves embedding invisible digital data as an electronic signature.
   (b) Less sensitive information is retained when verifying the sender's identity.
   (c) Electronically verified by the recipient.

16.30 Managing Keys in the Cryptosystem

In a cryptosystem, key management refers to the administration of cryptographic keys throughout their lifecycle. This includes key generation, replacement, storage, usage, and destruction (crypto-shredding). Effective key management ensures that the keys remain secure and are not exposed to unauthorized access during their lifespan. Properly managing keys is critical to maintaining the integrity and confidentiality of the cryptographic system. The key management process also involves establishing secure user procedures, implementing robust key servers, and designing an effective cryptographic protocol architecture. These protocols help regulate how keys are distributed, stored, and used within a system, ensuring that only authorized users have access. Additionally, it encompasses the implementation of key recovery mechanisms and disaster recovery protocols to ensure the availability of keys in case of system failures or breaches. Figure 16.11 is adapted from [65], which provides a visual overview of key management practices.

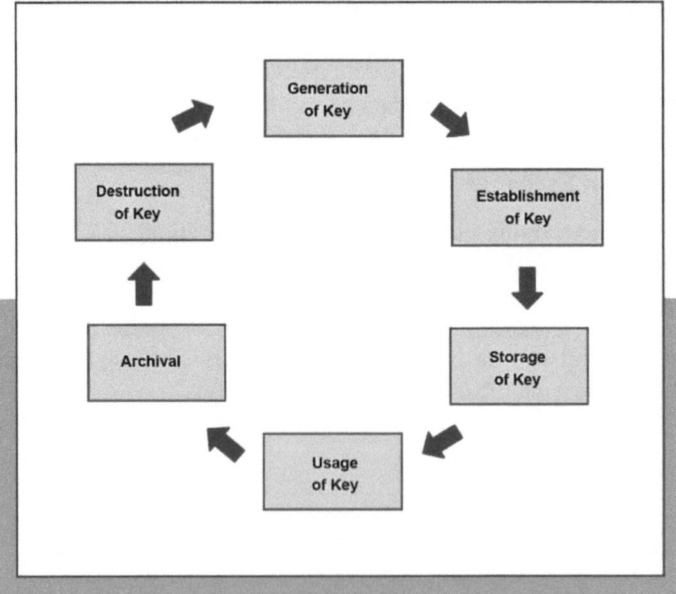

Fig. 16.11 Managing keys in the cryptosystem

Figure 16.11 depicts the key management process in a cryptosystem. It includes steps such as key generation, key establishment, key archival storage, key usage, and key destruction.

16.31   Public Key Infrastructure

Public Key Infrastructure (PKI) enables the creation of encrypted communication channels between components of an industrial control system. It ensures the confidentiality and integrity of data exchanged between controllers, sensors, and other devices by using digital certificates and encryption keys. PKI plays a critical role in securing industrial environments, where sensitive data and control signals must be protected from unauthorized access or manipulation.

By utilizing PKI, systems can authenticate identities, ensure data authenticity, and securely transmit information, thus preventing potential cyberattacks. This infrastructure provides a structured framework for managing public and private keys, certificate authorities, and other components essential for secure communications in distributed environments.

1. **Maintaining the Privacy of the Private Key:**

    (a) Only the owner of the private key should have access to it. It must remain inaccessible to anyone else, including unauthorized users or applications that could otherwise compromise its integrity.

2. **Assuring the Public Key:**

    (a) Public keys are accessible to everyone and are in the public domain, allowing others to encrypt data intended for the key's owner. It can be difficult to determine the validity and purpose of a key when it is widely accessible, highlighting the importance of verifying the key's authenticity through trusted sources such as certificate authorities. The function of a public key must be clearly defined, as improper handling or confusion about its use could result in miscommunication or security breaches.

16.32   Attacks on Industrial Control Systems

Industrial Control Systems are computer-based systems used to monitor and control industrial activities. These systems are used in many sectors, including energy (such as oil, gas, and electricity), manufacturing (such as food production and chemical plants), transportation, water treatment, mining, and healthcare.

Industrial Control System networks are different from regular computer networks because they run special software, have limited user access, and are not used for typical office tasks. These systems are mainly used to control equipment, collect data, and manage operations. Common components include Programmable

Logic Controllers, sensors, actuators, Supervisory Control and Data Acquisition systems, Programmable Automation Controllers, Remote Terminal Units, and Human-Machine Interface computers.

1. **Current Threats to Industrial Control Systems:** Many important sectors, such as nuclear energy, space programs, manufacturing, transportation, electric power, and building systems, have faced cyber threats. Attackers try to access and control system components, cause them to malfunction, or steal sensitive information using common software weaknesses.
2. **Security Challenges in Industrial Control Systems:** Research shows that many Industrial Control Systems rely on being hidden from public networks for security, which is no longer effective. These systems are at risk from both intentional attacks and accidental infections from viruses. Some systems are not updated regularly, which increases the risk. Common mistakes include using unsuitable general security tools and depending on secrecy as a form of protection.
3. **Attacks on Programmable Logic Controllers:** Programmable Logic Controllers are essential for monitoring and controlling machines, but many have security issues that have not been fixed. For example, some Programmable Logic Controllers made by Siemens are open to replay attacks, where recorded commands are used again by attackers. In secure places like prisons, these vulnerabilities could let attackers control systems such as automatic doors.
4. **Attacks on Sensors:** Sensors are important because they collect data for the system, but they can also be attacked. For example, smart meters used in modern electricity networks can be tricked into reporting false energy use. Attackers can also interfere with the communication between meters and utility centers. These weaknesses can lead to service disruptions or allow attackers to turn off power remotely.

16.33 Cybersecurity Strategies for Industrial Control Systems

Cyberattacks are becoming more advanced, and Industrial Control Systems need strong protection to defend against both known threats and new types of attacks.

Some cyber incidents are not reported or handled correctly, which increases the danger. Long-term and hidden attacks, often called advanced persistent threats, are especially difficult to detect and stop. Therefore, early action and careful planning are necessary to reduce damage.

Cyber Emergency Response Teams play a key role in defending these systems. They monitor cyber threats, issue warnings, and help organizations respond to attacks. These teams stay informed about new vulnerabilities and develop solutions to strengthen system security. Sharing information between the government, private companies, and industries also helps improve overall safety.

Fig. 16.12 A comprehensive countermeasure strategy for industrial control systems

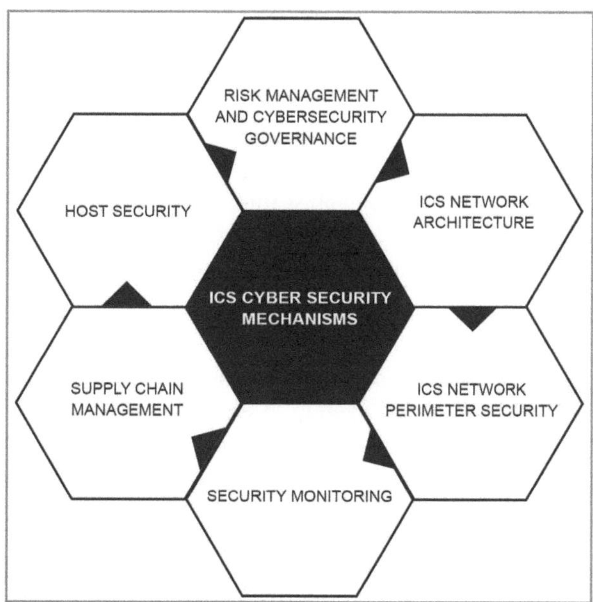

A complete security plan for Industrial Control Systems should cover both physical and digital risks. It should include modern security tools, risk assessment methods, and regular training to prepare for cyberattacks. The plan must protect both the equipment and the data that control critical operations.

Security teams must use several layers of defense. This includes dividing the network into smaller parts, using systems to detect intrusions, and constantly watching network activity. These defenses must act fast to find attacks, separate affected parts of the system, and stop any damage. This helps protect important operations and ensures safety and business continuity.

Figure 16.12 contains a diagram related to risk management, cybersecurity governance, host security, ICS network architecture, ICS cybersecurity mechanisms, supply chain, ICS network management, perimeter security, and security monitoring.

These protective techniques are being developed and improved largely due to the increasing cooperation among government agencies, cybersecurity experts, and industry participants. Proactive security measures, continuous information sharing, and threat intelligence analysis help maintain a strong cybersecurity posture for industrial control systems.

16.34 Defend and Govern: A Plan for Cybersecurity Strength

This section presents a clear and practical plan to help industries and organizations stay strong against cyberattacks and recover quickly if an attack happens. The plan focuses on defending against threats and managing cybersecurity with proper rules and actions. It includes steps for reducing risks, reacting to incidents, and improving protection over time.

1. **Finding Possible Risks:** Start by finding possible threats to the organization. Then follow a full process to manage these risks. It is important to list all system parts and use clear measures to understand each risk properly.
2. **Listing Control System Assets:** Make a complete list of all equipment and software used in industrial systems. This includes machines, tools, and the technologies that support them. This list helps identify the most important items to protect.
3. **Evaluating Impacts:** Evaluate what could happen if security goals like privacy, access, and accuracy are affected. This helps understand how serious each threat is and what damage it could cause.

Building Cybersecurity Rules and Actions
1. **Create and Follow Guidelines:** Develop and follow rules to keep systems safe, especially the most important ones. This includes writing clear instructions, giving training, and using tools from agencies such as the Cybersecurity and Infrastructure Security Agency.
2. **Stay Flexible:** Use rules that can change and improve over time. Instead of only depending on technology, use smart and adaptable methods to respond to new types of attacks.

Responding to Incidents
1. **Make and Practice Response Plans:** Create clear steps to follow during a cyberattack and practice them regularly. These plans should include how both information systems and control systems work together during a crisis to respond quickly and strongly.

Keep Improving
1. **Review and Update Regularly:** Check and update security rules, response plans, and risk assessments often. This helps the organization stay ready for new threats and fix any weak areas.

By using these steps, organizations can protect their most important systems, reduce risks, and act fast when cyber incidents occur. This plan focuses on always looking for new threats and updating protection rules to match. It also highlights the need to work together to protect control systems and keep essential services running safely.

16.35 Challenges in Recovering from Cyberattacks on Control Systems

Recovering from cyberattacks in industrial systems is often difficult because of how these systems work and the risk of serious physical damage. Below are the main challenges organizations may face:

1. **Complex System Design:** Industrial systems are often made of many connected parts. This makes it hard to find out how much damage an attack has caused and which operations are affected.
2. **Poor Visibility and Monitoring:** Many systems do not have good monitoring tools. Without proper visibility, it becomes harder to detect attacks and take action quickly.
3. **Limited Resources:** Some organizations do not have enough staff, money, or technology. This limits their ability to recover effectively after an attack.
4. **Difficult Integration:** Recovery steps must fit well with current systems. If not planned properly, recovery can cause further problems.
5. **No Standard Guidelines:** Without common rules and methods, it is hard to apply the same recovery steps in different places or situations.
6. **Following Legal Rules:** Laws about data and security reports must be followed during recovery. Meeting these legal needs can slow down the recovery process.
7. **Human Mistakes and Training:** A lack of proper training or mistakes by employees can slow down or weaken the response to an attack.
8. **Depending on External Vendors:** Some systems need help from outside companies for updates or repairs. This can delay the recovery process.
9. **System Redundancy and Backup:** Recovery works better when backup systems and extra equipment are in place. This requires planning and investment.
10. **New and Changing Threats:** Cyber threats keep changing. Recovery plans need regular updates to stay effective against new types of attacks.

16.36 Summary

The development and implementation of robust attack recovery strategies are vital for protecting industrial control systems (ICS) from cyberattacks. However, overcoming the challenges associated with attack recovery in ICS environments requires a multifaceted approach that addresses organizational, technological, and regulatory concerns.

To enhance their resilience to cyberattacks, organizations must prioritize investments in monitoring, detection, and response capabilities, despite the complexity of ICS systems and budget constraints. Recovery mechanisms must be integrated with existing ICS infrastructure and operations while ensuring regulatory compliance to minimize disruptions and maximize effectiveness.

Addressing human factors, such as through training and awareness campaigns, is also crucial for enabling staff to respond to cyber incidents and mitigate their impact on critical operations. Additionally, fostering strong relationships with vendors and industry partners can provide the necessary knowledge, support, and updates for rapid and effective recovery.

Finally, staying ahead of evolving cyber threats requires acknowledging the changing nature of the threat landscape and taking proactive steps to adjust and enhance recovery systems. In an increasingly digital world, organizations can strengthen their ability to recover from cyberattacks and maintain the integrity and reliability of industrial control systems by embracing resilience, redundancy, and continuous improvement.

Chapter 17
Some Case Studies of Industry Control System

Abstract This chapter provides a detailed review of case studies related to Industrial Control Systems, which are regarded as the backbone of industrial process management and its automation. ICS security and reliability are critical to operational success for the safety and efficacy of the distributed operations in many sectors. The case studies provide tangible learning points and demonstrate the practical application of ICS in various industries. The lessons learned are discussed, focusing on the challenges, approaches, and outcomes. Through these case studies, the chapter offers insight into best practices and methodologies for realizing optimal ICS performance and security across industrial domains. The results stress the need for a strong ICS structure and guide for the development of ICS in the future.

Keywords Industrial automation · ICS security · Operational efficiency · Best practices · ICS frameworks · Industrial processes · Performance optimization

17.1 Introduction

Industrial control systems are essential to today's industries. Typically, they are used for the control and monitoring of critical infrastructures in fields such as manufacturing, energy, transportation, or public infrastructure. These systems help ensure that essential services continue to run smoothly. These systems are relied on by so many people and the economy that it is a very great importance to keep them protected, resilient, and functional.

Case studies are helpful because they show real-life problems and how different companies have solved them. They share what worked, what failed, and what lessons were learned. By studying these examples, other organizations can better understand how to deal with risks, protect against threats, and reduce the chances of service interruptions.

These case studies also give reference areas that are systemically weak and require strengthening. They come bundled with wise remedies as many times as smart things such as new technological innovations or sophisticated resources to find and avoid cyber threats prior to their taking place.

M. A. Rahman et al., *Securing Industrial Control Systems*,
https://doi.org/10.1007/978-3-032-03018-4_17

By looking at case studies from many industries, experts and decision-makers can learn to better protect their own systems. These examples provide valuable insights for enhancing safety, mitigating risk, and preparing for future challenges. This introduction sets the stage for examining various case studies that illustrate the genuine issues and solutions in safeguarding industrial systems. They provide crucial information to engineers, researchers, and policymakers working on improving the safety and reliability of these systems.

17.2 Case Study

This book includes case studies in industrial control system security, which help organizations to understand their vulnerabilities, risks, and cyberattack outcomes. By studying these real-world examples, companies can learn more about how their infrastructure is being attacked, what defense mechanisms they have to prepare, and which response methods will be applied to protect themselves from dangers in the future. In aerospace, digital twin technology [100] plays a significant role in monitoring and predictive maintenance, including the virtual acknowledgment of sensor failures using virtual sensors. PHASE [72] secures the personalized smart health systems against the online fake data injection attacks with minimal data manipulation and is evaluated on Pima Diabetes and AIM-94 datasets. It also discusses ICS integration, data integrity, protocols, access control, anomaly detection, and the role for Industry 4.0 in the agri-food IoT and analytics [94], with the integration's benefits of optimized performance, unified asset management for predictive maintenance, and decision-making driven by data [50]. MCS becomes efficient and adaptable in telecom [87]. Chemical and petrochemical plants need to manage the security and efficiency cautiously [134], and an overview of the automation challenges in intelligent building is provided [165]. The significance of industrial cooling in temperature regulation is emphasized [130]. One approach to tackle this issue is to develop a Multivariable Iterative Learning Control (MILC) strategy for flexible feed drives [161] and to study load control in order to minimize energy consumption in fertilizer and chemical plants [11].

These case studies not only focus on the immediate consequences of cyberattacks but also emphasize the long-term strategic implications for industries that rely heavily on ICS. They illustrate the importance of proactive risk management, the role of advanced cybersecurity tools, and the need for continuous monitoring and updating of security protocols. By analyzing these real-world scenarios, stakeholders can gain a deeper understanding of how vulnerabilities in ICS can be exploited and what steps can be taken to mitigate these risks.

17.3 Stuxnet Attack (2010): Nuclear Facility Compromise [102]

The Stuxnet attack was such a landmark event for cyber warfare, and for industrial control system security in particular. The Stuxnet threat was created by American and Israeli intelligence services to slow down Iran's nuclear program, effectively shutting down the Natanz Uranium Enrichment Facility. The malware was originally believed to be limited to this facility, but its spread to other systems has raised alarms about broader implications. Instead, Stuxnet used several new and previously unknown Windows vulnerabilities, confirming its status as a state-sponsored cyber weapon created to target critical infrastructure. This specific case study is also applicable to discussions in Chaps. 1 and 2, providing the technical foundation required for grasping this case study's methods and results.

Development and Purpose
Stuxnet, which was created as part of the covert battle known as Operation Olympic Games, aimed to destroy Iran's nuclear program without a direct military intervention. The operation was started by the Bush administration and continued under Obama, and is thought to have required a team of very capable engineers. While it was never officially confirmed, Stuxnet is believed to have been part of a larger operation related to advanced malware tools, including Duqu and Flame.

Stuxnet's Design and Execution
Stuxnet was one very advanced malware that affected its uranium-enriching centrifuges at the Natanz facility. The original wiped commands for the specific PLCs used in the centrifuge control system, causing them to malfunction (changing rotational speeds). That simultaneously kept system monitors looking normal as well, so detection was difficult. The worm itself propagated via USB drives, which helped it to include most network defenses so that even isolated systems could become infected.

Exploitation of Vulnerabilities
Stuxnet exploited four zero-day vulnerabilities in Windows and industrial control systems, including flaws in print spoolers and privilege escalation mechanisms. It was written in multiple programming languages, including C and C++, underscoring its complexity. The attack successfully damaged over 2,000 centrifuges, significantly delaying Iran's nuclear enrichment program by at least two years. Various malware attack techniques, including those used in this case, are discussed extensively in Malware Attack Techniques Chap. 5.

Discovery and Legacy
Stuxnet was discovered in 2010 after spreading beyond its intended target. Anomalies observed in Iranian systems led to an investigation by the international cybersecurity community. Although the vulnerabilities it exploited have since been

mitigated, Stuxnet remains a landmark case in cyber warfare. It was the first publicly acknowledged cyberattack designed to cause physical damage, marking a turning point in the evolution of cyber threats. This incident exposed vulnerabilities in critical infrastructure and demonstrated the potential consequences of cyberattacks on industrial systems.

Stuxnet specifically targeted the Natanz Uranium Enrichment Facility using an advanced cyber-physical weapon. It was the first cyber weapon explicitly designed to inflict physical damage, distinguishing it from previous cyber intrusions. The malware manipulated programmable logic controllers to alter centrifuge spin rates, damaging nuclear equipment without triggering immediate detection. By spreading through USB devices and exploiting multiple security flaws, Stuxnet systematically disrupted Iran's nuclear infrastructure, marking a pivotal moment in cyber warfare history.

17.4   Ukrainian Power Grid Attack (2015) [167]

On December 23, 2015, a cyberattack targeted Ukraine's power grid, marking a significant event in cyber warfare. The attack compromised the information systems of three energy distribution companies, with the most severe impact on Prykarpattyaoblenergo, which serves Ivano-Frankivsk Oblast. As a result, approximately 230,000 people lost electricity for periods ranging from one to six hours. In addition to Prykarpattyaoblenergo, the energy companies Chernivtsioblenergo and Kyivoblenergo also experienced disruptions, though to a lesser extent.

Attack Details
The cyberattack was highly coordinated and involved multiple stages. Initially, hackers infiltrated corporate networks using spear-phishing emails that delivered malware, allowing remote control over industrial control systems. The attackers disabled key components of the power grid's infrastructure, including uninterruptible power supplies, modems, and remote terminal units, which led to the shutdown of substations. Additionally, malware erased critical files from both servers and workstations, further disrupting operations. A denial-of-service attack was also launched on call centers, preventing customers from receiving updates about the blackout. This incident highlights the need for strong cybersecurity measures to protect industrial control systems from cyber threats, as discussed in Chap. 2.

Impact and Consequences
The attack resulted in a substantial loss of electricity, disrupting up to 73 MWh of power, which accounted for approximately 0.015% of Ukraine's total daily consumption. The most severe consequences were felt by Prykarpattyaoblenergo, customers, though two other energy providers were also affected. This cyberattack not only disrupted Ukraine's infrastructure but also exposed vulnerabilities in critical systems, particularly those relying on outdated technology.

Vulnerabilities and Geopolitical Context
The attack on Ukraine's power grid must be understood within the broader context of the country's infrastructure vulnerabilities and geopolitical tensions. Ukraine's power grid, originally built during the Soviet era, had been modernized with components that left it susceptible to exploitation. The timing of the attack, during the holiday season, ensured minimal staff availability to respond to the breach, further exacerbating its impact. Additionally, the ongoing conflict and historical ties between the two nations made Ukraine particularly vulnerable to cyber infiltration, as attackers were already familiar with the systems and software used by Ukrainian operators.

17.5 Oldsmar Water Treatment Facility Attack (2021) [83]

On February 5, 2021, the Bruce T. Haddock Water Treatment Plant in Oldsmar, Florida, was targeted in a cyberattack. A malicious actor gained unauthorized remote access to an operator's workstation twice, first at 8:00 AM and again at 1:30 PM. During the second breach, the attacker attempted to increase the sodium hydroxide concentration in the water supply from a safe level of 100 parts per million to a dangerously high 11,100 parts per million. If successful, this could have posed a severe public health risk by contaminating the water supply for 15,000 residents. Fortunately, the attack was thwarted when the plant operator detected the changes in real time and immediately restored the correct settings, preventing any harm. This incident exposed vulnerabilities in control systems managing essential public services.

Significance of the Attack
The attack on the Oldsmar water treatment facility revealed critical weaknesses in public infrastructure security. Although the identity of the perpetrators remains unknown, their intent may have ranged from sabotage to testing the security of such systems. If the operator had not identified and reversed the unauthorized changes, the water supply could have been dangerously contaminated. Prolonged exposure to high levels of sodium hydroxide can cause severe health complications, highlighting the importance of effective cybersecurity measures in water treatment facilities.

System Redundancies and Rapid Response
The Oldsmar Water Treatment Plant had built-in safeguards designed to detect and alert staff to abnormal pH levels. According to city officials, it would have taken 24 to 36 hours for the altered sodium hydroxide concentration to affect the public water supply. Fortunately, the plant operator noticed the intrusion in progress and promptly corrected the settings, ensuring the safety of the community.

Lessons from a Previous Incident

Although the Oldsmar attack was mitigated, the potential consequences of an undetected cyberattack emphasize serious risks. A relevant historical example is the 2007 incident at the Spencer Water Treatment Plant in Massachusetts. Due to internal failures and notification system issues, excessive sodium hydroxide was inadvertently added to the water supply, resulting in over 100 individuals requiring hospital treatment for rashes, burns, and other health effects. Despite a high pH alarm being triggered, corrective actions were not taken in time to prevent harm. This underscores the necessity of robust monitoring systems, timely alerts, and swift responses to prevent similar incidents in the future.

17.6   Colonial Pipeline Ransomware Attack (2021)

On May 7, 2021, the DarkSide hacking group compromised the Colonial Pipeline system, which supplies fuel to the southeastern United States. The attackers gained access using a compromised virtual private network account that lacked multi-factor authentication. In response, the company halted operations and paid a ransom of 75 bitcoins (4.4 million USD) for a decryption tool, which proved slow in restoring full functionality.

This attack caused fuel shortages, panic buying, and significant disruptions, particularly in the Southeast. The US government responded with emergency measures, and the FBI later recovered a portion of the ransom. The incident highlighted vulnerabilities in critical infrastructure and raised concerns about cybersecurity [38]. Multi-factor authentication and ransomware threats are further discussed in Chap. 10, along with related topics in Chaps. 11, 12, 16, 7, and 8.

17.6.1   Impact of the Attack

Financial Losses

The attack resulted in substantial financial losses for both Colonial Pipeline and the broader economy. While the Department of Justice recovered 63.7 bitcoins (approximately 2.3 million USD at the time), the fluctuating value of Bitcoin and the unrecovered portion of the ransom contributed to financial setbacks. The six-day shutdown disrupted fuel supply chains across the southeastern United States, causing revenue losses for both the company and its clients. Businesses dependent on a stable fuel supply had to scale back operations, compounding economic impacts.

Beyond lost income, Colonial Pipeline incurred significant costs related to incident recovery. The company hired cybersecurity experts to assess vulnerabilities, replaced hardware and software, and implemented enhanced security measures. Business disruptions required system updates, reducing productivity. Additional costs included crisis management efforts, public relations, and communication with stakeholders. This attack underscored the widespread economic impact of cybersecurity breaches on critical infrastructure.

Reputational Damage

The ransomware attack severely damaged Colonial Pipeline's reputation. The decision to pay the ransom, despite law enforcement discouraging such actions, drew criticism from both the public and cybersecurity experts. Many argued that paying the ransom incentivized future attacks and exposed a lack of preparedness.

The company also faced scrutiny for shutting down the entire pipeline, even though operational systems remained intact. While intended as a precautionary measure, this decision was seen by some as an overreaction that worsened the crisis. Extensive media coverage amplified concerns, exposing vulnerabilities and eroding trust in Colonial Pipeline's cybersecurity measures. The loss of confidence among business partners and consumers posed long-term challenges for the company's reputation and market standing.

17.6.2 Pipeline Restart and Recovery

Operations resumed on May 12, following a six-day shutdown, marking one of the most significant disruptions to US infrastructure in recent history. Despite the pipeline reopening, intermittent disruptions continued as the system stabilized. Full operational normalcy was achieved by May 15, easing pressure on the fuel supply chain.

The shutdown had widespread consequences, particularly in Georgia, North Carolina, South Carolina, and Virginia, where fuel shortages were reported at numerous gas stations. Delays in resupply caused public frustration and logistical challenges. Gasoline prices surged to their highest levels since 2014 due to panic buying and limited availability.

Recovery efforts required coordination between Colonial Pipeline, federal and state agencies, and private-sector partners. Trucking companies and railroads were mobilized to transport fuel to affected areas. Additionally, cybersecurity experts and engineers worked to secure and restart operations, implementing safeguards to prevent further attacks.

This incident underscored the interconnectedness of critical infrastructure and the vulnerabilities of key supply chains to cyber threats. In response, Colonial Pipeline committed to reviewing its cybersecurity policies and strengthening defenses against future incidents.

17.6.3　Investigation and Ransom Recovery

The attack was attributed to DarkSide, a Russian-based cybercriminal group known for targeting high-profile organizations with ransomware. The group used a double extortion tactic—encrypting victims' systems while also threatening to release sensitive data unless payment was made.

While DarkSide had previously followed through with ransom demands, this attack prompted swift action from the US government. The FBI managed to recover 63.7 bitcoins (approximately 2.3 million USD at the time), which had been paid to the attackers. This marked a rare success in disrupting cybercriminal operations and reclaiming ransom payments.

Despite this partial recovery, the attack caused lasting financial and reputational damage. Investigations revealed gaps in coordination between government agencies, state authorities, and private sector organizations. The absence of a unified response system and clear communication channels during the attack exposed weaknesses in national cybersecurity strategies.

In response, the US government ramped up efforts to combat ransomware, including issuing sanctions and urging diplomatic pressure on Russia to crack down on cybercriminal groups operating within its borders. The attack also accelerated discussions on international collaboration to counter global cyber threats, emphasizing the need for stronger cybersecurity regulations and preventive measures across all industries.

17.7　German Steel Mill Blast Furnace Attack (2014)

In 2014, a German steel mill experienced a sophisticated cyberattack targeting its industrial control systems, specifically a blast furnace. The attack caused the furnace to shut down improperly, resulting in significant physical damage to the facility. This event marked only the second known instance of a cyberattack causing real-world physical destruction, following the Stuxnet malware attack in 2009. The incident drew widespread attention from cybersecurity experts due to its scale and implications for critical infrastructure security [52].

The identity and motives of the attackers remain unknown, though speculation ranges from industrial sabotage by competitors to a state-sponsored operation. The attack's complexity and similarities to Stuxnet and the Triton malware attack of 2017 have raised concerns about the vulnerabilities of interconnected operational technology systems. Despite the lack of attribution, detailed analyses of the incident have shed light on how the attackers exploited system weaknesses, offering valuable lessons on preventing similar threats. These include improving network segmentation, enhancing system monitoring, and adopting robust cybersecurity measures to safeguard critical infrastructure. This attack is directly relevant to Chaps. 1, 2, 5, 9, 10, 11, 14, and 16, as it encompasses foundational cybersecurity concepts.

Impact of the Attack
While the exact details of the incident's impact on the steel mill are not fully disclosed, several consequences can be inferred based on the available information. The most significant outcome was the physical damage to the blast furnace, which is typically used to melt metal at extremely high temperatures. An improper shutdown of such a furnace could lead to catastrophic damage not only to the furnace itself but also to surrounding machinery, equipment, and structural elements within the facility.

Fortunately, no employees or members of the public were harmed during the attack, but the property damage was substantial. The incident likely caused disruptions to the production process, leading to operational delays and financial losses. Additionally, recovery expenses were considerable. Repairing the blast furnace alone would have cost millions of dollars, making it financially challenging for the facility to recover. The cyberattack disrupted operational technology, leading to extended downtime and delays in restoring normal operations, further compounding the financial and operational impact.

Key Lessons Learned
This cyberattack provides several key lessons for organizations to enhance their cybersecurity practices and prevent similar incidents in the future:

- **Employees as the First Line of Defense:** The success of the cyberattack was largely due to employees unknowingly interacting with phishing emails. Cybersecurity awareness training for all employees is crucial to prevent such incidents. Workers should be educated to recognize phishing scams, avoid opening suspicious emails, verify senders' identities, and use strong, unique passwords.
- **Effective Security Software:** While employee training is vital, robust security software is equally important. The steel mill could have employed network monitoring systems, antivirus software, and endpoint detection tools to mitigate the attack. Regular updates and routine penetration testing of these systems are essential to identify and address vulnerabilities.
- **Physical Security in Cybersecurity Plans:** The incident highlights the need to assess physical vulnerabilities as part of a comprehensive cybersecurity strategy. Organizations should map out the relationship between IT systems and physical operations to identify potential risks and plan mitigation measures accordingly.
- **Adequate Insurance Coverage:** Proper insurance coverage is crucial to protect against digital and physical losses caused by cyberattacks. Organizations must work with insurance professionals to secure policies that cover data breaches and physical damage to infrastructure.

Case Highlights
The cyberattack on the German steel mill underscores the growing risks associated with cybersecurity breaches in industrial environments. The attackers initially gained access to the corporate network through a spear-phishing email, a common method used to exploit human vulnerabilities. Once inside, they moved laterally

through the IT systems, infiltrating the production network and targeting critical operational components. The hackers then manipulated the mill's control systems, preventing the proper shutdown of the blast furnace, which led to significant physical damage to the facility's infrastructure.

This attack demonstrated the dangerous potential of cybersecurity breaches to result in real-world physical destruction, highlighting the critical need for robust cybersecurity measures in industrial settings. The incident serves as a stark reminder that cyber threats to industrial infrastructure are constantly evolving. As cybercriminals become more sophisticated, organizations must not only secure their digital assets but also integrate cybersecurity strategies with their physical systems to avoid catastrophic consequences.

Key lessons from this attack emphasize the importance of comprehensive employee training to recognize and respond to phishing attempts, investing in effective security software, and performing regular risk assessments to identify potential vulnerabilities. Additionally, ensuring proper insurance coverage to mitigate financial losses and recovery costs is crucial. By implementing these best practices, organizations can strengthen their defense mechanisms and reduce the likelihood of devastating cyberattacks that threaten both their digital and physical assets.

Insights into LockerGoga

The LockerGoga ransomware infiltrated Norsk Hydro's systems through specific access vectors, significantly affecting both IT and operational technology systems. Broader trends were also observed as similar ransomware attacks targeted other industrial and manufacturing companies, revealing common vulnerabilities and exploitation patterns in critical infrastructure.

Refusing to Pay Ransom

The decision to refuse to pay the ransom was driven by ethical considerations, such as preventing the funding of criminal activities. The company's leadership expressed confidence in their ability to recover and rebuild infrastructure without complying with the attackers' demands. As part of the recovery process, the company restored systems from backups and implemented enhanced cybersecurity measures. Long-term investments were made to strengthen security protocols and prevent future attacks.

Restoring Operations

During the IT outage, the company shifted to manual production processes to maintain operational continuity. Employee expertise was crucial in adapting to the challenges posed by the disruption. Additionally, teams across departments and locations collaborated to restore operations. Retired employees were called back to assist with manual operations, ensuring a smooth transition and facilitating knowledge transfer during this challenging period.

Employee and Environmental Safety

The company prioritized employee safety and mitigated risks posed by operational disruptions. Preventive measures were implemented to reduce accident likelihood, and environmental safety was emphasized to prevent harm due to interrupted industrial processes. This focus on safety ensured minimal physical and environmental risks from the attack.

Lessons Learned and Recommendations

The company's response strategy underscored the importance of a robust incident response plan, including collaboration with external experts. The crisis highlighted the value of transparency and resilience in corporate culture. To enhance prevention and management, the company recommended increased investment in advanced cybersecurity tools and expertise, robust data backups, and regular recovery process testing. Additional recommendations included multi-factor authentication, stringent access controls, and employee training to raise awareness about phishing attacks and cybersecurity best practices.

17.8   Target Power Grid Attack (Brazil, 2005)

Massive power outages in Brazil in 2005 and 2007 that impacted millions were caused by cyber hackers attacking control systems, according to a report by a US television network. The report stated that the 2007 blackout in a Brazilian state, which affected over three million people, and a smaller incident in 2005, were perpetrated by hackers. The investigation included revelations as part of a broader examination of the threat of cyberattacks on critical infrastructure. A former official stated that a similar attack could potentially take place in the United States, emphasizing the nation's unpreparedness for such an event. If cyber hackers were able to infiltrate the US power grid, it was noted that the impact could be significant. Earlier this year, websites of major government entities were among those targeted in cyberattacks, amid suspicion of foreign involvement. It was also reported that nations had agreed to cooperate in addressing cyberattacks against their defense networks. An expert on strategic studies emphasized that US cybersecurity has been significantly targeted by foreign actors in recent years, including breaches of sensitive networks. Brazil has faced significant power outages over the years, with major events occurring in 2005, 2007, and 2009, highlighting vulnerabilities in the nation's electrical grid.

In January 2005 and 2007, power disruptions occurred in Rio de Janeiro and Espirito Santo State. These incidents were speculated to have been caused by cyberattacks, marking some of the earliest instances of potential interference with critical infrastructure. A major blackout in November 2009 left more than half of Brazil without electricity. This event was unprecedented in scale, exposing critical weaknesses in the energy grid. The Itaipu Dam, one of the largest hydroelectric plants in the world, was completely shut down. As a result, subway systems, traffic lights, and elevators across urban centers ceased operation, paralyzing essential services and disrupting millions of lives.

Initial Responses and Contradictions

In the wake of the 2009 blackout, conflicting explanations led to confusion and public concern.

The Brazilian government attributed the outage to adverse weather conditions and dismissed the possibility of a cyberattack. However, the explanation lacked compelling evidence to allay skepticism. The Brazilian National Space Research Institute challenged the government's claims, pointing out that the reported weather conditions were unlikely to cause simultaneous and widespread grid failures. This contradiction increased scrutiny of Brazil's infrastructure preparedness.

Cybersecurity Concerns in Brazil

The possibility of cyberattacks being involved in Brazil's blackouts has brought attention to the security of the country's critical infrastructure.

Supervisory Control and Data Acquisition (SCADA) systems, crucial for managing energy grids, are vulnerable to cyberattacks. These systems, often connected to the Internet, lack robust security measures. Compromised SCADA networks allow attackers to disrupt power supply or damage critical equipment. Cybersecurity experts have repeatedly raised alarms about Brazil's vulnerabilities. John Grines, the US Assistant Secretary of Defense, and Richard Clarke, a former White House cybersecurity adviser, emphasized Brazil's susceptibility to cyberattacks. Clarke noted that hackers had successfully disrupted Brazil's grid in previous incidents, warning that similar tactics could be employed elsewhere.

Implications for Critical Infrastructure

The vulnerabilities exposed by these blackouts extend beyond the energy sector.

Modern critical systems, such as healthcare, emergency response, telecommunications, and banking, rely on an uninterrupted power supply. A single grid failure can cascade across sectors, amplifying the impact. Blackouts disrupt millions of lives, cause economic losses, and erode public trust in infrastructure reliability. These vulnerabilities also pose national security risks, particularly during global events like the 2014 FIFA World Cup and 2016 Summer Olympics.

Brazil's Hacker Landscape

Brazil has become a hub for cybercriminal activities due to a combination of systemic and socioeconomic factors.

Brazil accounts for 30% of the world's top 50 website defacement groups, according to Safemode.org. High Internet connectivity and insufficient investment in cybersecurity foster an environment conducive to hacking. Portuguese has become the dominant language in cybercriminal networks. Brazil's laws against hacking are inadequate, requiring proof of financial fraud to prosecute offenders. This loophole emboldens cybercriminals and limits law enforcement's ability to act. Several high-profile cyberattacks illustrate the severity of Brazil's cybersecurity challenges. Hackers infiltrated a Brazilian government website, locking out over 3,000 employees for more than a day. A ransom demand of $350 million was refused. Despite data backups, it took over a week to restore functionality, highlighting operational risks. Brazilian hacking groups, such as Prime Suspectz and BHS, have targeted high-profile systems like US military computers and Microsoft servers, showcasing their global reach.

Recommendations for Cybersecurity Improvements
To strengthen the security of its critical infrastructure, Brazil must implement a comprehensive and multifaceted strategy. One of the first steps is the enactment of robust anti-hacking laws, which would not only close existing legal gaps but also provide law enforcement agencies with the tools and authority they need to investigate and prosecute cybercriminals. A well-defined legal framework would act as a deterrent to potential attackers, ensuring that the consequences of cybercrimes are clear and enforceable. This would contribute to a safer digital landscape and empower authorities to respond swiftly and effectively to cyber threats. Furthermore, Brazil should prioritize the development of public-private partnerships, where industries collaborate with government agencies to share intelligence, resources, and best practices for cybersecurity. Increased funding for cybersecurity research and development would also enable Brazil to stay ahead of emerging threats, ensuring its critical infrastructure remains protected against sophisticated cyberattacks.

Enhancing Grid Security
The security of Brazil's electrical grid is crucial for the stability of its economy and the safety of its citizens. To improve grid security, it is essential that security measures are embedded into the design and ongoing operations of the infrastructure. Regular security assessments, penetration testing, and system audits should be conducted to identify potential vulnerabilities and weaknesses in the grid. These assessments should not be limited to the technical aspects but should also evaluate human factors, such as the training and awareness of operators and other personnel. In addition to testing and assessments, continuous upgrades are necessary to keep pace with evolving cybersecurity threats. Significant investments must be directed toward modernizing Brazil's electrical grid, focusing on key components such as transmission systems, load management capabilities, and backup and redundancy planning. Upgrading these systems would not only increase the reliability and resilience of the grid but also make it more difficult for cybercriminals to disrupt critical services. Ensuring that the grid has sufficient redundancies and fail-safes in place would help mitigate the impact of any potential cyberattack, enhancing Brazil's ability to recover from such incidents quickly and efficiently. These efforts should be supported by the government and the private sector to guarantee the long-term security and sustainability of Brazil's energy infrastructure.

17.9   Havex Trojan Attack (Industrial Control Systems, 2014)

The Havex Trojan Attack, also known as the Dragonfly or Energetic Bear campaign, was a sophisticated cyber-espionage operation that targeted industrial control systems in various sectors, including energy, manufacturing, and transportation. Discovered in 2014, the attack is believed to have been carried out by a highly organized group with possible nation-state affiliations. This attack raised alarms within the cybersecurity community due to its advanced tactics and the critical nature of the systems it targeted. The Havex Trojan was a type of remote access trojan designed specifically to infiltrate industrial environments and provide attackers with unauthorized access to sensitive systems. The attackers employed a combination of strategies to penetrate their targets, including phishing emails, watering-hole attacks, and trojanized software updates. These methods allowed the attackers to establish a foothold in key infrastructure, steal data, and maintain access over extended periods. This attack is relevant to Chaps. 16 and 5.

17.9.1   Attack Tactics

Havex employs sophisticated attack tactics targeting cryptocurrency exchanges and financial institutions. The malware is designed to intercept user credentials, manipulate cryptocurrency transactions, and compromise financial systems.

Phishing Campaign
The attackers began by sending carefully crafted phishing emails to employees within the targeted organizations. These emails often appeared to be legitimate, containing attachments such as PDFs or Word documents that, once opened, triggered the installation of the Havex Trojan. The emails were designed to look relevant and trustworthy, increasing the likelihood that recipients would fall victim to the attack.

Watering-Hole Attacks
In addition to phishing, the attackers also used watering-hole tactics, compromising legitimate websites related to industrial control systems. These included websites for industry forums and software manufacturers. Malicious code was injected into downloadable software on these sites. When employees from targeted organizations visited these compromised sites and downloaded the software, the Havex Trojan was installed on their systems.

Trojanized Software Updates
Another method used by the attackers involved compromising trusted software vendors. They inserted the Havex Trojan into legitimate software updates. When customers of the affected vendors downloaded and installed the updates, the Trojan

was silently introduced into their systems, giving the attackers a backdoor to sensitive operations.

17.9.2   Remote Access and Data Exfiltration

Once the Havex Trojan was successfully installed on a victim's system, it granted the attackers remote access to the compromised machine, allowing them to monitor and control the infected device. The Trojan was capable of collecting a range of information, including details about the operating system, machine name, logged-in users, and network configuration. This data provided valuable insights into the infrastructure of the targeted organization.

The gathered information was then exfiltrated to command-and-control servers controlled by the attackers, giving them a comprehensive view of the targeted network. This allowed them to map the internal network, identify critical systems, and develop a strategy for executing further malicious actions, such as data theft, system manipulation, or sabotage.

The ability to stealthily exfiltrate valuable information and remain undetected for long periods highlighted the importance of continuous monitoring and strong cybersecurity measures within industrial environments.

17.9.3   Impact of the Attack

The Havex Trojan attack had a profound impact on the targeted industries, exposing major weaknesses in their cybersecurity infrastructure. One of the most critical outcomes was the unauthorized extraction of sensitive data related to ICS configurations and operations. This stolen information posed a serious threat, as it could potentially be used in future cyberattacks or for industrial espionage, giving adversaries insight into system vulnerabilities and internal processes.

The incident also raised serious concerns about the stability and security of industrial operations. Although no immediate physical damage was reported, the fact that such malware was present in key systems suggested that attackers had the potential to disrupt operations, manipulate processes, or even shut down entire facilities. This possibility alarmed both security experts and industry leaders, reinforcing the need for stronger defenses.

Most importantly, the Havex attack acted as a wake-up call for organizations relying on SCADA and ICS technologies. It made clear that cybersecurity is a critical component of operational safety—not just an IT issue. As a result, many companies began to invest more in security measures such as threat detection, network isolation, regular system audits, and staff training to better protect their infrastructure from future threats.

17.9.4   Response and Mitigation Measures

In response to the Havex Trojan attack, organizations and the cybersecurity community implemented several mitigation strategies:

Some Enhanced Security Protocols are Provided Below
Organizations strengthened their cybersecurity defenses by adopting more robust protocols such as regular software updates, network segmentation, and intrusion detection systems. These measures helped prevent malware spread and reduced the risk of future intrusions.

A major focus was placed on improving employee awareness, particularly regarding phishing and other social engineering attacks. Training employees to recognize suspicious emails and avoid opening untrusted attachments reduced the likelihood of successful compromises.

The attack underscored the need for better cooperation between stakeholders, including government agencies, cybersecurity firms, and industry participants. Sharing threat intelligence helped ensure a coordinated response to evolving cyber threats.

Lessons Learned
The Havex Trojan attack provided valuable lessons for organizations and cybersecurity professionals:

- Frequent software patching is essential to prevent attackers from exploiting known vulnerabilities.
- Continuous monitoring of industrial environments helps detect and respond to threats in real time.
- A multi-layered security approach, combining firewalls, intrusion detection, endpoint protection, and network segmentation, is key to mitigating cyberattack impact.

Collaboration and Information Sharing
The Havex attack demonstrated the importance of cross-sector collaboration. Sharing threat intelligence and vulnerability information between organizations and government entities is crucial for developing a comprehensive defense against advanced threats.

The Havex Trojan attack in 2014 served as a wake-up call to the vulnerabilities present in critical infrastructure. This incident emphasized the need for robust cybersecurity measures to protect industrial environments from cyber-espionage and other threats. By learning from this attack and implementing the lessons learned, organizations can better secure their systems and reduce the risk of similar incidents in the future.

17.10   Triton Malware Attack (Petrochemical Facility, 2017)

Triton is a sophisticated malware first discovered in 2017 at a petrochemical plant in Saudi Arabia. It is designed to compromise safety instrumented systems (SIS), critical components that ensure safe operations in industrial facilities. By disabling these systems, Triton can create conditions that lead to catastrophic plant failures or disasters. In December 2017, it was reported that an unidentified power station, believed to be in Saudi Arabia, had its safety systems compromised. The attack targeted Triconex industrial safety technology manufactured by Schneider Electric. The malware exploited a vulnerability in computers running the Microsoft Windows operating system. This incident is widely believed to have been a state-sponsored attack, raising concerns about the use of cyber tools to disrupt critical infrastructure. In 2018, cybersecurity firm FireEye linked Triton to Russia, specifically to the Central Scientific Research Institute of Chemistry and Mechanics (CNIIHM), a research entity believed to be involved in the development of the malware. This attribution underscored the growing role of nation-state actors in cyber warfare, particularly in targeting industrial control systems. Further reports indicated that Triton's attacks were not limited to Saudi Arabia. According to Wired, its activity has been observed in regions including North America and China, highlighting its global reach and the persistent threat it poses to industrial facilities worldwide. Triton's discovery has since prompted increased focus on securing industrial systems against sophisticated cyber threats. This chapter aligns closely with Chaps. 1, 8, 11, 14, 15, and 16, providing insights into cybersecurity threats targeting industrial control systems.

Trisis was first identified following an attack on a petrochemical facility in Saudi Arabia in 2017. The malware specifically targeted the Triconex SIS developed by Schneider Electric. These systems are essential for maintaining safety in industrial processes by detecting hazardous conditions and shutting down operations to prevent catastrophic failures.

Key Characteristics of Trisis Malware
The Trisis malware, also known as Triton, showcases a deliberate and targeted approach to attacking industrial control systems, particularly SIS. These systems play a vital role in ensuring industrial safety by automatically shutting down operations in the event of hazardous conditions. Trisis is specifically designed to reprogram or disable these safety systems, undermining their ability to perform crucial safety functions. This deliberate targeting not only poses a significant risk to industrial operations but also exposes vulnerabilities in the safety infrastructure of critical industries, such as petrochemical plants and power stations. The advanced technical capabilities of Trisis further highlight the sophistication of this cyber threat. The malware's complexity and design suggest that it was developed or deployed by state-sponsored actors or advanced hacking groups. These highly skilled threat actors possess the resources and expertise required to create such intricate and persistent malware, which makes Trisis particularly difficult to detect and mitigate. This level of sophistication is a clear indication of the growing

threat landscape within industrial cybersecurity, where advanced attacks can bypass traditional security measures. In addition to its technical sophistication, the malware's Python-based architecture provides it with a high degree of flexibility and adaptability. Python, known for its versatility, allows Trisis to easily integrate and interact with industrial control systems. This adaptability enhances the malware's potential to cause widespread damage, as it can modify system operations in subtle and complex ways, making it harder for security teams to identify and neutralize the threat. The combination of deliberate targeting, advanced technical capabilities, and a flexible architecture significantly elevates the threat posed by Trisis, emphasizing the need for robust and adaptive cybersecurity measures to protect critical infrastructure.

Summary of Triton Framework Attack

The Triton framework represents a highly advanced and sophisticated cyberattack that specifically targets SIS, with particular focus on the Triconex safety controller. These controllers are crucial for maintaining safety protocols in critical infrastructure sectors such as nuclear facilities, oil and gas refineries, and chemical plants. The attack, which demonstrated the significant capabilities of the attackers, began with the infiltration of the corporate network through spear phishing, a common technique used to exploit human vulnerabilities. Once inside the network, the attackers moved laterally, navigating through the corporate infrastructure to ultimately reach the industrial control systems network, where they aimed to manipulate and disable safety controls. The attack highlights the increasing risks to critical infrastructure, particularly the potential for cyberattacks to cause significant physical and operational disruptions. The early detection of the attack played a critical role in preventing a major industrial disaster, but the sophistication of the methods used underscores the growing challenges in securing vital systems against highly skilled threat actors.

Attack Methodology

The methodology behind the Triton framework attack reveals a highly organized and technical approach, leveraging advanced techniques to breach the targeted systems. The attackers utilized the proprietary TriStation communication protocol to establish communication with the Triconex controllers over port UDP/1502, which is specifically used for controller communication in industrial settings. Successfully exploiting this protocol required reverse engineering, a task that demands deep technical expertise and an understanding of the system's underlying architecture. By exploiting the knowledge of this communication protocol, the attackers were able to issue commands that manipulated the Triconex controllers, ultimately threatening to disable critical safety mechanisms that ensure the safe operation of industrial processes. The attackers' ability to reverse engineer the communication protocol demonstrates the advanced preparation and skill level of the threat actors, highlighting the need for robust cybersecurity measures and increased awareness of the vulnerabilities that can be exploited in industrial control systems.

Triton Framework Components

The Triton framework, which targeted the Triconex safety controllers, was composed of several sophisticated components designed to carry out the attack with stealth and precision. The malware used in the attack was disguised as a legitimate executable file named "trilog.exe," which was purported to collect logs for the system. However, upon closer inspection, it was discovered that "trilog.exe" was actually a Python script that had been compiled into an executable file. This deceptive technique allowed the attackers to remain undetected while infiltrating the system. The framework also included a set of additional Python scripts that were stored in a compressed file named library.zip, further obfuscating the attack's true nature. To gain deeper access to the targeted controllers, PowerPC shellcodes were used as part of the payload. The first shellcode injected a second-stage backdoor into the system, enabling the attackers to establish full control over the Triconex safety controllers. Despite the success of the attack in terms of infiltration, the full impact of the payload was not realized. The attack was detected early, which led to the interruption of the final stage of the assault, preventing any significant disruption to industrial operations. While the forensic investigation was able to trace the steps of the attackers, the ultimate consequences of a fully executed attack remain unclear, but it is likely that the attack could have had severe consequences for industrial safety and operations had it progressed.

Implications for Industrial Cybersecurity

The Trisis incident serves as a stark reminder of the growing vulnerabilities within critical infrastructure, particularly in industrial control systems that were once deemed too specialized or isolated to be targets for widespread cyberattacks. Historically, industrial systems, such as those found in power plants, petrochemical facilities, and manufacturing plants, were considered secure due to their isolated nature and lack of direct connection to public networks. However, the Trisis attack shattered this assumption by demonstrating that even highly specialized systems are susceptible to sophisticated, targeted cyberattacks. This shift in the threat landscape underscores the increasing complexity of cyber threats facing critical infrastructure today. The attack highlighted the vulnerability of Safety Instrumented Systems (SIS), which are responsible for ensuring the safe operation of industrial processes by shutting down operations when dangerous conditions are detected. By successfully compromising these systems, Trisis proved that even safety-critical systems can be exploited to cause physical damage, disrupt operations, and endanger lives. This realization emphasizes the urgent need for industrial cybersecurity practices to evolve in response to the growing sophistication of cyberattacks targeting the heart of industrial operations. The Trisis incident also points to the increasing convergence of information technology (IT) and operational technology (OT) networks. As industrial systems become more interconnected with corporate networks and the broader Internet, they present new attack surfaces for threat actors. While traditional IT security measures have been deployed to protect corporate data and systems, OT networks, which control physical processes, often operate with different protocols and face distinct security challenges. The Trisis attack revealed the need for a holistic cybersecurity approach that addresses both IT and OT concerns in tandem, rather than treating them as separate entities. Furthermore, the incident highlights the need for a proactive, multi-layered security strategy tailored specifically for industrial environments. Such a strategy should include the implementation of advanced intrusion detection systems, continuous monitoring, network segmentation, and robust access control mechanisms. Additionally, organizations must prioritize cybersecurity training for personnel, enabling them to recognize potential threats and respond effectively. Given the critical role that industrial systems play in the functioning of modern society, a failure to adequately secure them against evolving cyber threats could have devastating consequences, not only for the organizations involved but also for the safety and stability of entire industries and communities. The Trisis attack marks a pivotal moment in the evolution of cybersecurity for industrial control systems, demonstrating that no system, no matter how isolated, is immune to cyber threats. The implications of this attack are far-reaching, underscoring the need for comprehensive and adaptive cybersecurity measures to protect critical infrastructure from increasingly sophisticated threats.

Mitigation Strategies

To effectively defend against advanced cyber threats such as Trisis, organizations must adopt a comprehensive and multi-layered security strategy that addresses the complexities and unique challenges posed by industrial control systems. This approach should encompass a variety of interconnected tactics designed to minimize vulnerabilities and enhance the overall resilience of critical infrastructure.

One of the foundational strategies is **Network Segmentation**, which involves isolating industrial control networks from corporate IT networks. This isolation prevents the lateral movement of malware between different network environments, reducing the risk of a cyberattack spreading from less secure areas of the organization to the highly sensitive and operationally critical industrial control systems. By creating strong network boundaries, organizations can contain any potential breaches and limit their impact on essential systems.

Access Restrictions are also crucial in minimizing unauthorized access to industrial systems. Enforcing strict access control policies ensures that only authorized personnel with the necessary clearance can interact with critical infrastructure. This can be achieved through the implementation of secure authentication mechanisms such as multi-factor authentication (MFA) and role-based access control (RBAC), which significantly reduce the risk of credential theft and insider threats.

Routine Updates and Patching are vital components of any cybersecurity strategy. Industrial systems often rely on legacy software, which may have known vulnerabilities that cybercriminals can exploit. Regularly updating and patching systems ensures that these vulnerabilities are addressed, reducing the window of opportunity for attackers. This process should be part of a well-established maintenance schedule to ensure that all components of the system remain secure against emerging threats.

Deploying **Intrusion Detection Systems** further enhances an organization's defense by actively monitoring network traffic for unusual or suspicious activities.

IDS solutions can detect potential intrusions or signs of malware and alert security teams in real time, enabling prompt action to contain and mitigate the threat before it can cause significant damage.

Employee Training is another key element of an effective mitigation strategy. Ongoing cybersecurity awareness programs ensure that staff members are equipped with the knowledge to recognize phishing attempts, malware, and other social engineering tactics commonly used by cybercriminals. By fostering a culture of vigilance, organizations can strengthen their first line of defense and empower employees to act as active participants in the organization's cybersecurity efforts.

Incident Response Preparedness is essential to quickly respond to and mitigate the impact of any cyberattack. Organizations should develop and regularly update incident response plans tailored specifically for industrial environments. These plans should outline clear procedures for identifying, containing, and recovering from an attack, ensuring that responses are efficient and coordinated to minimize downtime and operational disruption.

Lastly, **Collaboration and Information Sharing** among industry stakeholders, government agencies, and cybersecurity experts play a critical role in strengthening collective defenses. By sharing intelligence about emerging threats and best practices, organizations can stay informed and better prepared to counteract evolving attack techniques. This collaborative approach fosters a broader cybersecurity community, creating a more resilient defense against global cyber threats.

Lessons Learned

The Triton attack highlights several crucial aspects for enhancing industrial cybersecurity. First, the use of spear phishing as an entry point emphasizes the need for thorough security training and awareness to prevent unauthorized access. The attack also underscores the importance of monitoring tools, such as Nozomi's Wireshark dissector, which can help detect advanced threats targeting industrial control systems. Additionally, the sophistication of the attack reinforces the necessity of implementing multi-layered security strategies to identify and prevent threats at multiple stages. The emergence of Trisis malware has transformed the understanding of cyber threats to industrial systems. It emphasizes the importance of prioritizing cybersecurity as a fundamental component of operational safety. By adopting rigorous measures and fostering continuous improvement in cybersecurity practices, organizations can protect critical infrastructure and ensure the reliability and safety of their operations in the face of evolving threats.

17.11   NotPetya Cyberattack (2017)

The NotPetya cyberattack, which began on June 27, 2017, is considered one of the most impactful and widespread cyber incidents in modern history. Initially targeting organizations in Ukraine, the attack quickly spread across the globe, affecting multiple industries and causing significant financial and operational damage. While it initially appeared to be a ransomware attack, demanding payment for the decryption of files, further investigation revealed that NotPetya was not designed to generate profits through ransom payments. Instead, it was a sophisticated wiper malware aimed at destroying data, disrupting operations, and inflicting long-term damage on affected organizations. This attack demonstrated the increasing use of cyberattacks as tools for geopolitical conflict, marking a significant escalation in the realm of cyber warfare. NotPetya is highly relevant to Chaps. 1, 10, 11, 12, and 16, as it covers key topics such as cybersecurity fundamentals, advanced threat methodologies, incident response strategies, attack analysis, and security policies.

17.11.1   Attack Methodology

Although NotPetya appeared to be a ransomware attack, it was primarily designed as wiper malware, focusing on destruction rather than financial extortion. It spread quickly through networks by exploiting security vulnerabilities and system weaknesses. The primary methods used in the attack included:

Initial Infection
The NotPetya attack began with a targeted supply chain attack, utilizing a widely used Ukrainian accounting software called M.E.Doc as the initial infection vector. The attackers compromised the software's update mechanism, ensuring the malware was delivered to all users who updated their version of M.E.Doc. By exploiting the trust placed in this commonly used software, the attackers were able to spread the malware throughout the targeted organizations, infecting a wide range of systems. This method, known as a supply chain attack, is particularly dangerous as it leverages legitimate and trusted processes, making detection and prevention more challenging.

Exploitation
After the malware was deployed, it used the EternalBlue exploit to spread across infected systems. EternalBlue was a critical vulnerability in Microsoft Windows, originally discovered by the US National Security Agency (NSA) and later leaked by the Shadow Brokers hacker group. This exploit allowed the malware to spread quickly within organizations, moving from system to system and affecting large portions of the network. Additionally, NotPetya used Mimikatz, a tool for harvesting user credentials from infected machines. With these stolen credentials, the malware was able to spread even further, compromising additional systems and amplifying the scale of the attack.

Payload Execution
The core objective of NotPetya was to disable and destroy systems. The malware overwrote the Master Boot Record (MBR) of infected systems, rendering them inoperable. Upon rebooting, the systems would display a ransom note demanding

payment in Bitcoin for file decryption. However, the malware was designed so that no recovery was possible, even if the ransom was paid. This made the ransom demand a mere facade, as the true goal of the attack was to inflict maximum damage by permanently erasing critical data. This wiper nature of the attack distinguished NotPetya from traditional ransomware, highlighting its role as a tool for destruction rather than financial gain.

17.11.2 Impact

NotPetya caused widespread disruption to businesses and organizations around the world. Its effects included crippling operations across various industries, inflicting billions in financial losses, and rendering critical systems inoperable. The attack's reach was amplified through supply chain vulnerabilities, spreading quickly and causing extensive data destruction.

Global Disruption

NotPetya had a profound and widespread impact on global operations, with several major corporations and organizations falling victim. Among the most notable victims were A.P. Møller-Mærsk, a global shipping and logistics giant, pharmaceutical company Merck, FedEx, and the Russian state-owned oil company Rosneft. The attack caused widespread disruption, halting production, crippling communication networks, and shutting down critical infrastructure. The total financial damage from the attack was estimated to exceed $10 billion worldwide, making it one of the most costly cyberattacks in history.

Operational Shutdowns

One of the most significant impacts was experienced by A.P. Møller-Mærsk, whose operations were severely disrupted. The company was forced to rebuild its entire IT infrastructure, leading to a complete halt in shipping activities and resulting in substantial financial losses. The estimated cost of this disruption was approximately $300 million. Similarly, Merck, a major player in the pharmaceutical industry, encountered serious operational setbacks. Its production lines were brought to a standstill, causing delays in drug manufacturing and losses exceeding $870 million. These incidents underscored the vulnerabilities within global supply chains, as the malware caused major disruptions across essential industries.

Public Sector Impact

The impact of NotPetya was not confined to private sector companies. In Ukraine, the attack caused widespread damage to critical public services, including government operations, banking systems, energy companies, and public transportation. The disruption of these services underscored the vulnerabilities in critical infrastructure and raised concerns about the security of national systems. The Ukrainian government, in particular, faced significant challenges in restoring operations and protecting against further attacks.

17.11.3   Response and Recovery

The global response to NotPetya involved immediate actions, such as isolating infected systems and deploying patches, as well as long-term measures like strengthening cybersecurity frameworks and improving response plans. Recovery was complex, requiring system rebuilds, data restoration, and infrastructure replacement, along with implementing stricter security measures and addressing legal and reputational impacts.

Immediate Response
In the immediate aftermath of the attack, organizations worldwide had to act quickly to contain the spread of the malware and prevent further damage. Affected organizations began by isolating infected systems from the network to prevent lateral movement. Incident response teams were deployed to assess the scope of the attack, identify its origins, and initiate recovery procedures. For instance, A.P. Møller-Mærsk used an unaffected domain controller in Ghana to help rebuild their entire network, demonstrating the importance of having isolated, unaffected systems for recovery.

Long-Term Recovery
For many organizations, the recovery process was long and difficult. Rebuilding IT infrastructures, restoring lost data from backups (if available), and addressing the vulnerabilities that allowed the attack to occur took months of effort. The attack emphasized the importance of having comprehensive disaster recovery and business continuity plans. Organizations learned that relying on regular data backups and maintaining isolated networks for critical infrastructure can help mitigate the effects of a cyberattack.

17.11.4   Lessons Learned

One of the most crucial lessons from the NotPetya attack was the need for effective network segmentation. The malware spread rapidly within organizations due to the lack of proper segmentation between different network segments. If organizations had segmented their networks effectively, the malware would have been confined to a smaller portion of the network, reducing its impact. Network segmentation is a key strategy for limiting the lateral movement of malware and reducing the overall damage caused by cyberattacks.

Patching and Vulnerability Management
The NotPetya attack exploited the EternalBlue vulnerability, which had been publicly disclosed and patched by Microsoft months before the attack. However, many organizations had not applied the patch, leaving their systems vulnerable to exploitation. This highlights the critical importance of regular patching and vulnerability management. Organizations must ensure they are aware of security patches and apply them promptly to protect their systems from known threats.

Incident Response Planning
The ability to respond quickly to cyber incidents is vital for minimizing damage. NotPetya underscored the importance of having a well-defined incident response plan. Organizations need to be prepared for the possibility of a cyberattack, with clear protocols for isolating affected systems, restoring data, and communicating with stakeholders during a crisis. The more prepared an organization is, the quicker it can mitigate the impact of an attack.

Cyber Hygiene and Awareness
Employee awareness and training play a crucial role in preventing cyberattacks. NotPetya, like many other cyber incidents, relied on social engineering tactics to gain initial access. Regular cybersecurity training and awareness programs can help employees recognize phishing attempts and other common attack vectors. Building a culture of cybersecurity within an organization can significantly reduce the risk of successful attacks [88].

The NotPetya cyberattack serves as a stark reminder of the destructive potential of modern cyber warfare. It highlights the importance of proactive cybersecurity measures, including network segmentation, timely patching, incident response planning, and employee awareness. By learning from the lessons of NotPetya, organizations can better prepare for the evolving threat landscape and take steps to protect their critical assets, infrastructure, and operations from future cyber threats. As cyberattacks continue to grow in sophistication and scale, the need for comprehensive cybersecurity strategies has never been more urgent.

17.12 Significance of ICS Frameworks

The study being presented emphasizes the critical importance of strong ICS frameworks across various industrial sectors. An effective ICS not only ensures operational efficiency but also enhances safety, ensures regulatory compliance, and provides resilience against cyberattacks. This is true across a wide range of industries, from aerospace to healthcare and beyond. For instance, digital twin technologies help in predictive maintenance within the aerospace industry by modeling sensor data and addressing potential issues early on. In healthcare, frameworks such as PHASE improve security by quickly identifying and preventing manipulation attempts, thereby safeguarding patient information and maintaining system integrity.

Furthermore, the integration of ICS maximizes operational reliability and enhances supply chain transparency in industries like chemical manufacturing and agriculture. Through data analytics, IoT-enabled agricultural systems—often referred to as "Agriculture 4.0"—improve market competitiveness and traceability. Automation and control systems, on the other hand, ensure process reliability in the chemical and petrochemical industries while minimizing the risk of human error, thus boosting operational effectiveness and safety.

17.13 Direction for Future Advancements

There are several key areas that require attention in the advancement of ICS frameworks. To protect vital infrastructure against evolving threats, improving cybersecurity safeguards remains imperative. This includes implementing proactive defense strategies, integrating threat intelligence, and maintaining continuous monitoring. Moreover, the integration of AI-powered technologies and advanced analytics can significantly enhance predictive maintenance and decision-making processes across industries. Leveraging big data for real-time anomaly detection and predictive modeling is essential for this advancement.

In addition, standardization and interoperability of ICS components are crucial for ensuring smooth integration and scalability across various industrial contexts. Open standards and protocols must be developed to enable secure data sharing and communication between systems, facilitating seamless integration.

17.14 Conclusion

In conclusion, this study highlights the transformative impact of Industrial Control Systems across various sectors, from improving operational efficiency and safety to enabling data-driven decision-making and enhancing resilience against cyber threats. By emphasizing the significance of robust ICS frameworks and outlining pathways for future advancements, the research underscores the pivotal role of technology in driving sustainable growth, innovation, and safety in industrial settings globally. As industries continue to evolve, prioritizing investments in resilient ICS frameworks will be essential for maintaining a competitive advantage and addressing the ongoing regulatory and operational challenges.

Appendix A
Important Definitions

A.1 Comprehensive Overview of Industrial Control Systems ICS: Evolution, Components, and Security Challenges

1. **Industrial Control System (ICS)** A system of devices, networks, systems, and controls that automate activities in a variety of industries. Designed to manage industrial processes.
2. **Programmable Logic Controllers (PLCs)** PLCs are industrial electromechanical process automation devices, which are digital computers. They are made to withstand vibration and impact, have different input and output configurations, wide temperature ranges, immunity to electrical noise, and all of these things. PLCs are utilized to operate machinery and processes in a wide range of industries.
3. **Distributed Control Systems (DCS)** A distributed control system (DCS) has controller elements dispersed throughout the system, with one or more controllers controlling each component subsystem. Centralized data gathering and control are made possible by the system's networking for communication and monitoring.
4. **Remote Terminal Units (RTUs)** RTUs are tools for gathering data from field sensors and other devices and sending it to a central control system. Additionally, the central system can send them commands to operate field equipment. Applications for RTUs include environmental monitoring, power distribution, and pipeline monitoring.
5. **Programmable Automation Controllers (PACs)** Industrial controllers known as PACs combine the functionality of a PC with the dependability and durability of a PLC. They are employed in intricate control applications that call for the integration of data management, motion control, and logic.

© The Author(s), under exclusive license to Springer Nature Switzerland AG 2026
M. A. Rahman et al., *Securing Industrial Control Systems*,
https://doi.org/10.1007/978-3-032-03018-4

6. **Intelligent Electronic Devices (IEDs)** IEDs are microprocessor-based tools for power system monitoring, control, and protection. They can exchange data and control in real time with other systems and devices through communication. IEDs are crucial parts of power system applications such as substation automation.

A.2   SCADA Systems in Industrial Control: Cloud Connectivity, Security Protocols, and Architectural Design

1. **Supervisory Control and Data Acquisition (SCADA) Systems:** Hardware and software systems that offer supervisory-level control, facilitating local or remote process operation and real-time data monitoring, gathering, and analysis.
2. **Infrastructure-as-a-Service, or IaaS:** Cloud-based Infrastructure as a Service (IaaS) offers resources on demand. Without having to handle their own data center infrastructure, organizations can access virtualization, networking, storage, and processing power. On the other hand, they are in charge of the virtual machines, middleware, operating system, and any data or apps.
3. **Platform-as-a-Service, or PaaS:** PaaS uses the cloud to supply and manage software and hardware resources for application development. By removing infrastructural specifics, it frees developers to concentrate on developing code. The underlying platform, which includes databases, runtime environments, and development tools, is managed by the cloud service provider.
4. **Software-as-a-Service, or SaaS:** SaaS provides cloud-based software applications on a subscription basis. Users don't need to worry about infrastructure, maintenance, or installation because they can access these programs online. Email services, teamwork tools, and customer relationship management (CRM) software are a few examples.

A.3   Understanding Communication and Protocols in ICS: Securing Network Infrastructure and Data Exchange

1. **Data Communication:** The process of moving data between two entities or from one location to another is referred to as data transmission. It makes it possible for digital and electronic data to travel across networks without restriction on format, content, or location. Examples include data transmission between devices or data transmission over Wi-Fi networks.
2. **Protocol:** A protocol is a set of guidelines or customs that control how computers and other devices communicate with one another. It guarantees appropriate communication, data interchange, and adherence to predetermined protocols. Protocols may be connected to security, networking, or other communication-related topics.

3. **TCP/IP (Transmission Control Protocol/Internet Protocol):** A widely used set of protocols for wide area networks (WANs) is TCP/IP. It was created by the US Department of Defense and allows for Internet-based communication between disparate platforms (such as Windows and UNIX). TCP/IP supports services like HTTP, DNS, and IP3 and consists of layers including application, transport, Internet, and network.

4. **Modbus:** A popular client/server data communication protocol in industrial settings is called Modbus. It was initially intended for use with programmable logic controllers (PLCs), but it is now widely accepted as the industry standard for industrial electronic device communication. Ethernet, serial connectivity, and the Internet protocol suite are all supported by Modbus.

5. **DNP3 (Distributed Network Protocol):** A group of communication protocols known as DNP3 is utilized by utilities (such as water and electricity companies), in particular for process automation systems. It is essential to SCADA systems because it links master stations to intelligent electronic devices (IEDs) and remote terminal units (RTUs). DNP3 is reliable, effective, and compatible.

6. **Ethernet/IP:** Ethernet/IP integrates industrial automation technologies with ordinary Ethernet connectivity. Because it enables device communication over Ethernet networks, industrial automation and control systems make extensive use of it.

7. **OPC UA (Open Platform Communications Unified Architecture):** OPC UA is an industrial automation machine-to-machine communication protocol. It offers dependable and safe data transfer across devices, supporting a range of platforms and data models.

8. **IEC 61850:** A standard for communication in power systems and electrical substations is IEC 61850. For devices in the smart grid to communicate with one another, it specifies data models and protocols.

9. **Virtual Private Network (VPN):** A Virtual Private Network (VPN) employs encryption to secure data transmitted over public networks.

A.4   Exploring Industrial Automation Systems: Security Strategies, Optimization of Control Mechanisms, and AI Integration

1. **Industrial Automation Systems:** Industrial automation is the process of carrying out operations with the aim of automating production using control systems, such as machines, actuators, sensors, processors, and networks. It includes many different kinds, including programmable automation (used in batch production), flexible automation (enabling customization), integrated automation (linking devices onto a single control system), and fixed automation for high-volume repetitive processes.

2. **Control Parameters:** Vital elements that work together to coordinate how machinery and procedures operate in industrial automation systems. They guarantee the automation system operates dependably and efficiently.
3. **Automated Integration:** Establishes connections between several automated systems inside a company to provide smooth, coordinated functioning. Guarantees effective coordination and communication amongst various components.
4. **Separate Automata:** Automates certain production processes, like assembly lines and quality control procedures. Frequently employed in sectors with discrete production phases.
5. **Autonomous Replication:** Utilized in industries like energy and chemicals that demand constant and continuous production. Ensures that production processes run smoothly and continuously.
6. **Sensors:** Sensors are devices that identify and quantify different physical attributes, including temperature, pressure, location, proximity, and others. They serve as the sensors of the automation system, collecting live data from the surroundings or the machines.
7. **Actuators:** Actuators respond to sensor data by triggering physical actions or modifications in the system. They convert control inputs into mechanical motion to adjust processes or conditions as needed by the automation system.
8. **Control System Elements:** Control system components are essential in coordinating the functioning of equipment and processes in industrial automation systems. These components provide the reliable and effective operation of the automation system.

A.5 Mitigating the ICS Attack Surface: Identifying Attack Vectors, Reducing Vulnerabilities, and Security Mapping Techniques

1. **Attack Surface:** An attack surface encompasses all potential vulnerabilities that an assailant could exploit to gain unauthorized access to a digital system or organization. It includes both weaknesses in digital and physical systems, as well as manipulative tactics used in social engineering. Think of it as the doors and windows of a house—increasing the number of entrance points also increases the likelihood of a break-in. Organizations proactively monitor their attack surface to detect and prevent possible threats, minimize it, and increase the difficulty for attackers to breach systems.
2. **Digital Attack Surface:** The digital attack surface is the total of all cyber threats' potential vulnerabilities and entry points into an organization's digital infrastructure. Software, networks, protocols, user access points, and more are included.
3. **Physical Attack Surface:** The physical attack surface encompasses all tangible assets and gadgets in an organization that enemies might attack. This includes

servers, PCs, mobile devices, and other hardware and physical security mechanisms.

4. **Attack vector:** An attack vector is a technique, or set of techniques, that hackers employ to compromise or compromise a victim's network. Adversaries gather a variety of attack methods that, in time, turn into virtual "calling cards." Threat intelligence analysts, cybersecurity vendors, and law enforcement can all detect various enemies with the use of these vectors. Organizations can fight against targeted assaults and get insight into adversaries' capabilities by identifying and monitoring attack vectors.

5. **Access Attack:** An access attack refers to any deliberate attempt to gain unauthorized access to a network, computer system, or digital device. Malicious actors use well-known vulnerabilities in authentication services, FTP services, and web services in order to get unauthorized access to web accounts, confidential databases, and other highly sensitive information. Essentially, it enables an individual to perceive information that they lack the legitimate authorization to access.

A.6   Network Segmentation in Industrial Operations: Enhancing ICS Security through Threat Mitigation and DNS Leak Prevention

1. **Network Segmentation:** Network segmentation is partitioning a computer network into smaller, separate segments or subnetworks. It is denoted as network isolation or network segregation. Each segment comprises a distinct set of devices or resources and is isolated from other segments by network devices such as routers, switches, or firewalls.

2. **Physical Segmentation:** Physical segmentation involves dividing networks into distinct physical portions or subnets. A firewall functions as a gateway and regulates the flow of incoming and outgoing network traffic, in conjunction with hardware components such as access points, routers, and switches.

3. **Logical Segmentation:** Logical segmentation involves dividing a network into smaller sections. A logical segmentation strategy utilizes VLANs or a network addressing system to leverage current network architecture ideas. VLANs facilitate automatic routing of traffic to the most suitable subnet, whereas a network addressing system is a more intricate and theoretical concept.

4. **DNS Leak:** A DNS leak is a security vulnerability that arises when requests are routed to an ISP's DNS servers even when a virtual private network (VPN) is being used by users. A VPN is intended to encrypt a user's Internet connection, maintaining their traffic in a private tunnel that conceals all browsing activity. A DNS leak happens when the user's DNS requests leave the encrypted tunnel and are visible to their ISP.

A.7   Comprehensive Overview of Field Devices in ICS: Protocol Management, Security Challenges, and Lifecycle Optimization

1. **Field Devices:** Field devices refer to a range of specialized hardware components and equipment used in industrial control systems (ICS) to coordinate, track, and manage physical operations in a variety of industries. These straightforward devices serve as a link between digital control systems and the real environment.
2. **Modbus Protocol:** The Modbus protocol is a request-response system that operates on a master-slave basis. The initiating device, known as the master, is in charge of starting every interaction in a master-slave relationship. Communication always takes place in pairs. One device must make a request and then wait for a response. A Human-Machine Interface (HMI) or Supervisory Control and Data Acquisition (SCADA) system is commonly used as the master, while sensors, PLCs, or PACs are commonly used as the slaves. The multiple layers of the protocol determine the content of these requests and responses as well as the network layers that these messages are transmitted across.
3. **Distributed Network Protocol:** A communications protocol used between components in process automation systems is called Distributed Network Protocol 3 (DNP3). It was created to facilitate communication between different kinds of control and data collection apparatus. It is essential to SCADA systems because SCADA Master Stations, Remote Terminal Units (RTUs), and Intelligent Electronic Devices (IEDs) use it. Communication between a master station and RTUs or IEDs is its main purpose.
4. **OPC UA (Unified Architecture Protocol):** OPC Unified Architecture is referred to as OPC UA. It is a platform-neutral, extendable standard that makes safe information sharing in industrial systems possible.
5. **PROFINET Protocol:** PROFINET is a communication protocol that lives at layer seven of the ISO/OSI model, the seven-layer model that generically describes the abstraction layers of a communication system. To ensure appropriate performance, PROFINET delivers data through the following communication channels: TCP/IP (or UDP/IP).

A.8   Exploring Supervisory Systems Security Threats: Legacy SCADA Vulnerabilities, Communication Protocols, and System Security Strategies

1. **Supervisory Systems:** Supervisory systems enable the tracking and monitoring of data from physical installations or production processes. Data collection devices are used to gather this kind of information, manage, process, analyze, store, and ultimately show it to the user.

A.9   Assessing Supervisory Systems Security Threats: Mitigating Sectoral Risks, Addressing Insider Threats, and Designing Human-Centric Security Solutions

1. **Cognitive Biases:** Cognitive biases are systematic patterns of decision-making divergence from rationality or objectivity caused by mental shortcuts.
2. **Advanced Persistent Threats (APTs):** Advanced Persistent Threats (APTs) are long-lasting, undiscovered cyberattacks with the intention of stealing confidential information, conducting cyberespionage, or compromising vital systems. An APT attack group aims to spread its presence throughout a target network while staying undetected, in contrast to other cyber threats like ransomware.

A.10   Controller Security Threats: Mitigating Advanced Persistent Threats (APTs), Enhancing Authentication, and Securing Control Architectures

1. **Data Manipulation and Tampering:** Data manipulation and tampering are the unlawful change or modification of data, whether deliberately or accidentally, in order to deceive, mislead, or gain an unfair advantage. This may happen in a variety of settings, including cybersecurity, data analysis, financial transactions, and more.
2. **Domain Controller:** A domain controller is a crucial server in a Windows domain that handles the management of authentication and authorization procedures. Domain controllers are crucial elements in the Microsoft Windows Active Directory framework, playing a vital role in the organization and management of people, computers, and network resources.
3. **Multi-factor Authentication (MFA):** MFA stands for "multi-factor authentication." This is a security method that asks users to verify their identity with more than one factor before they can access a system or app. The goal is to create a defense with many layers that makes it harder for attackers to get in and boosts security.

A.11   Building ICS Cyber Resilience: AI and Machine Learning Strategies, System Hygiene Practices, and Secure Smart Grid Frameworks

1. **ICS Cybersecurity Resilience:** ICS Cybersecurity Resilience is defined as the ability of industrial control systems (ICS) and operational technology (OT) environments to withstand, adapt to, and recover from cyber disasters.

2. **Active Directory:** Active Directory (AD) is Microsoft's directory and identity management solution for Windows domains. It was released in Windows 2000 and is bundled with the majority of Microsoft Windows Server operating systems. AD connects users to network resources and stores important information about your environment, such as user and computer information, access rights, and login credentials.

A.12   Strengthening ICS Attack Resiliency: Advanced Incident Response, Threat Intelligence, and Cyber-Physical Systems Monitoring

1. **Cyber Threat Intelligence:** Cyber threat intelligence, also known as threat intelligence, is a comprehensive awareness of cyber risks that may affect computer systems, networks, data, and digital infrastructure. To prevent and defend against cyberattacks, this intelligence gathers, analyzes, and interprets cybersecurity data.
2. **Cyber Threat Hunting:** The proactive cybersecurity approach known as "cyber threat hunting" entails cybersecurity specialists, sometimes referred to as "threat hunters," consistently and continuously scanning an organization's network or systems for security flaws and potential threats. This approach transcends conventional threat detection methods, which frequently rely on recognized patterns and indicators of compromise.
3. **Threat Intelligence Consumption:** Threat intelligence consumption is a key component of proactive cybersecurity methods and plays a vital role in the active cyber defense cycle. It involves the methodical collection, thorough analysis, and insightful interpretation of data related to current or potential cyber threats that endanger an organization's information systems and assets.
4. **Denial of Service (DoS):** A Denial of Service (DoS) attack is a malicious attempt to overwhelm a computer or network resource, rendering it unavailable to its intended users.
5. **Decoy Systems in Cybersecurity:** In cybersecurity, decoy systems are deliberately placed elements or entities within a network that imitate genuine resources, programs, or vulnerabilities.

A.13   ICS Security Requirements: Cybersecurity Frameworks, Incident Response Strategies, and IoT Device Compliance

1. **ICS Security Assessment:** ICS security assessment is a systematic and complete study of an organization's industrial control system security safeguards and vulnerabilities.
2. **Boundary Defense:** Boundary defense is the 12th control in the CIS Critical Controls and falls under the network category. This control consists of 11 subsections that include various aspects such as DMZ, firewalls, proxies, IDS/IPS, NetFlow, and remote access.
3. **Multi-layered Boundary Defense:** Multi-layered Boundary Defense in industrial control systems (ICS) is a comprehensive strategy that aims to safeguard critical infrastructure by implementing numerous levels of security controls at the network border. Industrial control systems (ICS) are often linked to the Internet or other external networks. Implementing a multi-tiered security system effectively mitigates the risk of unauthorized access, malware infections, and cyberattacks.
4. **Network Firewall:** A network firewall is a software or hardware security system that keeps an eye on and regulates all incoming and outgoing network traffic by pre-established security standards. It serves as a line of defense between untrusted external networks, like the Internet, and trusted internal networks.

A.14   Static Defense Strategies for ICS: Prioritizing Patch Management, Defense-in-Depth Approaches, and Regulatory Compliance

1. **Network Segmentation:** The process of breaking up a computer network into smaller parts is known as network segmentation.
2. **Physical Segmentation:** Networks are physically divided into several subnets or sectors through the process of physical segmentation. To do this, separate physical network equipment, such as switches or routers, can be used to form different segments.
3. **Logical Segmentation:** Dividing the network into smaller, easier-to-manage segments is most commonly accomplished by logical segmentation. Without the requirement for distinct physical devices, this method is achieved through logical setups, frequently utilizing virtual LANs (VLANs) or subnetting.

A.15 Intrusion Detection Systems (IDS) in ICS: Supervisory Frameworks, Signature vs Anomaly-Based Detection, and Architectural Design

1. **Intrusion Detection System:** An intrusion detection system (IDS) is a monitoring system that looks for unusual activity and, upon detection, sends out notifications. Based on these notifications, a security operations center (SOC) analyst or incident responder can look into the problem and take the necessary steps to eliminate the threat.

2. **Network Intrusion Detection System (NIDS):** Network intrusion detection systems provide uninterrupted network surveillance across both on-premise and cloud infrastructure to identify potential harmful actions such as policy breaches, lateral movement, or data extraction. NIDS security systems have a mostly passive character as opposed to an active one.

3. **Host-based Intrusion Detection System (HIDS):** A Host-Based Intrusion Detection System (HIDS) is a cybersecurity solution that actively monitors IT systems to identify any suspicious activity.

4. **Perimeter Intrusion Detection System:** A Perimeter Incursion Detection System (PIDS) refers to a sensor that is put on a fence and is designed to monitor and detect any kind of incursion occurring inside the perimeter.

5. **Virtual Machine-based Intrusion Detection System (VM-based IDS):** A virtual machine-based intrusion detection system (VMIDS) is a remote deployment method that utilizes a virtual machine (VM) to detect intrusions. It is comparable to one or a mixture of the three aforementioned IDSs.

6. **Physical Intrusion:** Physical intrusion is a phase of an attack where an attacker uses different techniques to enter a company's facilities. Alarms, control panels, and sensors are the usual components of physical IDS. The sensors, which may be installed on walls, doors, or windows, can pick up on environmental changes like sound and motion.

7. **Misuse Detection:** Abnormal system activity is defined as anything that is not considered normal, and this is how misuse detection, a technique for identifying computer threats, finds them. While misuse detection focuses on known attack patterns or illegal actions, anomaly detection finds deviations from a recognized typical behavior.

8. **Hybrid IDS:** Network-based intrusion detection system (NIDS) and host-based intrusion detection system (HIDS) elements are combined to create a hybrid IDS.

9. **Signature Detection for IDS:** Signature detection is used to discriminate between anomaly or attack patterns (signatures) and known intrusion detection signatures. It is a technique often used in Intrusion Detection Systems (IDS) and many anti-malware systems, such as antivirus, anti-spyware, etc.

10. **Statistical Anomaly-Based IDS:** Statistical anomaly detection involves the collection of data relating to the behavior of legitimate users over some time.

A.16   Attack Recovery Mechanisms for ICSs: Common Cyberattacks, Cryptographic Key Management, and Host-Based Mitigation Strategies

1. **Data Security Alliance:** The Data Security Alliance is a significant force in data protection. This collaboration combines experience, resources, and a shared commitment to defend digital landscapes from emerging dangers.
2. **Host-Based Man-in-the-Middle (MITM) Attack:** The methods used by attackers to control targeted processes and avoid detection at the host level are described in a Host-Based Man-in-the-Middle (MITM) assault.
3. **CRASHOVERRIDE Attack:** The first industrial virus, CRASHOVERRIDE, disrupted electric distribution. On the victim's ICS, novices may run its program. Credential theft, proprietary software deployment, and discovery were issues. Defense needed passive data aggregation.

References

1. Agarwal, T. Scada system: Architecture, components, types & its applications, October 31 2013. ElProCus - Electronic Projects for Engineering Students.
2. Ahmed, R. K. (2016). Overview of security metrics. *Software Engineering, 4*, 59–64.
3. Ajmal, A. B., Alam, M., Khaliq, A. A., Khan, S., Qadir, Z., & Mahmud, M. A. (2021). Last line of defense: Reliability through inducing cyber threat hunting with deception in scada networks. *IEEE Access, 9*, 126789–126800.
4. Alsalamah, M., Alwabli, H., Alqwifli, H., Ibrahim, D. M., et al. (2021). A review study on sql injection attacks, prevention, and detection.
5. An introduction to threat intelligence. Retrieved from https://www.ncsc.gov.uk/files/An-introduction-to-threat-intelligence.pdf.
6. Anomali. What is a threat intelligence platform (tip)? n.d. Retrieved from https://www.anomali.com/resources/what-is-a-tip.
7. Aqua Security. Fileless attacks in application security, Year of publication, e.g., 2022. Retrieved from https://www.aquasec.com/cloud-native-academy/application-security/fileless-attacks/.
8. Ara, A. (2022). Security in supervisory control and data acquisition (SCADA) based industrial control systems: Challenges and solutions.
9. Arraño-Vargas, F., & Konstantinou, G. (2022). Modular design and real-time simulators toward power system digital twins implementation. *IEEE Transactions on Industrial Informatics, 19*(1), 52–61.
10. Asghar, M. R., Hu, Q., & Zeadally, S. (2019). Cybersecurity in industrial control systems: Issues, technologies, and challenges. *Computer Networks, 165*, 106946.
11. Ashok, S., & Banerjee, R. (2000). Load-management applications for the industrial sector. *Applied Energy, 66*(2), 105–111.
12. ASMI Corporate Reporting. (2018). *Risk management approach.*
13. Automation Community. (2023). What is opc? - open platform communication architecture & benefits. Retrieved from https://automationcommunity.com/what-is-opc/.
14. AXA XL. (2020). The cyber incident response lifecycle. Retrieved from https://axaxl.com/fast-fast-forward/articles/the-cyber-incident-response-lifecycle.
15. Babayigit, B., & Abubaker, M. (2023). Industrial internet of things: A review of improvements over traditional scada systems for industrial automation. *IEEE Systems Journal, 18*(1), 120–133.
16. Balaban, D. 10 tips for creating a cyber resilience strategy, July 21 2023. Retrieved from https://cyberready.com/creating-a-cyber-resilience-strategy.

© The Author(s), under exclusive license to Springer Nature Switzerland AG 2026

M. A. Rahman et al., *Securing Industrial Control Systems*,

https://doi.org/10.1007/978-3-032-03018-4

17. Barbieri, G., Conti, M., Tippenhauer, N. O., & Turrin, F. (2021). Assessing the use of insecure ics protocols via ixp network traffic analysis.
18. Benjamin, A. (2022). What is system hardening? Standards and best practices. Retrieved from https://www.chef.io/blog/system-hardening-with-chef-for-a-secure-it-infrastructure.
19. Bitencourt, H. V. B., & Tavares, A. A. (2018). Desenvolvimento de um sistema scada aplicado a uma bancada didática com microgeração distribuída. *Revista Vincci-Periódico Científico do UniSATC, 3*(2), 98–133.
20. Blog, B. (2023). System intrusion: What, why, combat. Retrieved from https://blog.barracuda.com/2023/10/20/system-intrusion-what-why-combat.
21. Blumenthal, U., Marcovici, M., Mizikovsky, S., Patel, S., Sundaram, G. S., & Wong, M. (2002). Wireless network security architecture. *Bell Labs Technical Journal, 7*(2), 19–36.
22. Boero, L., Cello, M., Marchese, M., Mariconti, E., Naqash, T., & Zappatore, S. (2017). Statistical fingerprint-based intrusion detection system (SF-IDS). *International Journal of Communication Systems, 30*(10), e3225.
23. Bright Security. (2024). DNS tunneling: Techniques, tools, and prevention. https://brightsec.com/blog/dns-tunneling/#5_tech.
24. Butt, U. A., Amin, R., Aldabbas, H., Mohan, S., Alouffi, B., & Ahmadian, A. (2023). Cloud-based email phishing attack using machine and deep learning algorithm. *Complex & Intelligent Systems, 9*(3), 3043–3070.
25. Cao, Y., Zhang, L., Zhao, X., Jin, K., & Chen, Z. (2022). An intrusion detection method for industrial control system based on machine learning. *Information, 13*(7), 322.
26. Carden, F., Jedlicka, R. P., & Henry, R. (2002). *Telemetry systems engineering*. Artech House.
27. Check Point Software Technologies Ltd. Intrusion detection systems (IDS) vs firewalls, Year of publication, e.g., 2022. Retrieved from https://rb.gy/m06x62.
28. Chew, C. P. S., Logenthiran, T., & Woo, W. L. (2015). Smart centre of excellence: A tabletop demonstration kit.
29. Chipkin Automation Systems. AB-DF1 - overview, n.d. Retrieved from https://store.chipkin.com/articles/ab-df1-overview.
30. Chougule, D. ICS cybersecurity resilience and the importance of remote laboratory, October 10 2023. Retrieved from https://gca.isa.org/blog/ics-cybersecurity-resilience-and-the-importance-of-remote-laboratory.
31. Cisco Content Hub. Modicon communication bus (modbus), n.d. Retrieved from https://content.cisco.com/chapter.sjs?.
32. Clearwater Security. A multi-tiered approach to risk monitoring strategy, n,d. Retrieved from https://clearwatersecurity.com/blog/a-multi-tiered-approach-to-risk-monitoring-strategy/.
33. Cloudflare. (2024). How to prevent ddos attacks | methods and tools. Retrieved from https://www.cloudflare.com/learning/ddos/what-is-a-ddos-attack/.
34. Coalfire. Attack surface management, n.d. Retrieved from https://www.coalfire.com/services/threat-and-vulnerability-management/attack-surface-management.
35. Collantes, M. H., & Padilla, A. L. (2015). Protocols and network security in ics infrastructures. Technical Report. Access PDF: incibe.es.
36. Computer Weekly Staff. (2019). Norwegian aluminium producer hit by 'extensive cyber attack'. Accessed 11 Dec 2024.
37. Cooper, S. 8 best attack surface monitoring tools, 2021, July 12. Retrieved from https://www.comparitech.com/net-admin/best-attack-surface-monitoring-tools/.
38. CoverLink Insurance. (2021). Cyber case study: Colonial pipeline ransomware attack. Retrieved from https://coverlink.com/case-study/cyber-case-study-colonial-pipeline-ransomware-attack/.
39. Cross, C., & Gillett, R. (2020). Exploiting trust for financial gain: An overview of business email compromise (BEC) fraud. *Journal of Financial Crime, 27*(3), 871–884.
40. CrowdStrike. What is cyber threat hunting? [proactive guide], n.d. Retrieved from https://www.crowdstrike.com/cybersecurity-101/threat-hunting/.
41. CS Security. ICS security, n.d. Retrieved from https://www.slideshare.net/AhmedShitta/ics-security.

42. CyberSecOp Consulting Services. Security Audit Consulting | Compliance Audit | Data Security Audit | CyberSecOp Consulting Services, n.d. Retrieved from https://cybersecop.com/security-audits.

43. Cybersecurity & Infrastructure Security Agency (CISA). Seven steps to effectively defend industrial control systems, Date Accessed 10 Jan 2024. Retrieved from https://www.cisa.gov/sites/default/files/documents/Seven%20Steps%20to%20Effectively%20Defend%20Industrial%20Control%20Systems_S508C.pdf.

44. dacoso. Managed SIEM (security information and event management), n.d. Retrieved from https://www.dacoso.com/en/cyber-security/soc-as-a-service/managed-siem/.

45. Daneels, A., & Salter, W. (1999). What is scada? Retrieved from https://accelconf.web.cern.ch/ica99/papers/mc1i01.pdf.

46. Data Privacy Manager. (January 2023). 5 things you need to know about data privacy [definition & comparison]. Retrieved from https://dataprivacymanager.net/5-things-you-need-to-know-about-data-privacy/.

47. Department of Information Technology. Cybersecurity month: Multi-factor authentication (MFA), October 3 2022. Retrieved from https://www.hsph.harvard.edu/information-technology/2022/10/03/october-is-cybersecurity-month-week-1/.

48. Derhamy, H., Rönnholm, J., Delsing, J., Eliasson, J., & van Deventer, J. (2017). Protocol interoperability of opc ua in service oriented architectures.

49. Dini, P., Elhanashi, A., Begni, A., Saponara, S., Zheng, Q., & Gasmi, K. (2023). Overview on intrusion detection systems design exploiting machine learning for networking cybersecurity. *Applied Sciences, 13*(13), 7507.

50. Domingues, P., Carreira, P., Vieira, R., & Kastner, W. (2016). Building automation systems: Concepts and technology review. *Computer Standards & Interfaces, 45*, 1–12.

51. Dorjee, R. G. (2014). Monitoring and control of a variable frequency drive using PLC and SCADA. *International Journal on Recent and Innovation Trends in Computing and Communication, 2*(10), 3092–3098.

52. Dover Microsystems. Case study: German steel mill cyberattack, n.d. Accessed: 06 Dec 2024.

53. Dragos Inc. (2016). Reassessing the 2016 CRASHOVERRIDE cyber attack. Accessed 21 Jul 2025.

54. Drias, Z., Serhrouchni, A., & Vogel, O. (2015). Analysis of cyber security for industrial control systems.

55. Electronics Coach. Intelligent electronic devices, n.d. Retrieved from https://electronicscoach.com/intelligent-electronic-devices.html.

56. ElProCus. Scada system architecture & its working, n.d. Retrieved from https://www.elprocus.com/scada-system-architecture-its-working/.

57. Empowered Automation Solutions LLC. (2023). What are the 4 types of scada. Retrieved from https://www.empoweredautomation.com/what-are-the-4-types-of-scada.

58. Ferrari, P., Flammini, A., & Vitturi, S. (2006). Performance analysis of profinet networks. *Computer Standards & Interfaces, 28*(4), 369–385.

59. Fortinet. What is a dns leak? n.d. Retrieved from https://www.fortinet.com/resources/cyberglossary/dns-leak.

60. Fortinet. What is ICS (industrial control system) security? n.d. Retrieved from https://www.fortinet.com/resources/cyberglossary/ics-security?utm_source=blog&utm_medium=blog&utm_campaign=ics-security.

61. Fortinet Lab. Cyber attack surface, n.d. Retrieved from https://www.fortinet.com/resources/cyberglossary/attack-surface.

62. G. What is device hardening: Good explanation with 10 headings (September 14 2023).

63. Galinec, D., Možnik, D., & Guberina, B. (2017). Cybersecurity and cyber defence: national level strategic approach. *Automatika: časopis za automatiku, mjerenje, elektroniku, računarstvo i komunikacije, 58*(3), 273–286.

64. GeeksforGeeks. Intrusion detection system (IDS) - geeksforgeeks, April 8 2019. Retrieved from https://www.geeksforgeeks.org/intrusion-detection-system-ids/.

65. GeeksforGeeks. Types of security mechanism, September 2 2020. Retrieved from https://www.geeksforgeeks.org/types-of-security-mechanism/.

66. GeeksforGeeks. (June 2022). Difference between hids and nids. Retrieved from https://www.geeksforgeeks.org/difference-between-hids-and-nids/.

67. GE Fanuc Automation. Cimplicity® software, n.d.

68. Godoy, A. J. C., & Pérez, I. G. (2018). Integration of sensor and actuator networks and the scada system to promote the migration of the legacy flexible manufacturing system towards the industry 4.0 concept. *Journal of Sensor and Actuator Networks, 7*(2), 23.

69. Gunda, N. S. K., Dasgupta, S., & Mitra, S. K. (2017). Diptest: A litmus test for E. coli detection in water. *PLoS One, 12*(9), e0183234.

70. Guo, Y. (2023). A review of machine learning-based zero-day attack detection: Challenges and future directions. *Computer Communications, 198*, 175–185.

71. Gupta, M. (2015). Hybrid intrusion detection system: Technology and development. *International Journal of Computer Applications, 115*, 5–8.

72. Haque, N. I., & Rahman, M. A. (2022). Phase: Security analyzer for next-generation smart personalized smart healthcare system.

73. Hendry, G. (2007). Applicability of clustering to cyber intrusion detection. Available at: https://repository.rit.edu/theses/5477/.

74. Hive Pro. Threat exposure management can minimize attack surface, June 24 2022. Retrieved from https://www.hivepro.com/how-threat-exposure-management-can-minimize-attack-surface/.

75. HPE Juniper Networking. What is IDS/IPS? n.d. Retrieved from https://www.juniper.net/us/en/research-topics/what-is-ids-ips.html.

76. IBM. What is an attack surface? Retrieved from: https://www.ibm.com/think/topics/attack-surface.

77. ICS Cyber Resilience. (October 2021). Active defense & safety – part 2. Retrieved from https://www.youtube.com/watch?v=1zYKTovx_AI.

78. ICS Cyber Resilience. (September 2021). Active defense & safety – part 1. Retrieved from https://www.youtube.com/watch?v=QM5kHJfW4Zk.

79. Imperva. Man-in-the-middle (MITM) attack, Year of Access. Retrieved from https://www.imperva.com/learn/application-security/man-in-the-middle-attack-mitm/.

80. Imperva. Phishing attack and scam: What is it and how to prevent it, Year of publication (if available). Retrieved from https://www.imperva.com/learn/application-security/phishing-attack-scam/.

81. InAccess. (2022). *ICS architecture review process*.

82. Incident.io. How to improve incident triaging for better organization-wide incident response, n.d. Retrieved from https://incident.io/blog/understanding-incident-triage.

83. Inc. Logical Systems. (2021). Security incident: Oldsmar water treatment plant - lessons learned. Technical report, Logical Systems, Inc. Accessed 26 Nov 2024.

84. InstaSafe. What is domain controller: Everything you need to know, n.d. Retrieved from https://instasafe.com/blog/what-is-domain-controller-everything-you-need-to-know/.

85. International Risk Management Institute. Cyber-physical attack.

86. JumpCloud. What is a domain controller? n.d. Retrieved from https://jumpcloud.com/blog/what-is-a-domain-controller.

87. Jurkštiene, A., Darškuviene, V., & Dūda, A. (2008). Management control systemsand stakeholders' interests in lithuanian multinational companies: Cases from the telecommunications industry. *Journal of Business Economics and Management, 9*(2), 97–106.

88. Kalhoro, S., Rehman, M., Ponnusamy, V., & Shaikh, F. (2021). Extracting key factors of cyber hygiene behaviour among software engineers: A systematic literature review. *IEEE Access, 9*, 99339–99363.

89. Kaspersky. Zero-day exploit, n.d. Retrieved from https://www.kaspersky.com/resource-center/definitions/zero-day-exploit.

90. Katz, E. Top 10 cyber threat intelligence tools for 2022 - spectral, September 14 2022. Retrieved from https://spectralops.io/blog/top-10-cyber-threat-intelligence-tools-for-2022/.

91. Khan, B., Olanrewaju, R., Anwar, F., Mir, R., & Najeeb, A. (2019). A critical insight into the effectiveness of research methods evolved to secure iot ecosystem. *International Journal of Information and Computer Security, 11*, 332.

92. KIT IAI Webmaster. Hybrid intrusion detection system for smart grids, n.d. Retrieved from https://www.iai.kit.edu/english/464_4278.php.

93. KnowledgeHut. Intrusion detection system: What it is, how it works, and types, Year of Access. Retrieved from https://www.knowledgehut.com/blog/security/intrusion-detection-system.

94. Latino, M. E., Menegoli, M., Corallo, A., et al. (2018). From industry 4.0 to agriculture 4.0: How manage product data in agri-food supply chain for voluntary traceability, a framework proposed. *International Journal of Nutrition and Food Engineering, 12*, 146–150.

95. Lazarevic, A., Kumar, V., & Srivastava, J. (2005). Intrusion detection: A survey.

96. Macher, G., Messnarz, R., Armengaud, E., Riel, A., Brenner, E., & Kreiner, C. (2017). Integrated safety and security development in the automotive domain.

97. Majdalawieh, M., Parisi-Presicce, F., & Wijesekera, D. (2006). DNPSec: Distributed network protocol version 3 (DNP3) security framework. *Advances in Computer, Information, and Systems Sciences, and Engineering, 1*, 227–234.

98. Malik, M. I., Ibrahim, A., Hannay, P., & Sikos, L. F. (2023). Developing resilient cyber-physical systems: A review of state-of-the-art malware detection approaches, gaps, and future directions. *Computers, 12*(4), 79.

99. Manadhata, P., & Wing, J. (2011). An attack surface metric. *IEEE Transactions on Software Engineering, 37*, 371–386.

100. Manuja, A., Anilkumar, S., Varun, V. V., Mathew, A., Sureshkumar, S. P., & George, R. (2023). Multi-sensor failure recovery in aero-engines using a digital twin platform: A case study.

101. McGee, A. R., Vasireddy, S. R., Xie, C., Picklesimer, D. D., Chandrashekhar, U., & Richman, S. H. (2004). A framework for ensuring network security. *Bell Labs Technical Journal, 8*(4), 7–27.

102. McMillan, R. (2011). Stuxnet explained: The first known cyberweapon. Accessed 26 Nov 2024.

103. Messaoud, B. I. D., Guennoun, K., Wahbi, M., & Sadik, M. (2016). Advanced persistent threat: New analysis driven by life cycle phases and their challenges.

104. Mihret, E. T. (2021). Intrusion detection system-ids. *American Journal of Computer Science and Technology, 9*(8), 108.

105. Mindo solution. Introduction to modbus), n.d. Retrieved from https://www.csimn.com/CSI_pages/Modbus101.html.

106. Mohagheghi, S., Stoupis, J., & Wang, Z. (2009). Communication protocols and networks for power systems-current status and future trends.

107. Morris, C. (2020). An extensible big data software architecture managing a research resource of real-world clinical radiology data linked to other health data from the whole scottish population. *GigaScience, 9*(10), giaa095.

108. N-able Technologies. Intrusion detection system: A comprehensive guide, Unknown. Retrieved from https://www.n-able.com/blog/intrusion-detection-system.

109. Nankya, M., Chataut, R., & Akl, R. (2023). Securing industrial control systems: Components, cyber threats, and machine learning-driven defense strategies. *Sensors, 23*(21), 8840.

110. Normanyo, E., Husinu, F., & Agyare, O. R. (2014). Developing a human machine interface (HMI) for industrial automated systems using siemens simatic wincc flexible advanced software. *Journal of Emerging Trends in Computing and Information Sciences, 5*(2), 134–144.

111. OpenText Cybersecurity | CyberRes. (2023). What is privileged access management? Retrieved from https://www.microfocus.com/en-us/cyberres/what-is-privileged-access-management.

112. OWASP Cheat Sheet Series. Attack surface analysis. Retrieved from https://cheatsheetseries.owasp.org/cheatsheets/Attack_Surface_Analysis_Cheat_Sheet.html.

113. Pan, L., Zhizheng, W., & Fan, C. (2019). Design of scada system for CNC grinder workshop based on simatic net.

114. Paul, S., & Narang, S. (2015). Design of scada based wireless monitoring and control system using zigbee.

115. Pereira, J. M. D., Postolache, O., & Girao, P. S. (2003). Hart protocol analyser based in labview.

116. Pliatsios, D., Sarigiannidis, P., Lagkas, T., & Sarigiannidis, A. G. (2020). A survey on scada systems: secure protocols, incidents, threats and tactics. *IEEE Communications Surveys & Tutorials, 22*(3), 1942–1976.

117. PT Security. Pt security industrial control systems services, Accessed: 2024. Retrieved from https://www.ptsecurity.com/ww-en/services/ics/.

118. PurpleSec. How to reduce your attack surface (6 best practices for 2023), n.d. Retrieved from https://purplesec.us/learn/reduce-attack-surface/.

119. Qi, J., Hahn, A., Lu, X., Wang, J., & Liu, C.-C. (2016). Cybersecurity for distributed energy resources and smart inverters. *IET Cyber-Physical Systems: Theory & Applications, 1*(1), 28–39.

120. Raccomandino, L. (2022). Cyber security: Remote intrusion consulting. Retrieved from https://raccomandino.medium.com/cyber-security-remote-intrusion-consulting-aa74fb090c2e.

121. Radwan, F. A., & Martin, T. W. (2003). Real-time monitoring and controlling of an allen-bradley SLC 500 through the internet.

122. Raghvendra, N. (February 2019). Scada system - components, hardware & software architecture, types. Retrieved from https://electricalfundablog.com/scada-system-components-architecture/.

123. Ramachandran, A., Ramadevi, P., Alkhayyat, A., & Yousif, Y. K. (2023). Blockchain and data integrity authentication technique for secure cloud environment. *Intelligent Automation and Soft Computing, 36*(2), 2055–2070.

124. Risk Management. The benefits of a proactive cybersecurity program, n,d. Retrieved from https://www.rmmagazine.com/articles/article/2022/12/15/the-benefits-of-a-proactive-cybersecurity-program.

125. RiskOpticks. What are cybersecurity threats? July 24 2023. Retrieved from https://reciprocity.com/resources/what-are-cybersecurity-threats/.

126. Romm, D. Five major types of intrusion detection system (IDS), n.d. Retrieved from https://www.slideshare.net/davidromm/five-major-types-of-intrusion-detection-system-ids.

127. RT Automation. DNP3 overview, Year of Access. Retrieved from https://www.rtautomation.com/technologies/dnp3-overview/.

128. Rutkauskas, R., Lipnickas, A., Ramonas, Č., & Kubilius, V. (2008). New challenges for interoperability of control systems. *Elektronika ir Elektrotechnika, 83*(3), 71–74.

129. Sailawi, W. (2017). Scada system and industrial network for petroleum industry.

130. Samad, T., & Kiliccote, S. (2012). Smart grid technologies and applications for the industrial sector. *Computers & Chemical Engineering, 47*, 76–84.

131. SANS ICS Security. ICS cyber resilience, active defense, n.d. Retrieved from https://www.sans.org/cyber-security-courses/ics-scada-cyber-security-essentials/.

132. Santomero, A. M., & Babbel, D. F. (1997). Financial risk management by insurers: An analysis of the process. *Journal of Risk and Insurance, 64*(2), 231–270.

133. Schwartz, M., Mulder, J., Chavez, A. R., & Allan, B. A. (2014). Emerging techniques for field device security. *IEEE Security & Privacy, 12*(6), 24–31.

134. Seborg, D. E. (2009). Automation and control of chemical and petrochemical plants. In *Control systems, robotics and automation: Industrial applications of control systems II.* EOLSS Publishers.

135. SelectHub. (2023). What is endpoint monitoring? Comprehensive guide, September 29 2023. Retrieved from https://www.selecthub.com/endpoint-security/endpoint-monitoring/.

136. Selvarani, D. R., & Ravi, T. N. (2013). Issues, solutions and recommendations for mobile device security. *International Journal of Innovative Research in Technology & Science, 1*(5), 9–14.

137. Siddique, S., Rifat, R. H., Talukder, S., Alam, S. B., Gupta, K. D., et al. (2023). Challenges and opportunities of computational intelligence in industrial control system (ICS). Authorea Preprints.

138. Siddiqui, F., Hagan, M., & Sezer, S. (2019). Establishing cyber resilience in embedded systems for securing next-generation critical infrastructure.

139. Software Contributor. What is threat intelligence? Definition and types - dnsstuff, September 14 2020. Retrieved from https://www.dnsstuff.com/what-is-threat-intelligence.

140. Spiceworks. What is network traffic analysis? Definition, importance, implementation, and best practices, n.d. Retrieved from https://www.spiceworks.com/tech/networking/articles/network-traffic-analysis/.

141. Spurgeon, C. E. (2000). Ethernet: The Definitive Guide. https://www.researchgate.net/publication/234823320_Ethernet_The_Definitive_Guide.

142. Stouffer, K., Falco, J., Scarfone, K., et al. (2011). Guide to industrial control systems (ICS) security. *NIST Special Publication, 800*(82), 16–16.

143. Stuckman, J., & Purtilo, J. (2012). Comparing and applying attack surface metrics.

144. Sudhakar. (2021). Cybersecurity threat detection using machine learning and deep learning techniques.

145. Sulaiman, N. S., Nasir, A., Othman, W., Fahmy, S., Aziz, N., Yacob, A., & Samsudin, N. (2021). Intrusion detection system techniques: A review. *Journal of Physics: Conference Series, 1874*, 012042.

146. Tasmurzayev, N., Amangeldy, B., Baigarayeva, Z., Mansurova, M., Resnik, B., & Amirkhanova, G. (2022) Improvement of hvac system using the intelligent control system.

147. Techstrong. A complete guide to ICS security assessment, July 21 2025. Retrieved from https://securityboulevard.com/2023/08/a-complete-guide-to-ics-security-assessment/.

148. TechTarget. What is an attack surface and how to protect it? 2021, September 1. Retrieved from https://www.techtarget.com/whatis/definition/attack-surface.

149. TechTarget. How to recover from a ransomware attack, Year of publication, e.g., 2022. Retrieved from https://www.techtarget.com/searchsecurity/tip/How-to-recover-from-a-ransomware-attack.

150. Teece, D. J. (2016). Opec behavior: An alternative view. In *OPEC Behaviour and World Oil Prices* (pp. 64–93). Routledge.

151. Tekiner, E., Acar, A., Uluagac, A. S., Kirda, E., & Selcuk, A. A. (2021). Sok: cryptojacking malware.

152. Threat analysis and risk assessment. Accessed 3 January 2025. Retrieved from https://www.security-analyst.org/threat-analysis-and-risk-assessment/

153. Toderi, S., Gaggia, A., Balducci, C., & Sarchielli, G. (2015). Reducing psychosocial risks through supervisors' development: A contribution for a brief version of the "stress management competency indicator tool". *Science of the Total Environment, 518*, 345–351.

154. Trend Micro. Security patch management strengthens ransomware defense, October 24 2023. Retrieved from https://www.trendmicro.com/tr_tr/ciso/23/c/security-patch-management.html.

155. triangle microworks. ICCP tase.2, n.d. Retrieved from https://rb.gy/m06x62.

156. TXOne Networks. Protecting utilities: Strategies for mitigating common attack vectors, May 31 2023. Retrieved from https://www.txone.com/blog/protecting-utilities-strategies-for-mitigating-common-attack-vectors/.

157. U.S. Department of Energy. Defense in depth strategies, n.d. Retrieved from https://energy.gov/sites/prod/files/oeprod/DocumentsandMedia/Defense_in_Depth_Strategies.pdf.

158. Vaduva, B., Pop, I.-F., & Valean, H. (2022). One4all—a new scada approach. *Sensors, 22*(6), 2415.

159. Verhappen, I., & Pereira, A. (2008). *Foundation fieldbus.* ISA.

160. VISTA InfoSec. (2021). Benefits of implementing network segmentation, April 8. Retrieved from https://vistainfosec.com/blog/network-segmentation-benefits-for-businesses//.

161. Wang, Y., & Hsiao, T. (2024). Multivariable iterative learning control design for precision control of flexible feed drives. *Sensors, 24*(11), 87–105.

162. Wei, J., Zhang, R., Liu, J.-y., Niu, X., & Yang, Y. (2015). Defense strategy of network security based on dynamic classification. *KSII Transactions on Internet and Information Systems, 9,* 5116–5134.
163. Weiss, J. (2014). Industrial control system (ICS) cyber security for water and wastewater systems.
164. Yadav, G., & Paul, K. (2021). Architecture and security of scada systems: A review. *International Journal of Critical Infrastructure Protection, 34,* 100433.
165. Yang, Y., Zhu, Q., Maasoumy, M., & Vincentelli, A. (2012). Development of building automation and control systems. *Design & Test of Computers, IEEE, 29,* 45–55.
166. Yokogawa Electric Corporation. Field digital device diagnostics for effective use of device information, n.d. Retrieved from https://www.yokogawa.com/eu/library/resources/yokogawa-technical-reports/field-digital-device-diagnostics-for-effective-use-of-device-information/.
167. Zetter, K. (2016). Inside the cunning, unprecedented hack of ukraine's power grid. Accessed 26 Nov 2024.

Index